D1759918

CARDIAC RECEPTORS

CARDIAC RECEPTORS

Report of a symposium sponsored by the Commission on
Cardiovascular Physiology of the
International Union of Physiological Sciences
held in the
University of Leeds, 14–17 September 1976

Edited by

R. HAINSWORTH, C. KIDD & R. J. LINDEN

Department of Cardiovascular Studies, University of Leeds

CAMBRIDGE UNIVERSITY PRESS

CAMBRIDGE

LONDON · NEW YORK · MELBOURNE

Published by the Syndics of the Cambridge University Press
The Pitt Building, Trumpington Street, Cambridge CB2 1RP
Bentley House, 200 Euston Road, London NW1 2DB
32 East 57th Street, New York, NY 10022, USA
296 Beaconsfield Parade, Middle Park, Melbourne 3206, Australia

First published 1979

Printed in Great Britain at the Alden Press, Oxford

Library of Congress Cataloguing in Publication Data
Main entry under title:
Cardiac receptors.
Includes index.
1. Heart–Innervation–Congresses. 2. Muscle receptors–Congresses.
3. Neural receptors–Congresses.
1. Hainsworth, R. 2. Kidd, Cecil, 1933– 3. Linden, Ronald James.
4. International Union of Physiological Sciences.
Commission on Cardiovascular Physiology.
QP113.4.C37   612′.178   77-12404
ISBN 0 521 21853 5

CONTENTS

CHAIRMEN AND SPEAKERS

MAIN PAPERS

J. O. Arndt *Abteilung für Experimentelle Anaesthesiologie, Universität Düsseldorf, Moorenstraat 5, Düsseldorf, Germany*

D. G. Baker *University of California, Cardiovascular Research Institute, San Francisco, California 94143, USA*

A. M. Brown *Department of Physiology, University of Texas, Galveston, Texas 77550, USA*

H. M. Coleridge *University of California, Cardiovascular Research Institute, San Francisco, California 94143, USA*

J. C. G. Coleridge *University of California, Cardiovascular Research Institute, San Francisco, California 94143, USA*

D. E. Donald *Mayo Foundation, Rochester, Minnesota 55901, USA*

K. Floyd *Department of Physiology, University of Leeds, Leeds LS2 9 JT, UK*

B. Folkow *Institute of Physiology, University of Göteborg, Göteborg, Sweden*

A. G. Garcia-Aguado *Department Cardiopulmonar, Hospital Clinico de la Facultad de Medicina, Madrid, Spain*

O. H. Gauer *Physiologisches Institut, Der Freien Universität Berlin, 1000 Berlin 33, Germany*

J. P. Gilmore *Department of Physiology and Biophysics, University of Nebraska, Omaha, Nebraska 68105, USA*

K. L. Goetz *St Luke's Hospital, Kansas City, Missouri 6411, USA*

R. Hainsworth *Department of Cardiovascular Studies, University of Leeds, Leeds LS2 9JT, UK*

S. M. Hilton *Department of Physiology, University of Birmingham, Birmingham, UK*

C. T. Kappagoda *Department of Cardiovascular Studies, University of Leeds, Leeds LS2 9JT, UK*

C. Kidd *Department of Cardiovascular Studies and Physiology, University of Leeds, Leeds LS2 9JT, UK*

J. R. Ledsome *Department of Physiology, University of British Columbia, Vancouver 8, Canada*

R. J. Linden *Department of Cardiovascular Studies, University of Leeds, Leeds LS2 9JT, UK*

A. Malliani *Istituto di Ricerche Cardiovascolari, Dell Università di Milano, 20122 Milan, Italy*

A. L. Mark *University of Iowa College of Medicine, Iowa City, Iowa 52242, USA*

E. Neil *Department of Physiology, Middlesex Hospital Medical School, London W1P 6DB, UK*

B. Öberg *Department of Physiology, University of Göteborg, Göteborg, Sweden*

A. S. Paintal *Vallabhbhai Patel Chest Institute, University of Delhi, Delhi 7, India*

F. Perez-Gomez *Departmento Cardiopulmonar, Hospital Clinico de la Facultad de Medicina, Madrid, Spain*

J. T. Shepherd *Mayo Foundation, Rochester, Minnesota 55901, USA*

P. Sleight *Cardiac Department, Radcliffe Infirmary, Oxford OX2 6HE, UK*

P. Thorén *Department of Physiology, University of Göteborg, Göteborg, Sweden*

J. Tranum-Jensen *Anatomy Department C, University of Copenhagen, Copenhagen, Denmark*

Y. Uchida *Second Department of Internal Medicine, Faculty of Medicine, University of Tokyo, Tokyo, Japan*

D. Witteridge *Department of Physiology, University of Oxford, Oxford, UK*

A. Yamauchi *Department of Antaomy, Iwate Medical University, Morioka 020, Japan*

I. H. Zucker *Department of Physiology and Biophysics, University of Nebraska, Omaha, Nebraska 68105, USA*

OTHER PARTICIPANTS

Aars, H. *Oslo, Norway*
Abdalla, F.Z. *Mansoura, Egypt*
Abdul, M.L. *Zagazig, Egypt*
Abel, F.L. *Carolina, USA*
Ackermann, U. *Toronto, Canada*
Adigun, S.A. *Nottingham, UK*
Alabaster, C. *Sandwich, UK*
Al-Dulymi, R. *Leeds, UK*
Andrews, P. *Sheffield, UK*
Araneda, G. *Leeds, UK*
Baker, D.G. *California, USA*
Bannister, Sir Roger *London, UK*
Bennett, E.D. *London, UK*
Bie, P. *Copenhagen, Denmark*
Bishop, V.S. *Texas, USA*
Blaber, L.C. *Welwyn Garden City, UK*
Blair, M. *Framingham, USA*
Blount, D.H. *Bethesda, USA*
Boeck, V. *Copenhagen, Denmark*
Borgdorff, P. *Amsterdam, Holland*
Brooks, D. *Oxford, UK*
Brown, R.A. *Loughborough, UK*
Chance, E. *Nottingham, UK*
Chapple, D.J. *Beckenham, UK*
Chukwuemeka, A. *Basel, Switzerland*
Coleridge, H. *California, USA*
van Delft, A.M.L. *Oss, Holland*
Detloff, F.R. *Indiana, USA*
Diete-koki, K. *Benin, Nigeria*
Dobie, T.G. *Leeds, UK*
Dolamore, P.G. *Greenford, UK*
Donoghue, S. *Leeds, UK*

Downman, C.B.B. *London, UK*
Eckberg, D.L. *Iowa, USA*
Estavillo, J.A. *Illinois, USA*
Farjadazad, F. *London, UK*
Farmer, J.B. *Loughborough, UK*
Fitzgerald, J.D. *Alderley Park, UK*
Fox, D. *Leeds, UK*
Freedman, P.S. *London, UK*
Freyschuss, U. *Stockholm, Sweden*
Garnier, D. *Tours, France*
Gelb, I.J. *New York, USA*
Gillan, M.G.C. *Glasgow, UK*
Gilmore, J.P. *Nebraska, USA*
Gootman, P.M. *New York, USA*
Greenwood, P. *Leeds, UK*
Gupta, B.N. *Dusseldorf, Germany*
Gupta, P.D. *Delhi, India*
Hakumäki, M. *Kuopio, Finland*
Halinen, M. *Kuopio, Finland*
Hampton, I. *Leeds, UK*
Harley, A. *Liverpool, UK*
van Hell, G. *Haarlem, Holland*
Hemingway, A. *Leeds, UK*
Hick, V.E. *Leeds, UK*
Holton, A. *Leeds, UK*
Houpaniemi, T. *Helsinki, Finland*
Hutchinson, I. *Leeds, UK*
Jackson, D.M. *Loughborough, UK*
Joels, N. *London, UK*
Jones, S. *Sandwich, UK*
Jukes, M.G.M. *London, UK*
Kadiri, A. *Leeds, UK*

Kapteina, F.W. *Berlin, Germany*
Kapteina, U. *Berlin, Germany*
Karim, F. *Leeds, UK*
Knapp, M. *Leeds, UK*
Knibbs, A. *Leeds, UK*
Kunz, D.L. *Texas, USA*
Lawrenson, G. *Leeds, UK*
Linden, R. *London, UK*
McDonald, A. *London, UK*
Mackenna, B.R. *Glasgow, UK*
Mann, J. *Loughborough, UK*
Marillaud, A. *Poitiers, France*
Marin-Neto, J. *Oxford, UK*
Marley, E. *London, UK*
Mary, D. *Leeds, UK*
Maxwell, G.M. *Adelaide, Australia*
Merchant, S.J. *Leeds, UK*
Miller, M. *Oslo, Norway*
Mohammed, M. *Leeds, UK*
Morrison, G.W. *Leeds, UK*
Morrison, J.F.B. *Leeds, UK*
Motz, W. *Berlin, Germany*
Nady, A.H. *Mansoura, Egypt*
Nargeot, J. *Tours, France*
Newell, J. *Leeds, UK*
Newman, P.P. *Leeds, UK*
Norman, J. *Southampton, UK*
O'Connor, W.J. *Leeds, UK*
Octavio, J.A. *Caracas, Venezuela*
Opdyke, D.F. *Newark, USA*
Parameswaran, S.V. *Leeds, UK*
Pashley, M. *Leeds, UK*
Pearson, M.J. *Leeds, UK*
Perez-Gomez, F. *Madrid, Spain*
Peterson, D.F. *Texas, USA*
Potts, D. *Leeds, UK*
Pourrias, B. *Rueil-Malmaison, France*

Rahman, A.-R.A. *Leeds, UK*
Rai, U.C. *New Delhi, India*
Rao, P.S. *Oxford, UK*
Rasheed, B.M.A. *Tabriz, Iran*
Rayfield, K. *Leeds, UK*
Recordati, G. *Milan, Italy*
Reinhardt, H.W. *Berlin, Germany*
Reza, H. *Leeds, UK*
Ruch, A.-M. *Rueil-Malmaison, France*
Samodelov, L.F. *Dusseldorf, Germany*
Saum, W.R. *Texas, USA*
Saunders, D.A. *Sheffield, UK*
Saxton, C. *Sandwich, UK*
Schwartz, P.J. *Milan, Italy*
Seed, W.A. *London, UK*
Segall, H.N. *Montreal, Canada*
Shepherd, J.T. *Rochester, USA*
Siddiqui, J. *St Louis, USA*
Sivananthan, N. *Leeds, UK*
Slessor, I.M. *Watford, UK*
Sofola, O. *Leeds, UK*
Sreeharan, N. *Leeds, UK*
Summerill, R. *Leeds, UK*
Sutton, P. *London, UK*
Thames, M.D. *Rochester, USA*
Theilen, E.O. *Iowa, USA*
Thornhill, D.P. *Salisbury, Rhodesia*
Tidd, M. *Edinburgh, UK*
Ullmann, E. *London, UK*
Ward, S.A. *Liverpool, UK*
Weaver, L.C. *Michigan, USA*
Wennergren, G. *Göteborg, Sweden*
Whitaker, E. *Leeds, UK*
Whitwam, J.G. *London, UK*
Yokota, R. *London, UK*
Zoster, T.T. *Toronto, Canada*

The editors gratefully acknowledge the financial support of:

The Commission on Cardiovascular Physiology of the
    International Union of Physiological Sciences
The National Institute of Health, USA
The Physiological Society
The Wellcome Trust
Astra Chemicals Ltd
Boehringer Ingelheim Ltd
Ciba Laboratories Ltd
Fisons Limited
Imperial Chemical Industries Ltd
Reckett & Colman Ltd
G. D. Searle & Co. Ltd

PREFACE

Receptors in the heart have been described histologically since the turn of the century but it is only in the last 30 years that physiological studies have indicated that events in the heart were being signalled in terms of impulses in the vagal and sympathetic nerves. Much of the original work was completed by participants of this symposium, Whitteridge, Paintal and Neil, who made some of the first recordings of action potentials from receptors in the heart. Parallel studies have shown that stimuli applied to receptors in various chambers and on surfaces of the heart of animals and man have resulted in dramatic reflex responses. In this respect early studies by Dawes, who could not be present at the symposium, were important in defining the probable sites of action of drugs causing the Bezold–Jarisch reflex; and Gauer, who gave the closing address at this symposium, has been provocative in bringing cardiac receptors into the area of the control of extracellular volume.

We decided to hold the symposium in Leeds in 1976 because of our own interest and because we felt that the state of knowledge about cardiac receptors and their functions made this an opportune moment to gather together workers in the field for the benefit both of the individuals concerned and the overall research effort. We hoped that not only would knowledge of the basic mechanisms in animals be advanced but also that some evidence of the function of cardiac receptors in integrated responses in animals and in man would be presented and discussed. The reader will be able to judge whether to agree with us that, in the event, our hopes were realised.

The invited papers are printed in the order presented at the symposium which was arranged around a progression from the structure of the receptors, through their electrical and reflex activity, to an attempt finally to elucidate their contribution to integrated responses. We were fortunate indeed that this latter aspect was expertly reviewed in an Invited Lecture by Björn Folkow which was the culmination of the symposium. So as to provide a continuity of presentation and argument the short papers pertaining to each session have been included with the invited papers of that session. The discussion section following each paper has been edited so that only particular questions or points in the discussion are presented for their answer or comment by the speaker to whom they were addressed. We apologise if, in doing so, any remembered interesting comment has been omitted: we believe that the result is more coherent and informative in print.

We must thank those who, without exception, accepted our invitation to talk about their research and submit it to the challenge of a critical audience, and the chairmen who skilfully, ruthlessly and rigidly held the symposium to its right time schedule. We would also like to thank those scientists from all over the world who attended the meeting and contributed to the criticism and discussion.

Thanks must go to the sponsors who so kindly helped us financially, some of whom by contributing to the social programme, helped the physiological discussion to be prolonged long after the formal sessions had ended.

We should like to thank all our helpers in Leeds who contributed in very many ways to the preparation and minute to minute running of the meeting, particularly the other members of the Department of Cardiovascular Studies who selflessly gave up their time and Mr J. T. Gleave, Director of Special Courses in the University of Leeds for his exceptional control of the overall organisation.

Lastly, we must thank our publishers who patiently waited while we collated the delivered manuscripts.

R. Hainsworth
C. Kidd
R. J. Linden

Histology of cardiac receptors

1. Light microscopy of nerve endings in the atrial endocardium

K. FLOYD

Summary

The atrial endocardium contains a variety of sensory nerve endings which arise from the subendocardial plexus. The available evidence indicates that most of these fibres are of vagal origin, and the majority of nerve endings appear to be concentrated in the region of the vein–atrial junctions. The nerve endings present are discussed under headings of compact unencapsulated endings, diffuse unencapsulated endings and end-nets, all of which are derived (but not necessarily exclusively) from myelinated axons. It is considered that the morphological characteristics of the various 'types' of nerve ending serve to emphasise their similarity – terminal expansions of various sizes derived from myelinated axons. In addition, interconnections between unencapsulated endings and end-nets, and the presence of terminal expansions arising from end-nets are interpreted as further indications that the unencapsulated endings and end-nets should not be regarded as distinct types of nerve ending. Reports of encapsulated endings require confirmation. The contribution, if any, from non-myelinated fibres is not known.

The endocardium of the atrial appendage contains small numbers of nerve endings which are morphologically similar to those found in the atrial endocardium.

Introduction

It is widely agreed that the atrial endocardium is profusely innervated and many sensory terminations have been described, including encapsulated endings, unencapsulated endings and end-nets. Since the work of Holmes (1957) and Miller & Kasahara (1964) it has become accepted that the atrial endocardium is innervated by two main types of nervous structure, 'complex unencapsulated endings' and 'end-nets', the latter being described as a syncytial structure of fibres and cells (Holmes, 1957). Miller & Kasahara (1964) emphasised that the division of complex unencapsulated endings into two groups, compact and diffuse, was artificial because intermediate forms could always be found. I hope to present evidence to suggest that the distinction between unencapsulated endings and end-nets may be equally artificial.

All illustrations are taken from preparations of the left atrial endocardium of the dog, stained with methylene blue (see Appendix I, p. 21) and were made in collaboration with R. J. Linden and D. A. Saunders.

Subendocardial plexus

The nerve supply to the atrial endocardium is derived from the nerve plexuses of the epicardium and myocardium and appears as a coarse-meshed plexus which lies directly above the myocardium. This subendocardial plexus (Smirnow, 1895) is composed of large and small nerve trunks which appear to divide and rejoin, thereby forming the meshes of the plexus. The nerve endings in the endocardium are all derived from this plexus. Two examples of the subendocardial plexus (Fig. 1.1) also show single myelinated fibres which are derived from the nerve trunks.

Encapsulated endings

The presence of encapsulated endings ('Eingekapselte Nervenknäuel') in the atrial endocardium of the horse has been reported by Michailow (1908). He described the endings as oval, rounded, or irregular in shape, and surrounded by a lamellated connective tissue capsule which gradually merges with the surrounding tissue. Within the capsule is an inner bulb ('Innenkolben') which contains a terminal expansion consisting of dense, entangled and anastomosed nerve filaments. The encapsulated endings were derived from one or more myelinated nerve fibres, and were found either singly or in groups which may lie in a row along a fine nerve trunk. Nonidez (1941) also mentioned the presence of encapsulated endings in the inferior vena cava of the cat, but unfortunately gave no description.

Ábrahám (1969) who has also studied the atrial endocardium of the horse did not refer to encapsulated endings but drew attention to loose coil-like end systems, composed of numerous delicate smooth fibres, in the right atrium. However, Ábrahám's illustration (Ábrahám, 1969, Fig. 89) resembles the 'sensory nodal points' – areas where fibres of the nerve

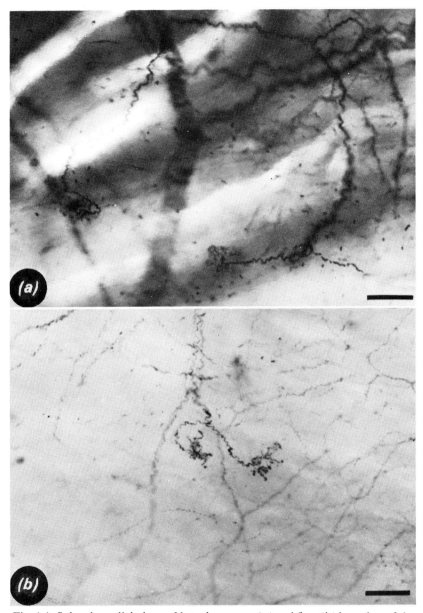

Fig. 1.1. Subendocardial plexus. Note the coarse *(a)* and fine *(b)* branches of the plexus from which single myelinated axons are derived. *(a)* was counterstained with 1% O_sO_4. Scales: 100 μm.

Table 1.1. *Reports of unencapsulated endings and end-nets in the atrial endocardium. Divided squares indicate the author and species studied. Symbols above the diagonal refer to unencapsulated endings, those below the diagonal refer to end-nets. Symbols: +, present; (+), probably present; ?, possibly present; −, not described. Ag = silver impregnation; Mb = methylene blue staining*

Author	Stain	Cat	Cow	Dog	Fox	Goat	Guinea pig	Hedgehog	Horse	Human	Monkey	Mouse	Pig	Rabbit	Rat	Sheep
Ábrahám (1969)	Ag	+/?	+/?	+/?	+/?				?/?				+/?			+
Coleridge et al. (1957)	Mb			+												
Holmes (1957)	Mb			+/+												
Holmes (1958)	Ag/Mb	+/+		+/+			+/+							+/+	−/+	
Johnston (1968)	Mb									+/+						
Khabarova (1963)	Ag	+/?				+/?				+/?						+
Meyling (1963)	Mb		−/+												−	
Michailow (1908)	Mb								+/+							
Miller & Kasahara (1964)	Mb	+/+		+/+							+/+					+/+
Nettleship (1936)	Mb	+/(+)														
Nonidez (1937)	Ag	+/−		+/−										−		
Nonidez (1941)	Ag	+/−		+/−												
Pannier (1940)	Ag	+/−		+/−												
Sato (1954)	Ag			+/−												
Smirnow (1895)	Mb	+/(+)		+/(+)			+/(+)	+/(+)				+/(+)		(+)/(+)	−/(+)	
Tcheng (1951)	Ag			+/−												
Woollard (1926)	Mb	+/(+)		+/(+)										+/(+)		

trunks split, interweaved, coiled and twisted upon themselves – described by Nettleship (1936) and which may be sites of nerve branching rather than terminations. These discrepancies could perhaps be resolved by a further study.

Unencapsulated endings

Unencapsulated endings with a wide variety of shapes and forms have been described in many species, and a list of authors and species is presented in Table 1.1.

Compact unencapsulated endings

Quantitative estimates of the sizes of the myelinated axons and their endings have been made by several authors. Coleridge, Hemingway, Holmes & Linden (1957), Holmes (1957) and Miller & Kasahara (1964) gave similar descriptions of endings of up to 50×350 μm supplied by myelinated axons of diameter 3–14 μm, while Khabarova (1963) described compact endings of area 0.005–0.008 mm^2 which arose from fibres 5–8 μm in diameter. These descriptions embrace the terminal apparatus commonly known as the 'compact complex unencapsulated ending'; examples of such endings from the left atrial endocardium of the dog are illustrated in Figs 1.2, 1.3 and 1.4. From these illustrations and those of the authors mentioned previously, it can be seen that there is extensive morphological variation within one species, and the only feature common to all the endings is that the myelinated axons which innervate them retain their myelin sheath until they branch extensively within the terminal arborisation.

The sizes of the axons (2.5–8.5 μm) and terminal expansions (0.003–0.1 mm^2) illustrated are typical of the range observed in the present study of the left atrial endocardium of the dog, and large diameter axons up to 14 μm (Holmes, 1957; Miller & Kasahara, 1964) have not been encountered. This may reflect, in part at least, the difficulty in measuring internodal diameters of axons which take such a sinuous course through the endocardial tissues. A further complication may arise because measurements are made at different points along the course of an axon, and the values quoted here refer to measurements from the internodes just before the terminal expansion. If the axons become smaller as they approach their terminal expansion there may be substantial decreases in conduction velocity, such as those encountered in the peripheral course of splanchnic mechanoreceptors (Floyd & Morrison, 1974*a, b*), which would

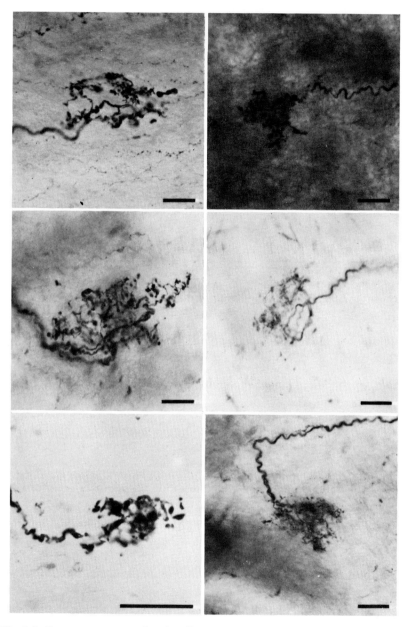

Fig. 1.2. Compact unencapsulated endings. Note the differences in axon diameter and the size of the terminal expansions, and compare with Figs 1.3 and 1.4. The preparation in the lower right corner is counterstained with 1% OsO₄.
Scales: 50 μm.

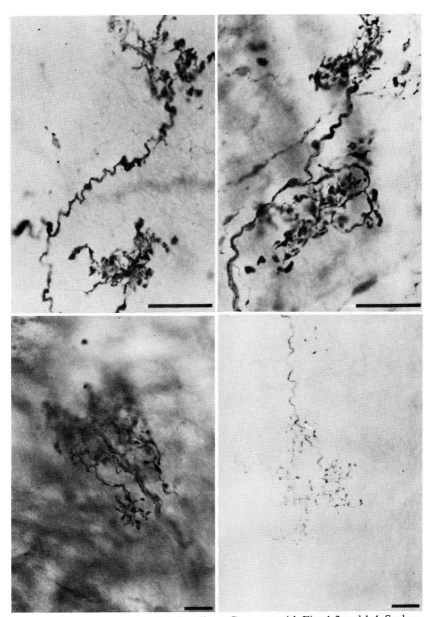

Fig. 1.3. Compact unencapsulated endings. Compare with Figs 1.2 and 1.4. Scales: 50 μm.

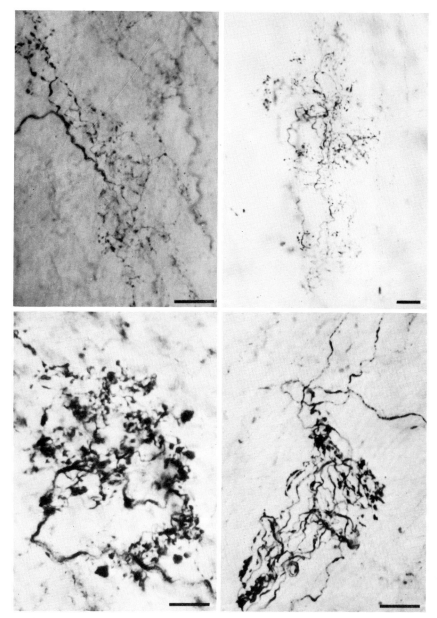

Fig. 1.4. Compact unencapsulated endings. Note the size of the terminal expansions compared with Figs 1.2 and 1.3. Scales: 50 μm.

lead to further difficulties in comparing axon diameters and overall conduction velocities.

Diffuse unencapsulated endings

Nettleship (1936) described nerve endings of enormous size and elaborate branching, often with varying structure but all derived from myelinated axons of about 6 μm in diameter, which gradually lost their myelin sheaths before entering the endings. Holmes (1957) referred to 'terminal areas' which were formed from repeated branching of small myelinated fibres to produce a complex network of fine terminal branches. Khabarova (1963) described complex arborisations of area 0.2–0.8 mm² arising from myelinated fibres 2.1–4.5 μm in diameter, and Miller & Kasahara (1964) show diffuse endings which occupy several square millimetres and arise from myelinated axons of 4–6 μm in diameter. Both Holmes (1957) and Miller & Kasahara (1964) stress that there are no anastomoses between the terminal branches. Examples of 'diffuse complex unencapsulated endings', similar to those depicted by the above authors are illustrated in Fig. 1.5. Such endings arose from myelinated axons of diameter 2–6.5 μm and extend through areas from 0.23 to 0.8 mm².

Fig. 1.6. shows an extreme example of the problems associated with attempts to classify endings into 'types': the parent axon, of diameter *c*. 5 μm, divides several times giving off both myelinated and non-myelinated branches, some of which end as unencapsulated endings while others appear to form free endings, and the total area innervated extends to over 2 mm². As pointed out previously there seem to be no grounds for a distinction between the compact and diffuse unencapsulated endings because intermediate forms exist (Miller & Kasahara, 1964). In any event, the morphological similarities between the two 'types' do not lend support to any distinction, but morphological similarity must not obscure possible functional differences such as the sites and modes of initiation of nerve impulses, or the nature of the adequate stimuli.

Number and distribution of unencapsulated endings

Nettleship (1936) reported that the complicated branched endings which he observed were most abundant at the region of insertion of the great vessels, around the atrio-ventricular orifices, and at the base of the inter-atrial septum. Like Nettleship, Nonidez (1937) did not draw any distinction between compact and diffuse terminal arborisations and his Fig. 1, which

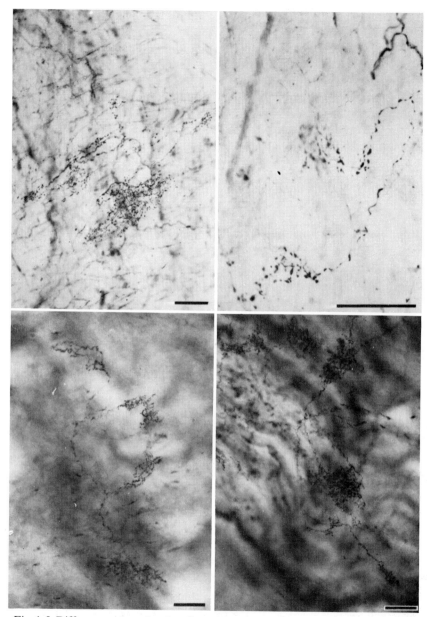

Fig. 1.5. Diffuse unencapsulated endings. Note the diffuse terminal expansions and the apparent loss of myelin sheath towards the endings. Compare with Figs 1.2, 1.3 and 1.4. Scales: 100 μm.

Fig. 1.6. Myelinated axon which branches several times and gives rise to both unencapsulated endings and free endings. Scale: 250 μm.

has been widely quoted, shows receptor areas at the insertions of the venae cavae and the pulmonary veins and he also observed them on the posterior wall of the left atrium and at the opening of the coronary sinus. Nonidez also noted that the regions of the veins where receptors were found were invested by a continuation of the myocardium forming a transitional zone. In a later article Nonidez (1941) distinguishes between diffuse and compact endings in the pulmonary veins of the dog and reported that the compact type was more scattered and less numerous (10–15 in each vein), and that the myelinated fibres which form them branch two or three times.

Coleridge *et al.* (1957), Holmes (1957) and Miller & Kasahara (1964) all confirm the concentration of unencapsulated endings at the vein–atrial junctions, and these authors give estimates of 100–300 for the numbers of compact unencapsulated endings from both sides of the heart. In addition, Holmes (1957) estimated that the distribution ratio of the endings between the left and right atria varied from about 2:1 to 4:1.

In contrast, Tcheng (1951) stated that there were more sensory fibres in the right atrium than the left. However, in his paper he also states that Nonidez did not find sensory fibres in the atria, and this suggests that Tcheng did not include in his estimate the endings which occur in the transitional zone of the vein–atrial junctions. Meyling (1953) and Mitchell (1956) both appear to argue that unencapsulated endings are artifacts, and describe only the end-nets.

End-nets

Recent authors refer frequently to the description of the end-net (terminal nervous apparatus) given by Holmes (1957) after Meyling (1953) and Mitchell (1956): a network of nerve fibres and associated cells forming a syncytium of variable appearance, but always in the form of a mesh, composed of thick strands and single filaments, the latter beaded in appearance. Examples of end-nets are illustrated in Fig. 1.7. Holmes (1957), Miller & Kasahara (1964) and Johnston (1968) also describe myelinated fibres which contribute to the formation of end-nets (Fig. 1.7*a*) and there can be little doubt that these descriptions refer to the end-nets described by Michailow (1908).

The paucity of descriptions of end-nets in the earlier literature is attributed in part by both Holmes (1957) and Miller & Kasahara (1964) to the use by some authors of silver impregnation techniques in contrast to methylene blue (Table 1.1). However, although Smirnow (1895), Woollard

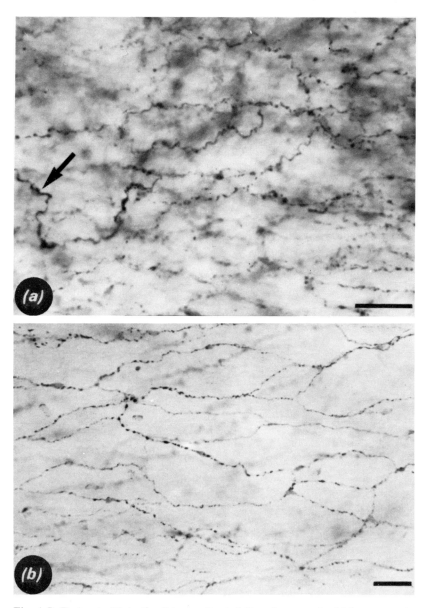

Fig. 1.7. End-nets. Note the thin myelinated fibre (*a,* arrow) which appears to contribute to the formation of the end-net, and the meshwork of coarse and fine beaded strands of which the net is composed *(b)*. Scales: 50 μm.

(1926) and Nettleship (1936), who used methylene blue, and Khabarova (1963) and Ábrahám (1969), who used silver techniques, did not give specific descriptions of the end-net, they did describe the subendocardial plexus and extensions of the plexus which lie at all levels of the endocardium, including the subendothelium, and anastomose repeatedly to form what Smirnow (1895) described as 'a fine close-meshed plexus out of fine varicose nerve fibres and fibre bundles'. This description is consistent with that of end-net in all but name.

End-nets appear to be found more extensively within the atrial endocardium than are unencapsulated endings (Coleridge *et al.*, 1957; Holmes, 1957) but are more frequently found together with the latter (Miller & Kasahara, 1964). The axons which form the end-net are said to be finely myelinated (Michailow, 1908; Holmes, 1957) or of various sizes (Miller & Kasahara, 1964).

It must be pointed out, however, that these descriptions of the end-net serve only to identify it. The composition of the fine fibres and beaded filaments is unlikely to be resolved by light microscopy, and there is at present no information available to assess the possible contribution to the end-net from independent non-myelinated fibres.

Origin of endocardial nerve endings

A number of authors have attempted to define the source of the nerve endings in the atrial endocardium by nerve section and degeneration studies. Bilateral vagotomy has been performed by Ábrahám (1969) who reported degeneration of the majority of sensory systems in the endocardium, and by Nettleship (1936) who observed degeneration of both the large myelinated fibres and the endocardial plexus. Nettleship (1936) also performed bilateral vagotomies proximal to the nodose ganglion, and his study provides evidence that both unencapsulated endings and end-nets are sensory and of vagal origin. Unilateral vagotomy (Smirnow, 1895; Holmes, 1957) leads to degeneration of some unencapsulated endings but the end-net is apparently unchanged. Holmes (1957) also noted that degeneration of unencapsulated endings was not limited to the side of nerve section, and the apparent absence of change in the end-net may be explained by postulating a bilateral origin for the fibres which contribute to it. Khabarova (1963) reported degeneration of a variety of nerve endings but did not distinguish between results obtained from uni- and bilateral vagotomy.

Experiments involving partial or complete removal of sympathetic ganglia and trunks have been reported by Nonidez (1941) and Khabarova (1963), who reported degeneration of some endings, and by Woollard (1926) and Nettleship (1936) who observed no change in either the endocardial plexus or the unencapsulated endings. Nonidez (1941) rightly pointed out the possibility of anastomoses between vagal and sympathetic branches which may occur extensively (Pick, 1970) and evidence of spinal origin requires section of nerve roots within the spinal cord. Bilateral dorsal root ganglionectomy (T1–T5) is said to have no effect upon either the unencapsulated endings or the endocardial plexus (Nettleship, 1936). Thus there is no unequivocal evidence for or against a spinal contribution to the atrial endocardium, but, as mentioned above, there appear to be difficulties in detecting degeneration within the end-net, and it is possible that fibres of spinal origin contribute some nerve endings which Ábrahám (1969) reports to be present following bilateral vagotomy. On the other hand, the 30–150 myelinated fibres (diameter 1–9 μm) in the inferior cardiac nerve of the cat, and perhaps some of the 25 000–40 000 non-myelinated fibres are thought to be afferent (Emery, Foreman & Coggeshall, 1976). There remains the further possibility that, as in the sacral spinal cord, some afferent fibres may enter via the ventral roots (Clifton, Coggeshall, Vance & Willis, 1976; Floyd, Koley, Juthika & Morrison, 1976).

It should also be noted that there is little information from light microscope studies on the possible contribution to the innervation of the endocardium by non-myelinated fibres of either vagal or spinal origin.

Relationship between unencapsulated endings and end-nets

Coleridge *et al.* (1957) and Holmes (1957) concluded that connections probably existed between the discrete endings and end-nets. Fig. 1.8*a* shows such a connection in which fine branches appear to arise from the terminal expansion of an unencapsulated ending and pass directly into an adjacent end-net. It must be noted, however, that morphological evidence of continuity between these structures may be apparent rather than real. What is observed is an apparent continuity within the focal plane of the microscope objective, and such continuity may be the result of failure to detect discontinuity or due to the capricious nature of the staining techniques. Much stronger evidence would be the demonstration of axoplasmic continuity by electron microscopy.

A further example of interconnection (Fig. 1.8*b*) shows two myelinated

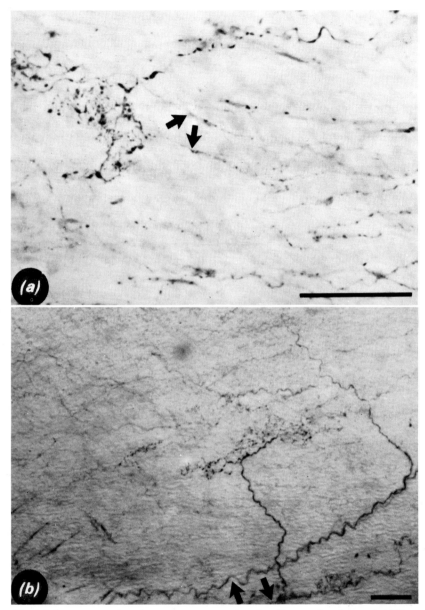

Fig. 1.8. Interconnections between unencapsulated endings and end-nets. *(a)* Fine branches (arrows) appear to arise from an unencapsulated ending and pass into an adjacent end-net. *(b)* Two myelinated axons (arrows) which are branches of a single parent axon supply an end-net (upper arrow) and two unencapsulated endings (lower arrow). Scales: 100 μm.

axons, which are branches of a single fibre, one of which divides and terminates as unencapsulated endings while the other divides twice, progressively loses its myelin sheath and appears to divide and anastomose to form an end-net.

Discrete endings which arise from fibres which are branching to form an end-net, or arise directly from the meshes of an end-net are illustrated in Figs 1.9 and 1.10. The shape of these endings is variable, some of them resembling unencapsulated endings, and they are formed as the terminal expansions of non-myelinated fibres. The existence of such endings arising from the end-net is not mentioned by recent authors, but both Woollard (1926) and Nettleship (1936) describe simple endings ('brush-like' and 'dot-like', respectively) which arise from the fine branches of subendocardial plexus.

These results, which present evidence of interconnection and common innervation and demonstrate the existence of discrete terminal expansions arising from the end-net, indicate some common features in the morphology of unencapsulated endings and end-nets, and such common

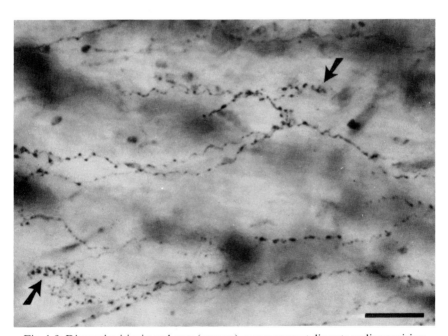

Fig. 1.9. Discontinuities in end-nets (arrows) may represent discrete endings arising from the meshes of an end-net. Compare with the continuous meshwork shown particularly clearly in Fig. 1.7*b*. Scale: 50 μm.

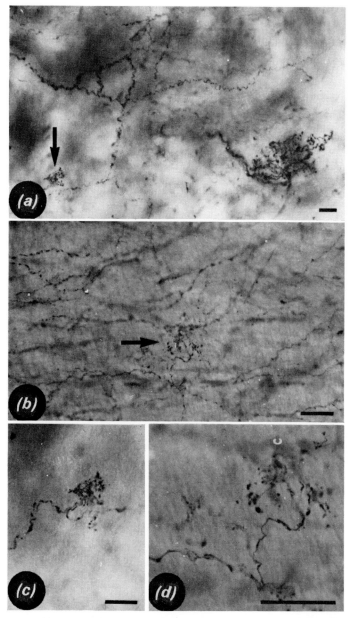

Fig. 1.10. Discrete endings arising from end-nets. The structures indicated by arrows in *(a)* and *(b)* are shown at higher magnification in *(c)* and *(d)*, respectively. Compare the structure of the endings with the unencapsulated endings shown in Figs 1.2, 1.3, 1.4 and 1.5, and note that these endings arise from non-myelinated fibres. Scales: 50 μm.

features do not support the rigid distinction commonly drawn between these two forms of nerve ending in the atrial endocardium. Indeed the differences that appear to exist, such as axon size and extent of terminal arborisation, are quantitative differences and do not provide an adequate basis for a qualitative distinction between unencapsulated endings and end-nets. Thus there seems to be no morphological basis to regard end-nets as anything other than extensive unencapsulated endings.

Appendix I: Methods

The illustrations are taken from preparations from the hearts of 18 adult dogs. The hearts were removed immediately after death and the left side of the heart was filled with a solution of methylene blue in 0.9% saline for 30 min. Eleven experiments involved the use of a 0.5% solution of vital methylene blue (G. Gurr & Son Ltd) but more recent experiments have employed a 0.01% solution of methylene blue GPR (Hopkin & Williams Ltd, No. 571400) which appears to give more consistent and reproducible staining. The left atrium and appendage were then dissected out, pinned out in oxygenated 0.9% saline, and observed under a dissecting microscope. At this stage unencapsulated endings were clearly visible in the region of the pulmonary vein–atrial junctions. When the stain appeared to have reached its maximum selectivity for the observed nerve endings the stain was fixed overnight in 8% ammonium molybdate at 4 °C.

The tissue was prepared for examination in one of three ways. In nine hearts, blocks of tissue from both atrium and appendage were placed, endocardium down, on a freezing microtome and myocardial tissue was trimmed off in the manner described by Holmes (1957). The remaining tissue was floated on to slides coated with 3% gelatin, dehydrated in *n*-butanol, cleared in xylene and mounted in DePeX (G. Gurr & Son Ltd). In a few preparations the myelinated nerve fibres were counter-stained with 1% OsO_4 (Gray, 1957). In six hearts the endocardium and myocardium were separated by fine dissection, but both of these procedures have severe limitations when applied to the appendage with its highly trabeculated structure. In an attempt to overcome these limitations the appendages from three hearts were embedded in paraffin wax and serially sectioned at 50 μm; a piece of tissue from a vein–atrial junction in each heart was similarly treated and used as a staining control (Floyd, Linden & Saunders, 1972).

The preparations were examined by normal transmitted light microscopy. The 'whole mounts' of dissected endocardium were rather

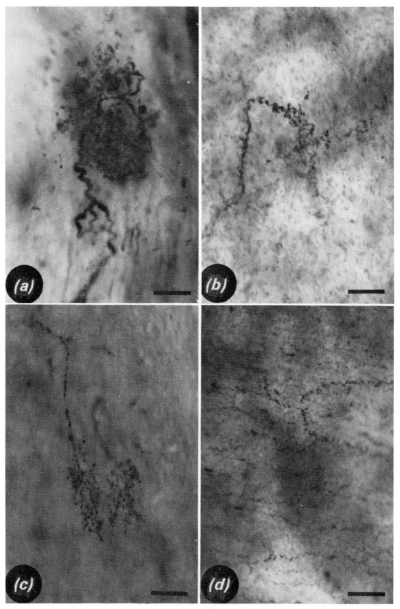

Fig. 1.11. Nerve endings from the endocardium of the left atrial appendage of the dog. Unencapsulated endings from 'whole mount' preparations *(a)*, *(b)* and 50-μm paraffin sections *(c)* resemble those observed in the remainder of the atrium (cf. Figs 1.2, 1.3, 1.4 and 1.5). An example of an end-net is shown in *(d)* (cf. Fig. 1.7). Scales: 50 μm.

thick and were examined at high magnification with an oil immersion objective (45 ×, N.A. 0.85), modified by Ealing Beck Ltd, to give a working distance of approximately 1.3 mm.

Appendix II: Nerve endings in the endocardium of the atrial appendage

The endocardium of the atrial appendages has been less extensively studied than that of the atria. Michailow (1908) refers to the presence of a nerve plexus in the endocardium of the appendage of the horse, and reports of the presence of end-nets are given by Coleridge *et al.* (1957), Holmes (1957), Miller & Kasahara (1964) and Johnston (1968), but these authors suggest that unencapsulated endings are rare, absent or undetected.

Floyd *et al.* (1972) re-investigated the endocardium of the left atrial appendage of the dog and reported that both unencapsulated endings and end-net were present. In serial sections of each of three appendages (see Appendix I: Methods) they observed between 8 and 12 unencapsulated endings which were derived from myelinated axons up to 10 μm in diameter, and were located towards the apex of the appendages. Illustrations of unencapsulated endings and end-net from whole mount preparations and 50-μm sections are shown in Fig. 1.11 and resemble the endings observed in the remainder of the atrium.

References

Ábrahám, A. (1969). *Microscopic innervation of the heart and blood vessels in vertebrates including man.* Pergamon Press: London.

Clifton, G.L., Coggeshall, R.E., Vance, W.H. & Willis, W.D. (1976). Receptive fields of unmyelinated ventral root efferent fibres in the cat. *J. Physiol.,* **256**, 573–600.

Coleridge, J.C.G., Hemingway, A., Holmes, R.L. & Linden, R.J. (1957). The location of atrial receptors in the dog: a physiological and histological study. *J. Physiol.,* **136**, 174–97.

Emery, D.G., Foreman, R.D. & Coggeshall, R.E. (1976). Fiber analysis of the feline inferior cardiac sympathetic nerve. *J. comp. Neurol.,* **166**, 457–68.

Floyd, K., Koley, Juthika & Morrison, J.F.B. (1976). Afferent discharges in the sacral ventral roots of cats. *J. Physiol.,* **259**, 37–8P.

Floyd, K., Linden, R.J. & Saunders, D.A. (1972). Presumed receptors in the left atrial appendage of the dog. *J. Physiol.,* **227**, 27–8P.

Floyd, K. & Morrison, J.F.B. (1974a). Interactions between afferent impulses within a peripheral receptive field. *J. Physiol.,* **238**, 62–3P.

Floyd, K. & Morrison, J.F.B. (1974b). Splanchnic mechanoreceptors in the dog. *Q. Jl exp. Physiol.,* **59**, 359–64.

Gray, E.G. (1957). The spindle and extrafusal innervation of a frog muscle. *Proc. R. Soc., B,* **146**, 416–30.

Holmes, R.L. (1957). Structures in the atrial endocardium of the dog which stain with methylene blue, and the effects of unilateral vagotomy. *J. Anat.*, **91**, 259–66.

Holmes, R.L. (1958). Nervous structures in the mammalian atrial wall. *J. Physiol.*, **142**, 46–7P.

Johnston, B.D. (1968). Nerve endings in the human endocardium. *Am. J. Anat.*, **122**, 621–9.

Khabarova, A.Y. (1963). *The afferent innervation of the heart.* Consultants Bureau: New York.

Meyling, H.A. (1953). Structure and significance of the peripheral extension of the autonomic nervous system. *J. comp. Neurol.*, **99**, 495–543.

Michailow, S. (1908). Die Nerven des Endocardiums. *Anat. Anz.*, **32**, 87–101.

Miller, M.R. & Kasahara, K. (1964). Studies on the nerve endings in the heart. *Am. J. Anat.*, **115**, 217–33.

Mitchell, G.A.G. (1956). *Cardiovascular innervation.* E. & S. Livingstone: Edinburgh.

Nettleship, W.A. (1936). Experimental studies on the afferent innervation of the cat's heart. *J. comp. Neurol.*, **64**, 115–33.

Nonidez, J.F. (1937). Identification of the receptor areas in the venae cavae, and pulmonary veins which initiate reflex cardiac acceleration (Bainbridge's reflex). *Am. J. Anat.*, **61**, 203–31.

Nonidez, J.F. (1941). Studies on the innervation of the heart. II. Afferent nerve endings in the large arteries and veins. *Am. J. Anat.* **68**, 151–89.

Pannier, R. (1940). Contribution a l'étude de l'innervation presso- et chemo-sensible des oreillettes et des vaisseaux de la base du coeur. *Archs int. Pharmacodyn. Thér.*, **64**, 476–84.

Pick, J. (1970). *The autonomic nervous system.* Lippincott: Philadelphia.

Sato, H. (1954). Innervation of heart in dog. *Tohoku J. exp. Med.*, **59**, 343–56.

Smirnow, A. (1895). Ueber die sensiblen Nervenendigungen im Herzen bei Amphibien und Säugetieren. *Anat. Anz.*, **10**, 737–49.

Tcheng, K.T. (1951). Innervation of the dog's heart. *Am. Heart J.*, **41**, 512–24.

Woollard, H.H. (1926). The innervation of the heart. *J. Anat.*, **60**, 345–73.

Discussion

(a) Floyd was asked whether he had any histological evidence of smooth muscle in the endocardium close to the nerve terminals since a number of previous histological studies had shown smooth muscle in this position. Floyd replied that he had not consistently observed this in his preparations.

(b) The presence of histologically and electrophysiologically identifiable atrial receptors in the rabbit was questioned. Mary indicated that he had unequivocable histological and electrophysiological evidence for the presence of atrial receptors in this species (Kappagoda, Linden & Mary, *J. Physiol.* **272**, 709–815, 1977).

(c) The close relationship between the end-net and the diffuse, complex unencapsulated nerve terminals raised the question of the fundamental structure of the end-net and its relationship to other fibres. It was suggested

that there may be axoplasmic continuity between branches but, during the discussion, it was pointed out that there was no evidence for such continuity and that a combined light and electron microscopic analysis would be required to examine this.

(*d*) There was a suggestion that the end-net might be similar to the autonomic ground plexus found generally in tissues but now accepted as not having a neural origin even though its staining properties appear to be influenced by denervation. However, it was pointed out that the end-net appears to disappear permanently with division of the vagal nerves distal to the nodose ganglia.

(*e*) Floyd was asked whether the apparent branching of axons in the net, which was described under relatively low magnifications, might represent divergence of axons from a common pathway. He accepted that this could be so for coarse strands of the end-net, which may also have discrete endings on them, or for strands apparently diverting from a stem axon. It was agreed that a combined light and electron microscopic study would be required to define adequately the nature of the fine strands and whether they consisted of a number of fine nerve strands running together and then diverging.

(*f*) One participant indicated that earlier histological studies, notably those of Nonidez, defined a restricted distribution for the complex unencapsulated endings to the pulmonary vein–atrial junctions; the end-net, on the other hand, was extensive and extended to the ventricles. It was suggested, therefore, that the conclusion that the end-net functioned as a very diffuse, complex unencapsulated ending was not in keeping with the earlier histological and electrophysiological studies. Floyd pointed out that the distribution of the end-net was only a little more extensive than that of the complex unencapsulated endings and there were large areas of endocardium devoid of any structure other than the subendocardial plexus. Furthermore, the presumed relationship of the histological evidence to the electrophysiologically defined discrete areas of location is false, since it is possible to explain the electrophysiological findings of a discrete region by apportioning specific electrical properties to the nerve terminals.

(*g*) There was some discussion as to whether the end-net contained an efferent motor component. Floyd made the point that the previous histological evidence indicated light staining for cholinesterase and the general conclusion seemed to be that there may be a sparse motor component to the net. Tranum-Jensen pointed out that the methylene blue technique does not distinguish between afferent and efferent nerves and indicated that his studies showed that adrenergic fibres form a very considerable proportion of the plexus in the endocardium. In his opinion the end-net was, to a large extent, an adrenergic plexus (see Chapter 2 of this volume). The structures seen after methylene blue staining are probably composed of several fibres, of which a high proportion are to be regarded as adrenergic.

2. Ultrastructural studies on atrial nerve-end formations in mini-pigs

J. TRANUM-JENSEN

Our current knowledge about the fine structural organisation of afferent cardiac receptors is sparse, though of obvious importance for the advancement of detailed models for their function.

For an electron microscopic study of cardiac receptors, structures which are widely dispersed in the cardiac tissues, it is an almost indispensable prerequisite that the structures can be localised, identified and selected for study in the light microscope. Thus, the uncertainties and limitations relating to the fine structural identification of these receptors are closely linked with the limitations at the light microscope level, a fact which must be clearly recognised. In the following, therefore, I have made an attempt to outline briefly and discuss the major problems relating to: *(a)* the available morphological methods; *(b)* the correlation of morphological findings with the physiology of the receptors; and *(c)* the identification of atrial endocardial receptors, which is the main object of our studies.

(a) Specific staining methods enabling one to see afferent neurites or their endings are not available. The traditional methods utilising methylene blue or metal impregnations were developed on a purely empirical basis, and the methods stain both afferent and efferent fibres and other tissue components, such as connective tissue fibrils and cells, may be stained as well. It is also well known that the reproducibility of the staining is not satisfactory. The histochemical methods for cholinesterase give more reproducible results, but are also beset with problems concerning specificity (Koelle, 1955). By means of the fluorescence histochemical methods for biogenic amines it is possible specifically to visualise the adrenergic system; it is difficult, however, at the same time to identify other fibres (Björklund, Falck & Owman, 1972).

Given the above conditions, the identification, by light microscope, of sensory nerve fibres and their endings become to a large extent dependent on morphological criteria: e.g. fibre diameter, presence or absence of a myelin sheath and the shape of the terminal ramification. On this basis, and from studies on the degeneration following experimental nerve transections (rostral or distal to the nodose ganglia, sympathectomies and ablation of spinal ganglia (Lawrentjew, 1929; Nettleship, 1936; Nonidez, 1941; Holmes, 1957a; Chabarowa, 1959; Semenov, 1963) our knowledge about the structure of cardiac receptors has been compiled.

The mentioned techniques for light microscopic studies are unfortunately accompanied by a marked deterioration of fine structural detail. Thus, to obtain an optimal ultrastructural preservation with the techniques presently available, indirect approaches must be used (Tranum-Jensen, 1975a), but these, on the other hand, raise the problems of ensuring identity between structures observed with different methods.

(b) Few attempts have been made to correlate a physiological identification of cardiac receptors with morphological studies (Coleridge, Hemingway, Holmes & Linden, 1957). Afferent cardiac impulses are conveyed by myelinated as well as unmyelinated fibres both in the vagus nerves and in the cardiac sympathetic nerves (Malliani, Recordati & Schwartz, 1973; Uchida, Kamisaka, Murao & Ueda, 1974; Uchida & Murao, 1974; Uchida, 1975). Together with various patterns of discharge in relation to the cardiac cycle (Paintal, 1972) several functionally different types of endings may be discerned. Larger and distinct endings may be studied in tiny biopsies excised after a very precise probing and recording of a characteristic receptor discharge (Coleridge *et al.*, 1957). This attractive direct approach can hardly be used for small receptors arising from unmyelinated fibres, and perhaps mixed with other types of sensory endings, among the multitude of unmyelinated efferent fibres present in all cardiac tissues and at present only vague suggestions can be made about the morphological substrates for most of the physiologically different types of receptors.

(c) The epi- and endocardium are the most extensively studied parts of the heart with respect to afferent innervation because they are easily studied with whole-mount techniques. Therefore, given the limitations mentioned above, our studies have been confined to the atrial endocardium.

Two main types of terminal nervous ramifications have been described in the atrial endocardium of several mammalian species: (1) plexus formations, covering large areas and with contributions from several fibres; and (2) circumscribed end formations arising from single fibres (Smirnow,

1895; Lawrentjew, 1929; Nettleship, 1936; Nonidez, 1941; Holmes, 1957a; Miller & Kasahara, 1964; Johnston, 1968).

The structural components and physiological significance of the *plexus formations* are ill defined. They occur in the entire atrial endocardium but exhibit differences in morphology in different regions (Holmes, 1957a). Neither observations on the degenerations following selective denervation nor direct morphological studies have so far allowed definite conclusions about its constituents, and it is possible that a large proportion of the fibres are efferent (Johnston, 1968), a possibility which is surprisingly rarely considered. Another component of the plexus is thick fibres *en route* to circumscribed endings (Holmes, 1957a). The question whether sensory functions are inherent in the plexus *per se* or not (Malliani *et al.*, 1973) remains open.

Circumscribed endings arise from single fibres which are usually thick and myelinated. They arborise within small areas and are often observed to have a cellular component forming distinct end-organs. Observations following selective denervation have shown that the afferent fibres of circumscribed endings are conveyed centrally by the vagus nerves as well as the cardiac sympathetic nerves (Chabarowa, 1959; Semenov, 1963). Circumscribed endings are scattered in the atria (they are rarely observed in the auricles (Floyd, Linden & Saunders, 1972)), but they show a preferential occurrence in certain locations in the atria, especially around the venous entrances. This appears to be the case particularly for the endings with vagal afferent fibres (Nonidez, 1937; Coleridge *et al.*, 1957; Holmes, 1957a; Coleridge, Coleridge & Kidd, 1964), while the endings with 'sympathetic' afferent fibres seem to be more scattered (Chabarowa, 1959; Semenov, 1963).

Combined physiological and morphological studies (Coleridge *et al.*, 1957) have provided strong direct evidence that distinct, circumscribed end-organs located around the atrio-venous junctions are mechano-sensitive receptors, and such end-organs, arising from thick or medium-sized myelinated fibres and taken at the sites of their preferential occurrence, have been the main object of our studies.*

Materials and methods

Fifteen mini-pigs of both sexes (Corselitze stock, 6–10 weeks of age, 3.5–8 kg body weight) were used in the investigation.

* Parts of these studied have been reported previously (Tranum-Jensen, 1975a,b).

Staining for cholinesterase was performed on whole-mount preparations from eight hearts fixed by vascular perfusion with a mixture of 2.5% gluteraldehyde and 1% formaldehyde in 0.1 M phosphate buffer (Tranum-Jensen, 1975a) as for the electron microscopical study, but with a shorter total fixation time (0.5–1 h). Two types of specimen were used: (1) Large pieces of *intact atrial wall*. After staining for cholinesterase these specimens were studied through Leitz Ultropak immersion objectives while in buffer. Following photography, single end-organs or small groups of these were excised and embedded in Epon as for electron microscopy. Serial 2-μm sections were cut from these blocks, which allowed correlation of the whole-mount portrait with the histological appearance of the end-organs. (2) Thin *sheets of endocardium* were obtained by dissection in a cleavage plane at the junction with the myocardium. By simultaneous processing of the corresponding pieces of underlying myocardium, it was found that in successful dissections nearly all the end-organs were contained in the endocardial sheets. These specimens were cleared and mounted on slides as ordinary histological preparations. The staining procedure for the two types of specimen was essentially the same. Following a wash to remove fixative and phosphate (several changes of 0.07 M acetate buffer, pH = 5.7, made isotonic with sodium chloride), the specimens were incubated on an agitator at 0–4 °C in a modified Karnovsky–Roots medium (Karnovsky & Roots, 1964) containing 2 mM substrate (acetyl-thiocholine or butyryl-thiocholine iodide), 2.8 mM copper sulphate, 8.3 mM sodium citrate, and 0.95 mM potassium ferricyanide, all dissolved in 40 mM acetate buffer (final concentration), pH = 5.5–5.7. The mixture was made isotonic with sodium chloride. The incubations were stopped when the reaction product was just visible in the dissecting microscope (usually after 1–2 h), and the specimens were washed in the isotonic acetate buffer. Some specimens were treated with an 0.05% solution of 3,3-diaminobenzidine in the isotonic acetate buffer for 10–20 min at room temperature, which caused a marked intensification of the staining (Hanker, Thornburg, Yates & Moore, 1973). Controls were incubated in substrate-free medium. The specificity of the enzyme was further characterised by the addition of 0.1 mM tetra-isopropyl-pyrophosphoramide or 0.05 mM eserine to the incubation medium. Specimens incubated in the presence of these inhibitors were first pre-incubated with the same concentrations of the inhibitors dissolved in 0.1 M phosphate buffer at pH = 7.2 (Eränkö, 1973). They were then washed in the isotonic acetate buffer containing the same concentration of inhibitors before the final incubation with the inhibitors.

Fluorescence histochemical methods for catecholamines were applied to preparations from four hearts. During sodium pentobarbitone anaesthesia (about 50 mg·kg^{-1} body weight i.p.) the pigs were given heparin (10 mg·kg^{-1} i.v.) and the heart was removed and placed in cold, calcium-free Tyrode solution containing 1 μg·cm^{-3} of noradrenaline. Two types of preparations were made from these hearts: (1) Small *Blocks of endocardium* together with the adjacent myocardium were cut from regions around the venous junctions where end-organs were likely to occur. The blocks were frozen in Freon 22 cooled by liquid nitrogen, freeze dried at -85 °C, and processed for catecholamine fluorescence by the method of Falck and Hillarp (Björklund *et al.,* 1972). They were embedded in epoxyresin (Araldite, Ciba) according to Hökfelt (1965). These blocks were serially sectioned at 4 μm, parallel to the plane of the endocardium, and end-organs were found in about one-third of the blocks. Every second section was mounted in fluorescence-free immersion oil (Leitz) and used for fluorescence microscopy, the intervening sections were stained with toluidine blue. (2) Thin *sheets of endocardium* were dissected from the atria of the unfixed hearts, and slightly stretched over small circular coverslips. They were quickly dried in a stream of dry nitrogen, stored in a desiccator over phosphorus pentoxide until they were processed by the same method as in (1). The sheets were then mounted together with the coverslip in fluorescence-free immersion oil and observed by fluorescence microscopy and with phase contrast optics.

Specimens for electron microscopy were obtained by a method previously reported (Tranum-Jensen, 1975*a*). Electron microscopy was also performed on end-organs localised in the whole mounts stained for cholinesterase. The ultrastructure of these preparations was found to be unsatisfactory for high resolution, probably because of the prolonged incubations at low pH, and it was not possible to determine the precise origin of the cholinesterase activity at the fine structural level. However, these preparations served as controls to ensure the identity with the end-organs localised by the indirect approach used for optimal ultrastructural preservation.

Observations and discussion

In whole-mount preparations of pig atria stained for cholinesterase accumulations of distinct, circumscribed end-organs were found in the endocardium around the entrances of the pulmonary veins (Fig. 2.1), in the

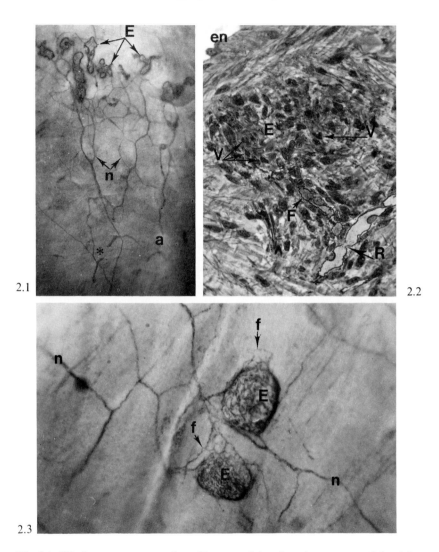

Fig. 2.1. Whole-mount preparation of intact atrial wall at the entrance of the right inferior pulmonary vein, showing an accumulation of irregularly contoured, distinct end-organs (E) together with nerves (n) forming a coarse-meshed plexus in the endocardium. At (a) a nerve enters the endocardium. At the point marked * a nerve takes a stitch in the myocardium. Cholinesterase-staining. Magnification: 12:1.

Fig. 2.2. Epon section (2 μm thick) of an end-organ (E) supplied by a thick myelinated fibre (F) departing from a stem fibre at a node of Ranvier (R). The dense bodies (V) seen among the cells of the end-organ are identified in the electron microscope as mitochondria-filled varicosities of the branches of the thick fibre.

endocardium of the superior caval vein and the adjacent atrial endocardium, on the right side of the inter-atrial septum opposite to the root of the aorta and in the anterior circumference of the mitral ostium a few millimetres above the valve. They were found less frequently around the inferior caval vein. In addition to these preferential locations, end-organs were found scattered over the entire atrial endocardium, with the exception of the auricles. In the coronary sinus a few end-organs were present. The end-organs are often assembled in small groups around a common stem fibre (Fig. 2.2). Individual end-organs are usually irregular and oblong with a largest diameter of 100–400 μm, and the largest end-organs often consist of two or three apposed endings.

The observed distribution is, except for the fewer end-organs around the inferior caval vein, largely in accordance with the distribution of circumscript end formations observed in other species (Smirnow, 1895; Dogiel, 1898; Lawrentjew, 1929; Nettleship, 1936; Nonidez, 1937, 1941; Holmes, 1957a; Chabarowa, 1959; Semenov, 1963; Miller & Kasahara, 1964; Johnston, 1968; Floyd et al., 1972).

The end-organs of the pig heart reacted strongly when acetyl-thiocholine was used as substrate for the whole-mount incubations, whereas very little reaction occurred when butyryl-thiocholine was used, and then only after prolonged incubation. Tetra-isopropyl-pyrophosphoramide did not inhibit the reaction in the acetyl-thiocholine medium, whereas reaction with the butyryl-thiocholine medium was completely inhibited. Eserine blocked the acetyl-thiocholine reaction almost completely. Accordingly, the observed cholinesterase activity is very likely to be of the so-called specific type (Eränkö, 1973) in contrast to the cholinesterase activity observed in the dog and cat heart, where the activity has been identified as being largely non-specific (Holmes, 1957b). This difference is puzzling, because one would expect the cholinesterase to exert identical functions in these

---

(caption continued)

The end-organ was localised and excised from a specimen similar to Fig. 2.1 and sectioned parallel to the plane of the endocardium. Tangentially sectioned endothelium (en) at upper left. Toluidine blue stain. Magnification: 160:1.

Fig. 2.3. A specimen similar to Fig. 2.1 seen at higher magnification. Two end-organs (E) are seen among nerves (n) in the endocardium. Thin cholinesterase-positive fibres (f) form a plexus at the periphery of the end-organs. The dotted staining seen over the end-organs is located at their surface and does not represent nerve fibres. Magnification: 80:1.

structures in the different species, whatever the function may be (Koelle, 1955; Holmes, 1957b).

In Epon sections obtained from the whole-mount specimens, an end-organ appears as a disc-like aggregate of cells into which a thick nerve fibre enters (Fig. 2.2). The thickness of the end-organ is usually about 20 μm and it may appear as a single layer of cells. The position of the end-organ in the endocardium seems to vary considerably. Some are located very close to the myocardium, occasionally even bridged by a thin strand of myocardium, while others are located entirely in the endocardial connective tissue. Collagenous fibrils are found in large amounts inside the end-organ, while there is a marked tendency for the elastic fibrils not to enter the end-organ (Figs 2.4a,b, 2.8). Although the majority of the end-organs studied possessed numerous collagenous fibrils, end-organs containing very few were occasionally found. Strands of smooth muscle cells are frequently seen in the atrial endocardium, but they do not seem to have any special relation to the end-organ.

It seems likely that such differences in the micro-topographical location in the endocardium of the end-organs and their relations to connective tissue fibrils, as well as their macro-topographical location in the atria, may influence their moment of activation during the cardiac cycle (Arndt, Brambring, Hindorf & Röhnelt, 1974).

The thick nerve fibres supplying the end-organs were found to be myelinated in most cases. When myelination was observed, it always ended before the entrance of the fibre into the end-organ at a distance varying from a few to more than 100 μm. Thus, the stem fibre supplying a group of end-organs may have lost the myelin sheath before branching, or it may retain the sheath on to the branches, as seen in Fig. 2.2. Taking the model for receptor function (Paintal, 1972) into consideration, such differences may possibly be of importance for phenomena of summation of stimuli.

In the whole mounts stained for cholinesterase, the end-organs were often associated with thin, strongly cholinesterase-positive fibres. This could be observed in intact whole-mount preparations (Fig. 2.3) and was particularly easy to see in the sheet preparations (Fig. 2.5). In these specimens the thick fibres were nearly always seen accompanied by thin fibres to the entrance into the end-organ. Thin fibres joining the end-organ from other directions were also frequently observed. In sheet preparations of the endocardium, processed for the demonstration of adrenergic nerves, the course of the thick fibres could be easily followed with phase contrast

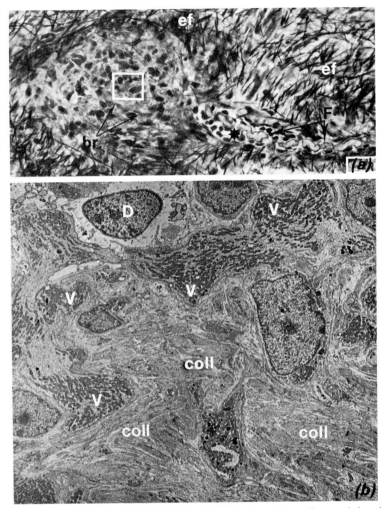

Fig. 2.4 *(a)* Epon section (1 μm thick) cut parallel to the endocardium and showing an end-organ (located in the anterior circumference of the mitral ostium a few millimeters above the valve) supplied by a thick myelinated fibre (F). The myelin sheath is lost at the point marked * before the entrance into the end-organ, inside which a major branch (br) may be discerned. The densely stained elastic fibrils (ef) in the endocardium do not enter the end-organ. Toluidine blue stain. Magnification: 255:1. *(b)* Low power electron micrograph corresponding to frame in *(a)*. A large profile of a nerve branch with mitochondria-filled varicosities (V) is seen across the upper half of the picture. A varicosity with several dense residual bodies is seen among the numerous collagenous fibrils (coll) which intersect the end-organ. One of the Schwann cells (D) of the end-organ is apparently undergoing degeneration. Magnification: 3200:1.

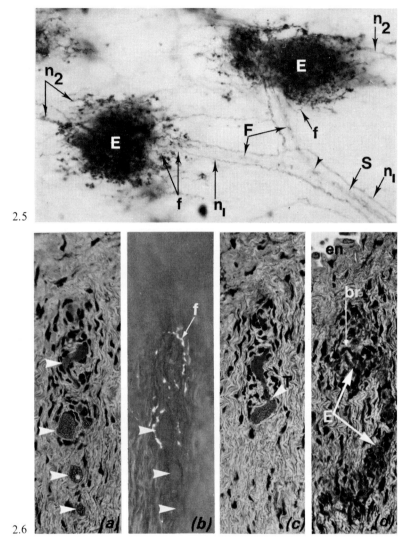

2.5

2.6

Fig. 2.5. Sheet preparation of atrial endocardium taken at the entrance of the superior caval vein and stained for cholinesterase. A thick nerve fibre (S) divides (at arrowhead), and each branch (F) gives rise to circumscribed end-organs (E). The thick nerve fibres are accompanied by thin cholinesterase-positive nerves (n_1) on to the end-organs where they split into fine fibres (f). Several thin fibres are seen around the end-organs, and nerves (n_2) coming from other directions also join the end-organs. Magnification: 160:1.

Fig. 2.6*a–d*. Series of 4 μm sections cut from a block taken at the entrance of the right inferior pulmonary vein, freeze dried, and processed for the demonstration

optics,* particularly when fibres were myelinated. Microscopy of the same field in phase contrast and fluorescence clearly showed that the thick fibres were always accompanied by thin, fluorescent fibres. Where an end-organ could be localised (identified by the presence of a cellular aggregate at the termination of a thick fibre) a plexus of thin fluorescent fibres could be seen to surround the terminal portion of the thick fibre (Fig. 2.7). This observation was confirmed when sections cut from blocks processed for the demonstration of catecholamines and serially sectioned were studied (Fig. 2.6*a–d*). In the fluorescent preparations it was also frequently observed that thin, fluorescent fibres, approaching from different directions, came into close relationship with the periphery of the end-organs.

Electron microscopic examination of the terminal portion of the thick fibre close to the end-organ frequently showed that the thick fibre 'shared' a Schwann cell with thin, varicose fibres containing accumulations of small vesicles, which strongly suggests that the thin fibres are efferent axons (Fig. 2.9). In addition, small separate nerves containing several similar thin axons were nearly always found in the immediate vicinity of the thick fibre. Single, vesicle-containing axon profiles were also regularly found at the periphery of the end-organ.

Although synapses have not been observed between the thick fibre or its ramification within the end-organ and the thin fibres, the observations are suggestive of an efferent component of the end-organ. Based on the fluorescence microscopic observations it is clear that at least some of the thin fibres are adrenergic. Because their relation with the terminal portion of the thick fibre seems to be the most constant, it would – on a purely

---

* Such preparations of fresh endocardium where the course of myelinated fibres may be followed for distances up to more than a millimetre, together with their end-organs, might be used for physiological studies *in vitro* on isolated, morphologically identifiable receptors.

(caption continued)

catecholamines. *(a)–(c)* are cut in series and show fragments of a thick fibre (arrowheads) surrounded by several thin fluorescent fibres (f). At (T) the myelin sheath of the thick fibre terminates. A few sections further revealed an end-organ (E) at the end of the thick fibre. A larger branch (br) of the thick fibre can be seen inside the end-organ. For correct topography the lower edge of *(d)* should be placed on top of *(c)*. Tangentially sectioned endothelium (en) at upper left. The felt-work structure in the myelinated part of the thick fibre is a freeze artifact. The endocardium was slightly curved and the end-organ could not been seen full face in one section. Magnification: 305:1.

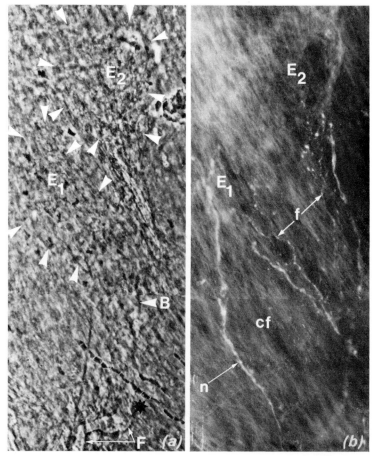

Fig. 2.7. *(a)* Phase contrast micrograph of a sheet preparation of intact endocardium taken at the anterior circumference of the mitral ostium half a centimetre above the valve. A thick myelinated fibre (F) enters the field from below and loses its myelin at the point marked *. The fibre proceeds unmyelinated (hardly seen on photograph), surrounded by cells. At (B) the fibre could be seen branching in the microscope, each branch giving rise to a circumscript end formation (E_1 and E_2) identified as a cellular aggregate (outlined by arrowheads) in the connective tissue of the endocardium. Magnification: 320:1. *(b)* Fluorescence micrograph of the same field as *(a)*. The two divisions of the thick fibre is seen to be surrounded by a plexus of thin fluorescent fibres (f). A fluorescent nerve (n) entering the field together with the thick fibre takes a separate course to join E_1. The orientation of the fibrils (cf) in the connective tissue of the endocardium can be seen in weak autofluorescence, and a slight disturbance of the fibrils is noted in the position of E_2. Magnification: 320:1.

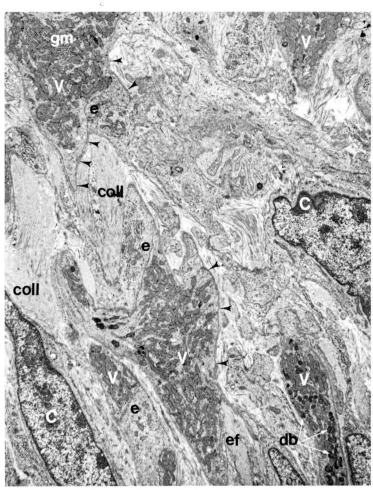

Fig. 2.8. Low power electron micrograph of the interior of an end-organ located on the right side of the inter-atrial septum. The bulky varicosities (V) contain numerous small mitochondria, and are partly surrounded by extensions (e) of the modified Schwann cells (C) of the end-organ, but long stretches of the nerve membrane (arrowheads) are exposed to the surrounding collagen fibrils (coll). A single elastic fibril (ef) is seen at lower right. A glycogen-membrane (gm) complex is seen in the varicosity at upper left, and the varicosity at lower right contains several dense residual bodies (db). Magnification: 5200:1.

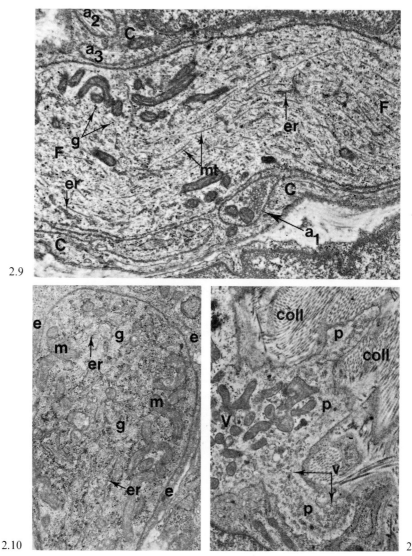

Fig. 2.9. The terminal portion of a thick nerve fibre (F) just before its entrance into an end-organ (located on the right side of the inter-atrial septum). An axon profile (a_1) containing a cluster of small vesicles is seen in the Schwann cell (C) covering the thick fibre. Two other small axon profiles (a_2, a_3) are seen at upper left. The thick fibre contains numerous microtubules (mt), irregular profiles of smooth endoplasmic reticulum (er), and some densely stained glycogen particles (g). Magnification: 14 400:1.

Fig. 2.10. A nerve branch in the interior of an end-organ is surrounded by

morphological basis – be expected that, if these fibres have a functional significance, they are likely to exert their influence on the regeneratory region of the receptor.

The observation of efferent fibres in relation to the end-organs in the heart is reminiscent of observations on thin fibres entering Paccinian corpuscles, in which adrenergic fibres have recently been identified in the inner core region (Santini, 1969). In studies on Golgi tendon organs the presence of thin fibres of unknown destination has been noted along the thick afferent fibre (Schoultz & Swett, 1972). Previous investigators of atrial nerve-end formations have also sometimes noted a relationship between circumscript end formations of thick fibres and surrounding thin fibres (Lawrentjew, 1929; Holmes, 1957a).

It has been demonstrated on different mechanoreceptors like the Paccinian corpuscles (Loewenstein & Altamirano-Orrego, 1956) and carotid sinus receptors (Zapata, 1975) that catecholamines influence the responses of the receptors. Physiological studies did not, however, demonstrate a direct efferent effect on atrial mechanoreceptors (Zucker & Gilmore, 1974; Wahab, Zucker & Gilmore, 1975).

The thick nerve fibre splits into branches shortly after it has entered the end-organ, forming a dense arborisation of very irregular fibres with bulky varicosities containing numerous mitochondria (Figs 2.4b, 2.8). Morphologically, all the cells of the end-organ are alike. No structures suggestive of a primary receptor function have been observed in these cells, and they are considered to be modified Schwann cells. They ensheath the branches of the thick fibre except at the varicosities, which are partly or completely free of cellular covering. These exposed areas of the nerve membrane are covered by an external (basal-) lamina and they come into direct contact with connective tissue fibrils, most of which are collagenous. The varicosities of the branches of the thick fibre are often furnished with small protrusions which extend between the collagen fibrils (Fig. 2.11). These protrusions

---

(caption continued)

extensions (e) of a Schwann cell. The nerve branch contains a profusion of glycogen granules (g) most of which are partly eluated due to low section thickness. Among the glycogen is seen several tubular profiles of smooth endoplasmic reticulum (er) and mitochondria (m). Magnification: 14 400:1.

Fig. 2.11. A mitochondrion-filled varicosity (V) in an end-organ is furnished with small protrusions (p) containing small clear vesicles (v), and is in intimate contact with the surrounding collagenous fibrils (coll). Magnification: 12 800:1.

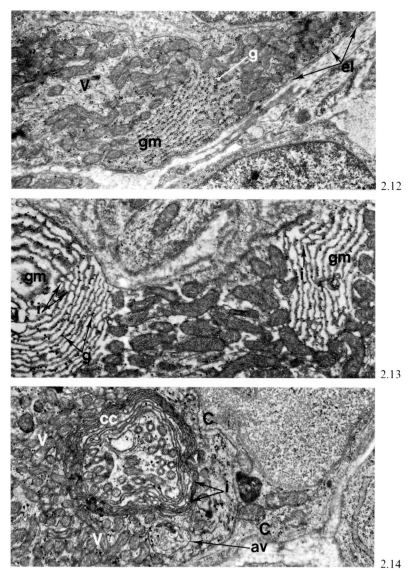

2.12

2.13

2.14

Fig. 2.12. A mitochondria-filled varicosity (V) is seen containing a small, parallel aligned glycogen-membrane complex (gm) with numerous densely stained glycogen particles (g). A large stretch of the nerve membrane is exposed to connective tissue fibrils. The external (basal-) lamina (el) of the varicosity is clearly seen. Magnification: 13 600:1.

Fig. 2.13. A very thin section showing two glycogen-membrane complexes (gm). Most of the glycogen particles (g) have been eluated and the cisterns of smooth

contain a population of small, clear vesicles and occasionally a few larger vesicles with dense cores but are devoid of larger organelles. Because these protrusions have the most intimate relation to the collagen fibrils, they are, on a morphological basis, considered an important part of the generator membrane of the receptor. The morphology of the protrusions and their relations to collagenous fibrils bear a close relation to some other mechanoreceptors, e.g. Golgi tendon organs (Schoultz & Swett, 1972).

Accumulations of numerous mitochondria are a prominent feature in all the varicosities. Similar accumulations of mitochondria have been found so frequently in several other types of mechanoreceptors (Knoche & Schmitt, 1964; Rees, 1967; Schoultz & Swett, 1972; Hanker, Dixon & Moore, 1973; Spencer & Schaumberg, 1973; Castano, 1974), that it appears to be a characteristic of such endings. A high level of cytochrome oxidase activity has been demonstrated by cytochemical methods in mitochondria of other sensory endings (Hanker, Dixon & Moore, 1973), suggesting that they are metabolically very active. The work required to restore a receptor potential may require much energy, and this would apply particularly to those cardiac receptors which discharge in every cardiac cycle.

Another prominent feature of many of the varicosities, likely to be related to metabolism, is that they contain a large number of glycogen granules. This has also often been observed in other mechanoreceptor endings. Accumulations of glycogen are not restricted to the varicosities (Fig. 2.10), but may be seen in the thick fibre already at its entrance into the end-organ. The glycogen granules were often found closely associated with profiles of smooth endoplasmic reticulum, which may appear as complexes of concentric or parallel cisterns with numerous glycogen particles between the cisterns (Figs 2.12, 2.13). Glycogen granules are often present in large quantities around such complexes. Similar complexes, but without associated glycogen (Fig. 2.14), have been observed together with large lamellated bodies (Fig. 2.15); it is very likely that they represent a degenerative end stage of these complexes. Complexes without associated

---

(caption continued)

endoplasmic reticulum stand out clearly. Interconnections (i) between the cisterns can be seen. Magnification: 17 600:1.

Fig. 2.14. A concentric cistern complex (cc) without associated glycogen is seen in a varicosity (V) which is partly covered by a modified Schwann cell (C). An autophagic vacuole (av) is seen below the complex. Interconnections (i) between cisterns are seen. Magnification: 16 000:1.

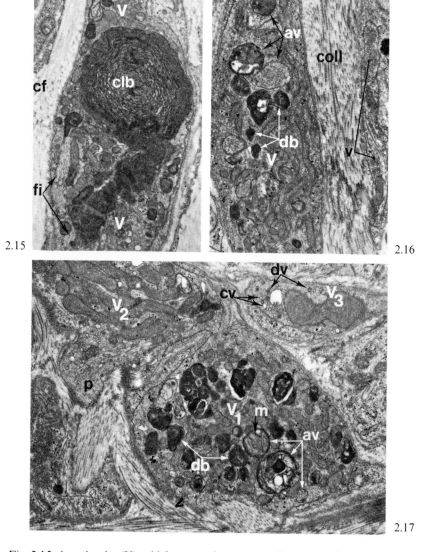

2.15

2.16

2.17

Fig. 2.15. A varicosity (V), widely exposed to surrounding connective tissue fibrils (cf), contains a large, dense, concentric lamellated body (clb). This type of inclusion as well as the concentric complex of Fig. 2.14 are thought to represent degeneratory stages of glycogen-membrane complexes; (fi) are bundles of filaments. Magnification: 12 400:1.

Fig. 2.16. A varicosity (V) containing several autophagic vacuoles (av) and dense residual bodies (db). The varicosity is broadly exposed to the surrounding

glycogen and with no signs of degradation have occasionally been observed at some distance from the end-organ (Tranum-Jensen, 1975a). These findings perhaps indicate that the complexes are transported from the perikarya of the thick fibre to the terminals where they exert a function in glycogen synthesis and then degenerate. Glycogen associated with profiles of smooth endoplasmic reticulum was found in all the end-organs studied, but highly ordered complexes were found in only a few. Morphologically identical membrane complexes with associated glycogen particles have been found in perikarya in spinal ganglia (Pannese, 1969), and very recently in cardiac ganglia too (Ellison & Hibbs, 1976). This type of complex is not very common anywhere and has not been observed in other nerve endings, and one might speculate whether the findings in spinal ganglia, cardiac ganglia and atrial end-organs are in some way related.

Degeneration processes seem to be constant phenomena in the end-organ. Autophagic vacuoles, containing recognisable mitochondria and fragments of endoplasmic reticulum in the main, together with residual bodies, were constant findings in the varicosities of the thick fibre (Figs 2.16, 2.17). Often varicosities were found which were completely filled with residual bodies and what appeared as entire, fully degenerated nerve branches were occasionally seen. Likewise, single scattered cells of the end-organ had a degenerated appearance (Fig. 2.4b). Similar degenerative phenomena have frequently been observed in other mechanoreceptors (Cauna, 1959). The above findings in the end-organs of hearts of young animals suggest that the receptors are subject to a continued replacement of neural material throughout life, and that a slow remodelling of the end-organs may take place continuously.

_____

(caption continued)

collagenous fibrils (coll). A nearby axon profile contains clusters of vesicles (v) suggestive of it being an efferent axon. Magnification: 12 800:1.

Fig. 2.17. A varicosity (V_1) with numerous autophagic vacuoles (av) and dense residual bodies (db). Recognisable mitochondria (m) can be seen in one of the vacuoles. Varicosities of this appearance are likely to be in a process of degeneration. A nearby varicosity (V_2) furnished with a protrusion (p) shows no signs of degeneration. A fragment of a varicosity (V_3) contains an unusually large mitochondrion and some dense cored vesicles (dv) together with clear vesicles (cv). Magnification: 12 800:1.

References

Arndt, J.O., Brambring, P., Hindorf, K. & Röhnelt, M. (1974). The afferent discharge pattern of atrial mechanoreceptors in the cat during sinusoidal stretch of atrial strips *in situ. J. Physiol.*, **240**, 33–52.

Björklund, A., Falck, B. & Owman, Ch. (1972). Fluorescence microscopic and microspectro-fluorometric techniques for the cellular localization and characterization of biogenic amines. In *Methods of investigative and diagnostic endocrinology*, vol. 1, ed. S. A. Benson; *The thyroid and biogenic amines*, ed. J. E. Hall & I. J. Kopin, pp. 318–68. North Holland Publishing Company: Amsterdam.

Castano, P. (1974). Further observations on the Wagner–Meissner's corpuscle of man, an ultrastructural study. *J. submicrosc. Cytol.*, **6**, 327–37.

Cauna, N. (1959). The mode of termination of the sensory nerves and its significance. *J. comp. Neurol.*, **113**, 169–209.

Chabarowa, A.J. (1959). Die Afferente Innervation des Herzens. *Z. mikrosk. anat. Forsch.*, **66**, 236–50.

Coleridge, H.M., Coleridge, J.C.G. & Kidd, C. (1964). Cardiac receptors in the dog, with particular reference to two types of afferent ending in the ventricular wall. *J. Physiol.*, **174**, 323–39.

Coleridge, J.C.G., Hemingway, A., Holmes, R.L. & Linden, R.J. (1957). The location of atrial receptors in the dog: a physiological and histological study. *J. Physiol.*, **136**, 174–97.

Dogiel, A.S. (1898). Die sensiblen Nervenendigungen im Herzen und in den Blutgefässen der Säugethiere. *Arch. mikrosk. Anat.*, **52**, 44–70.

Ellison, P.E. & Hibbs, R.G. (1976). An ultrastructural study of mammalian cardiac ganglia. *J. mol. cell. Cardiol.*, **8**, 89–101.

Eränkö, L. (1973). Effect of pH on the activity of nervous cholinesterases of the rat towards different biochemical and histochemical substrates and inhibitors. *Histochemie*, **33**, 1–14.

Floyd, K., Linden, R.J. & Saunders, D.A. (1972). Presumed receptors in the left atrial appendage of the dog. *J. Physiol.*, **227**, 27–8P.

Hanker, J.S., Dixon, A.D. & Moore, H.G., III (1973). Cytochrome oxidase activity of mitochondria in sensory nerve endings of mouse palatal rugae. *J. Anat.*, **116**, 93–102.

Hanker, J.S., Thornburg, L.P., Yates, P.E. & Moore, H.G., III (1973). The demonstration of cholinesterases by the formation of osmium blacks at the sites of Hatchett's brown. *Histochemie*, **37**, 223–42.

Hökfelt, T. (1965). A modification of the histochemical fluorescence method for the demonstration of catecholamines and 5-hydroxy-tryptamine, using araldite as embedding medium. *J. Histochem. Cytochem.*, **13**, 518–19.

Holmes, R.L. (1957*a*). Structures in the atrial endocardium of the dog which stain with methylene blue, and the effects of unilateral vagotomy. *J. Anat.*, **91**, 259–66.

Holmes, R.L. (1957*b*). Cholinesterase activity in the atrial wall of the dog and cat heart. *J. Physiol.*, **137**, 421–6.

Johnston, B.D. (1968). Nerve endings in the human endocardium. *Am. J. Anat.*, **122**, 621–30.

Karnovsky, M.J. & Roots, L. (1964). A 'direct-coloring' thiocholine method for cholinesterases. *J. Histochem. Cytochem.*, **12**, 219–21.

Knoche, H. & Schmitt, G. (1964). Beitrag zur Kenntnis des Nervengewebes in der Wand des Sinus Caroticus. *Z. Zellforsch.*, **63**, 22–36.

Koelle, G.B. (1955). The histochemical identification of acetyl-cholinesterase in cholinergic, adrenergic and sensory neurons. *J. Pharmac. exp. Ther.*, **114**, 167–84.

Lawrentjew, B.J. (1929). Experimentell-morphologische Studien über den feineren Bau des autonomen nervensystems. I. Die Beteiligung des Vagus and der Herzinnervation. *Z. mikrosk.-anat. Forsch.*, **16**, 383–411.

Loewenstein, W.R. & Altamirano-Orrego, R. (1956). Enhancement of activity in a Paccinian corpuscle by sympathomimetic agents. *Nature (Lond.)*, **178**, 1292–3.

Malliani, A., Recordati, G. & Schwartz, P.J. (1973). Nervous activity of afferent cardiac sympathetic fibres with atrial and ventricular endings. *J. Physiol.*, **229**, 457–69.

Miller, M.R. & Kasahara, M. (1964). Studies on the nerve endings in the heart. *Am. J. Anat.*, **115**, 217–34.

Nettleship, W.A. (1936). Experimental studies of the afferent innervation of the cat's heart. *J. comp. Neurol.*, **64**, 115–33.

Nonidez, J.F. (1937). Identification of the receptor areas in the venae cavae and pulmonary veins which initiate reflex cardiac acceleration (Bainbridge's Reflex). *Am. J. Anat.*, **61**, 203–31.

Nonidez, J.F. (1941). Studies on the innervation of the heart. II. Afferent nerve endings in the large arteries and veins. *Am. J. Anat.*, **68**, 151–89.

Paintal, A.S. (1972). Cardiovascular receptors. In *Handbook of sensory physiology*, vol. III/I, *Enteroceptors*, ed. E. Neil, pp. 1–45. Springer-Verlag: Berlin.

Pannese, E. (1969). Unusual membrane-particle complexes within nerve cells of the spinal ganglia. *J. Ultrastruct. Res.*, **29**, 334–42.

Rees, P.M. (1967). Observations on the fine structure and distribution of presumptive baroreceptor nerves at the carotid sinus. *J. comp. Neurol.*, **131**, 517–48.

Santini, M. (1969). New fibres of sympathetic nature in the inner core region of Paccinian corpuscles. *Brain Res.*, **16**, 535–8.

Schoultz, T.W. & Swett, J.E. (1972). The fine structure of the Golgi tendon organ. *J. Neurocytol.*, **1**, 1–26.

Semenov, S.P. (1963). Experimental-morphological study of afferent endings of different cardiac nerves. *Arkh. Anat. Gistol. Embriol.*, **45**, 72–83.

Smirnow, A. (1895). Ueber die sensiblen Nervenendigungen im Herzen bei Amphibien und Saugetieren. *Anat. Anz.*, **10**, 737–49.

Spencer, P.S. & Schaumburg, H.H. (1973). An ultrastructural study of the inner core of the Paccinian corpuscle. *J. Neurocytol.*, **2**, 217–35.

Tranum-Jensen, J. (1975a). The ultrastructure of the sensory end-organs (baroreceptors) in the atrial endocardium of young mini-pigs. *J. Anat.*, **119**, 255–75.

Tranum-Jensen, J. (1975b). The fine structure of afferent nerve endings in the atria. *10th Int. Congr. Anat.*, Tokyo, Japan, p. 40.

Uchida, Y. (1975). Afferent sympathetic nerve fibres with mechanoreceptors in the right heart. *Am. J. Physiol.*, **228**, 223–30.

Uchida, Y., Kamisaka, K., Murao, S. & Ueda, H. (1974). Mechanosensitivity of afferent cardiac sympathetic nerve fibres. *Am. J. Physiol.*, **226**, 1088–93.

Uchida, U. & Murao, S. (1974). Afferent sympathetic nerve fibres originating in left atrial wall. *Am. J. Physiol.*, **227**, 753–8.

Wahab, N.S., Zucker, I.H. & Gilmore, J.P. (1975). Lack of a direct effect of efferent cardiac vagal nerve activity on atrial receptor activity. *Am. J. Physiol.*, **229**, 314–17.

Zapata, P. (1975). Effects of dopamine on carotid chemo- and baroreceptors *in vitro*. *J. Physiol.*, **244**, 235–51.

Zucker, I.H. & Gilmore, J.P. (1974). Evidence for an indirect sympathetic control of atrial stretch receptor discharge in the dog. *Circulation Res.*, **34**, 441–6.

Discussion

(a) When asked whether information was available from the electron microscopic study which would indicate why some receptors fire only during atrial systole and others do not, Tranum-Jensen replied that he had found discrete nerve endings at all levels in the endocardium. Some endings were closely opposed to the underlying myocardium, some lay in the middle area and some existed in a subendothelial position. Collagen fibres, principally in the lower part of the endocardium, were in close connection with cells of the underlying myocardium. The fibres were arranged in a parallel manner and entered into the structure of the end-organs. Elastic fibres have a marked tendency to be located away from the region of the receptors. Occasionally he had noted that the receptor apparatus was bridged by a thin strand of myocardium and suggested that it was possible that a receptor in this position would be stimulated when the myocardium contracted; it would clearly be stimulated early during contraction. Basically he felt that the structure of the circumscribed endings throughout the atrium was fundamentally the same.

(b) The question was asked whether it was possible to make quantitative estimates of the extent of the fluorescent staining and was it possible to decide whether the observation that noradrenaline-containing fibres coursing with the myelinated fibres was a chance phenomenon? Tranum-Jensen indicated that the plexus could contain afferent and efferent fibres and he had never seen thick myelinated fibres which were not accompanied by the fluorescent fibres. However, fluorescent fibres may join the plexus and leave it, so that the relationship between individual fibres was possibly a chance one. He could make no comment whether there was any functional influence of the adrenergic fibres upon the end-organs.

(c) The speaker was asked about the proportion of end-organs lying against the muscle and if this suggested that the proximity of the muscle to the receptors is the event which controls the discharge that occurs during systole. In addition, he was asked how the conversion of one discharge into another was achieved. Tranum-Jensen replied that he thought that most receptors lay close to the myocardium and it was exceptional to observe

them close to the endothelium. He accepted that his technique would, to a large extent, determine the observed distribution of the receptors since, as soon as an end-organ was recognised using light microscopic techniques for making the sections, he transferred to the electron microscopic analysis. Therefore, he did not feel that he had made a true population study across the endocardium. He did not know how the conversion of one pattern of discharge into another might be expected to occur.

(d) When asked to speculate on the character of the acetyl-cholinesterase staining on the end-organs, Tranum-Jensen replied that in the pig he had concluded that most of cholinesterase activity was specific according to the usual criteria. On the other hand, in the dog heart, Holmes had found that the activity was largely caused by non-specific cholinesterase activity. Tranum-Jensen had additionally worked on the dog heart and also found it to be non-specific in character. He did not wish to imply that the enzyme had the same functional role in the two animals and the reasons for the differences were not at all clear. When asked whether there were cholinergic vesicles in the efferent axons he replied that, when using normal microscopic preparations it was very difficult to distinguish with any real certainty between cholinergic and adrenergic fibres. The presence of small-core vesicles was reasonable evidence that the axon was adrenergic. However, if they were not present the axon could still be adrenergic and, if one really wanted to solve the problem, it required an approach using other techniques, e.g. 5-hydroxydopamine labelling.

(e) There was some discussion as to whether there may be some form of re-setting of atrial receptors by efferent fibres. The comment was made that, in the dog, data were not consistent with there being an efferent control from either vagal or the sympathetic fibres. Gilmore commented that in his laboratory they had looked at type A atrial receptors in the dog and had been unable to observe any change in activity when the sympathetic nerves were stimulated, noradrenaline infused, or large alterations made in blood volume. Recordati talked of experiments in the cat which indicated that stimulation of efferent sympathetic or vagal fibres, or infusion of acetylcholine always affected the discharge of type A atrial receptors. In discussion, it became clear that it was difficult to differentiate between direct effects of efferent stimulation upon the receptors themselves and any influence of stimulation upon the muscle, e.g. an alteration in 'contractility' of the atrial muscle.

(f) There was some discussion as to whether nerve terminals were labile and, whether there was some substantial 'turnover' of nerve terminals.

Tranum-Jensen indicated that there was evidence in the literature that in a variety of the mechanoreceptors, e.g. Meissner's corpuscles, there was considerable turnover of the neural material within nerve terminals and he suggested that not only was there a turnover of organelles but also there was a 'turnover' of terminals. In reply to a question as to whether there was evidence of any branching or terminal sprouting from axons, the speaker replied that he had had no real opportunity to investigate this aspect. The point was made that the ultrastructure studies had been performed on very young animals, aged between 5 and 8 weeks, and if degenerative activities were occurring in such young pigs it might suggest that it is a process which takes place throughout life. Evidence from studies by other workers on ganglion cells in new-born mice which showed changes in the perikaryon of the ganglion cells supported the notion that degenerative activity is part of a normal cyclical process.

(g) The question was raised as to whether there was any correlation between the incidence of smooth muscle fibres and the close proximity of afferent and efferent fibres or whether the presence of smooth muscle was correlated to the receptor apparatus. The point was made that the field of observation under the electron microscope was too narrow to be certain whether this was a constant relationship. Tranum-Jensen had the impression that the smooth muscle was not consistently related to the receptor apparatus.

3. Comparisons of the fine structure of receptor end-organs in the heart and aorta

A. YAMAUCHI

The receptor end-organs in the heart and aorta have been intensively studied by light microscopic examination of neurohistological preparations from a variety of vertebrates (Smirnow, 1895; Dogiel, 1898; Woollard, 1926; Nonidez, 1935, 1937, 1941; Seto, 1935; Nettleship, 1936; Hollinshead, 1939, 1940; Mitchell, 1956; Coleridge, Hemingway, Holmes & Linden, 1957; Muratori, 1962; Miller & Kasahara, 1964; Ábrahám, 1969). These studies are of great significance since they established the gross outlines of the structure and innervation of the aortico-cardiac receptors: evidence from degeneration experiments was presented, e.g. that receptor end-organs in the aortic region are supplied by sensory axons whose cell bodies are located in the inferior vagal (nodose) ganglion, and those in the heart by vagal axons of the same origin in addition to numerous spinal sensory axons derived from the perikarya located in the upper five thoracic root ganglia.

Baroreceptor end-organs in the heart, aorta and other vascular beds, including the carotid sinus, were also revealed by light microscopy to have the common morphological feature of being represented by extensive terminal branching of a thick myelinated nerve fibre accompanied by numerous Schwann-like cell nuclei. It is to be noted that although some of these morphological findings have been well confirmed and further studied by electron microscopic examination of the baroreceptor end-organs in the carotid sinus (Knoche & Schmitt, 1964; Rees, 1967; Chiba, 1972) and in the heart atrium (Tranum-Jensen, 1975), the corresponding ultrastructural work on the aorta has been lacking. One of the main objectives of this chapter was, therefore, to present the results of our recent studies on the fine structure of the baroreceptor end-organ in the rat aorta, and to make comparisons with the intracardiac baroreceptors.

Concerning the chemosensory apparatus, there have been several electron microscopic studies of its end-organs located in the aortic region, the aortic bodies* (Knoche & Schmitt, 1963; Abbott & Howe, 1972; Hansen & Yates, 1975; Yamauchi, 1976), which supplemented the previous finding by the light microscopists that the end-organs are usually unmyelinated terminals of nerve fibres intimately associated with the paraganglionic epithelioid cells mass and are thus quite comparable to the carotid bodies (see also Heymans & Neil, 1958; Comroe, 1964; Purves, 1975). However, in view of the recent progress in synaptic morphology of the carotid bodies (cf. Hess & Zapata, 1972; Butler & Osborne, 1975; King, King, Hodges & Henry, 1975; McDonald & Mitchell, 1975), it seemed to us necessary to do further observations, at least on normal aortic bodies, in search of evidence that the paraganglionic cells in them were in reciprocal synapse with sensory axon terminals. We hoped also to try to distinguish the chemosensory axons from the efferent axons in the aortic body, in view of the previous finding that sympathetic preganglionics also innervate a minority of the paraganglionic cells in the carotid body (McDonald & Mitchell, 1975).

It has been widely recognised that paraganglionic cells exist also in the heart. In common with the cells of similar type which constitute carotid and aortic bodies, they show the property of storing catecholamines within their cytoplasm (Jacobowitz, 1967; Ehinger, Falck, Persson & Sporrong, 1968; Ellison, 1974; van der Zypen, 1974), and they are in general richly innervated. The intracardiac paraganglionic cells at times are closely intermingled, as in the carotid and aortic bodies, with blood capillaries forming tiny glomera which may be considered as being the cardiac chemoreceptor end-organs (Ellison & Hibbs, 1974; Papka, 1974). On the other hand, at least some of the intracardiac paraganglionic cells have been shown to have direct contacts and/or reciprocal synapses with postganglionic neurons of the cardiac ganglia, suggesting that these cells act as an adrenergic interneuron involved in the efferent autonomic pathways to the heart (Jacobowitz, 1967; Yamauchi, Fujimaki & Yokota, 1975; Yamauchi, 1976).

The latter part of this chapter will be devoted to detailed descriptions of

---

* The terminology proposed by Howe (1956) and revised by Coleridge, Coleridge & Howe (1970) is adopted in this chapter. These authors classified the aortic bodies into four groups according to their position, blood supply and innervation: groups 1 and 2 are located at the roots of the right and left subclavian arteries, respectively; group 3 comprises the bodies on the ventral surfaces of the arch of aorta, ductus arteriosus and pulmonary bifurcation, and group 4 bodies are distributed between the ascending aorta and the pulmonary trunk.

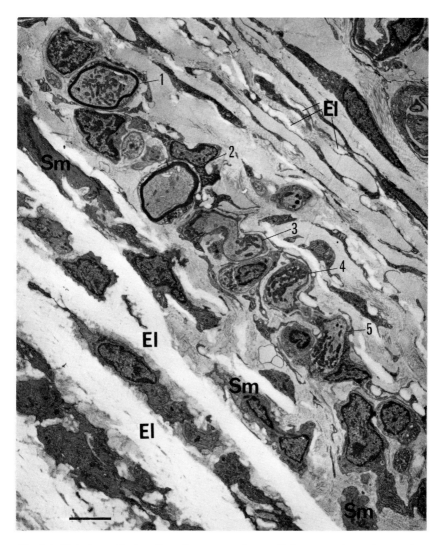

Fig. 3.1. This shows the location of a presumptive aortic baroreceptor end-organ in the rat in three dimensions. The tunica media and the tunica adventitia of the arch of the aorta are to the lower left and to the upper right in the figure, respectively. Note that numerous mitochondria are contained within both the myelinated and non-myelinated segments of the sensory axon terminal, and that eight profiles of Schwann-like cells with nuclei are shown to accompany the axon terminal. El: elastic fibres and fibrils. Sm: smooth muscle cell. The numbers 1–5 indicate axon profiles. Scale: 5 μm.

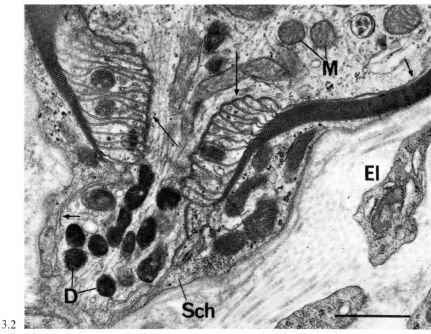

3.2

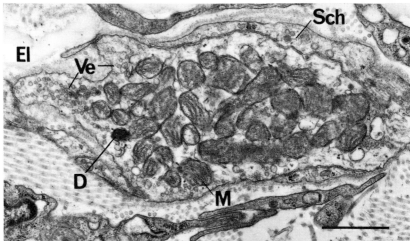

3.3

Fig. 3.2. This shows a hemi-node formed by the last myelin sheath of a sensory axon terminal in the rat aorta. An undercoating of dense axoplasmic material (at long arrows) occurs in the hemi-node region, but is absent elsewhere (short arrows). M: mitochondria, D: dense lysosomic bodies in the axoplasm. Sch: Schwann-like cell cytoplasm, El: elastic fibril. The scale represents 1 μm.

Fig. 3.3. An unmyelinated segment of the presumptive aortic barosensory axon

the ultrastructure of dog aortic bodies, and to discussion of similarities and dissimilarities between the chemoreceptor end-organs in the heart and those in the aorta.

Ultrastructure of the presumptive baroreceptors in the rat aorta

Fig. 3.1 shows a low-power electron micrograph of the terminal segment (3–7 μm thick) of a myelinated axon located at the medio-adventitial border in the wall of the arch of the aorta of a young rat. We consider this axon terminal to be one of the aortic baroreceptors, in view of its massive content of mitochondria and its being in a connective tissue space rich in collagenous and elastic fibrils without any obvious relation to the blood capillaries or to the autonomic effectors in the wall of the aorta, such as smooth muscle cells and postganglionic nerve cell bodies. Also, it is to be noted that the axon is accompanied by a remarkably higher concentration of the Schwann-like cells (in a row) than is usual for the typical Schwann cells investing the terminal segments of the autonomic efferent axons.

In order to study the fine structure of this complex of baroreceptor axon and Schwann-like cells (baroreceptor end-organ), we performed a serial sampling at 0.5 ± 0.4 μm intervals of the tissue block for a distance of 50 μm approximately, in which 500 thin sections were cut serially and mounted five per Formvar-coated 150-mesh grid. Examination of the 100 recordings of the relevant profiles of the baroreceptor end-organ, made at 8500 times magnification from each of the successive grids, showed the axon profiles labelled 1–5 in Fig. 3.1 all to be continuous with each other, and showed the maximum length of an unmyelinated segment of sensory axon terminal detectable by this method to be about 57 μm. At the hemi-node formed by the last myelin sheath, the axonal plasma membrane had a dense undercoating of axoplasmic material (Fig. 3.2) which was similar to that found previously in the transitional zone between the myelinated and ummyelinated segments of 1A axons that supply human muscle spindles (Kennedy, Webster & Yoon, 1975).

The presumptive baroreceptor axons in the rat aorta contain, besides a

---

(caption continued)

terminal is shown to have a mass of mitochondria (M), a considerable number of small vesicles (Ve) of rather uniform size, as well as a dense lysosomic body (D). Sch: Schwann-like cell process with numerous pinocytotic vesicles, El: elastic fibril. The scale represents 1 μm.

large accumulation of mitochondria, many of which show longitudinal cristae, dense lysosomic bodies, smooth-surfaced endoplasmic reticulum, moderately high numbers of the small vesicles (40–60 nm in size) with contents of low to medium electron opacity, and a few larger, dense-cored vesicles. In the sectional level of sensory axon that is depicted in Fig. 3.3, the total axoplasmic area as measured by means of a planimeter was 9.6 μm^2 and concentrations per μm^2 axoplasm of the mitochondria, small vesicles, and large dense-cored vesicles were 3.5, 24.1 and 1.4, respectively. Within the axoplasm, a tendency was noted that small vesicles are especially numerous near the axonal surface which is not covered by Schwann-like cells and is exposed to the connective tissue space at the medio-adventitial border in the aortic wall. Basal lamina substance surrounds the axonal surface devoid of Schwann-like cells, except where collagenous fibrils are directly attached to a portion of the naked axolemma. The plasma membranes of a naked sensory axon containing a vesicle cluster and a smooth muscle cell located in the outermost layer of the media in the wall of the aorta were observed to come as close to each other as 130 nm.

No efferent autonomic axons were identified in the periphery of the baroreceptor end-organ in the rat aorta. Although there occurred small axonal profiles containing rather diffusely scattered vesicles, these often turned out to be tips of the naked part (containing a vesicle cluster) of typical sensory axons. However, the presumed cholinergic and adrenergic efferent axons, which were much thinner than the baroreceptor axons and contained dense aggregations of small granular or agranular vesicles, did occur often in pairs within the adventitial connective tissue of the aorta, but did not show any intimate relation to the baroreceptor end-organ in the aorta.

The Schwann-like cells in the baroreceptor end-organ resemble typical Schwann cells in that they invest sensory axon terminals, but the former show the peculiarity of more perikaryal cytoplasm than the latter. This cytoplasm contains the well-developed Golgi apparatus, numerous free and membrane-bound ribosomes, and the dilated cisterns of the ergasto-plasm. Also, the pinocytotic vesicles are remarkably numerous within the attenuated cytoplasmic processes from these cells that embrace the sensory axon terminals (Fig. 3.3).

The above observations on the aortic baroreceptor end-organ in the rat appear to indicate that its fine structural features are nearly identical to those of the baroreceptor end-organ in the carotid sinus of the rabbit and dog (Knoche & Schmitt, 1964; Rees, 1967; Chiba, 1972). However, a remarkably high content of small vesicles of fairly uniform size was notable especially in the denuded portions of the sensory axons found in the present material. It is to be added in this respect that a similar finding has been made in this laboratory in the mechanosensory axon terminals supplying the rat muscle spindle, which are of a thickness comparable with the aortic baroreceptive axon terminals and are naked, being devoid of Schwann cell coverings for most of their paths along the surface of intrafusal muscle fibres. The significance of the small vesicles contained within the mechano-receptor axon terminals in general is to be clarified by future studies.

Comparisons between the baroreceptor end-organs in the heart and those in the aorta

The work of Tranum-Jensen (1975; also see Chapter 2 of this volume) has provided us with a detailed knowledge of the ultrastructure of endocardial baroreceptor end-organs in the mini-pig atrium. According to this author, the end-organ is a flattened structure in a plane parallel to the surface of the endocardium and is composed of a 4–9 μm thick axon which loses its myelin sheath before it enters and arborises inside an aggregate of Schwann-like cells to form a large number of mechanosensitive terminals. The latter have been shown to contain a profusion of mitochondria, glycogen granules associated with cisterns of smooth-surfaced endoplas-mic reticulum, microtubules, filaments, dense lysosomic bodies, as well as a few small clear and large dense-cored vesicles. A point was made also that small axonal profiles containing dense accumulations of vesicles of uniform size were regularly present both around the thick sensory axon and in the periphery of the end-organs, which suggests a double innervation of the atrial baroreceptor.

In order to confirm this information about the cardiac baroreceptor, we made an attempt to find similar structures in the rat and dog. Three examples (Figs 3.4–3.6) of a thick, unmyelinated axon containing many mitochondria are from the subepicardial connective tissue adjacent to the sino-atrial node region in three different animals. These large axons loaded with mitochondria are interpreted, together with those previously reported to occur in the guinea pig atrium (Chiba, 1973, cited by Yamauchi, 1973) and the human atrium and ventricle (Chiba & Yamauchi, 1970), to be a portion of the cardiac baroreceptor because of their characteristic intra-axonal content and the large size of these axons. It seems notable that, whereas in the aorta there was no evidence for the presumptive barorecep-

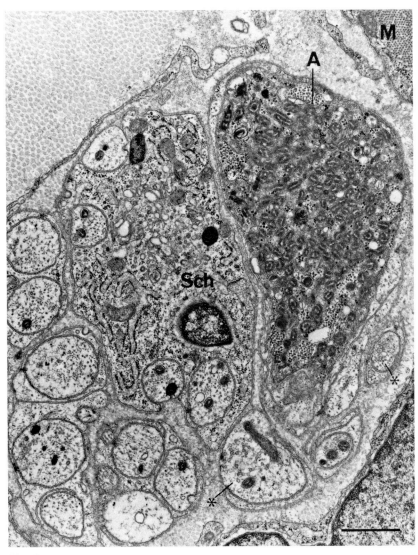

Fig. 3.4. A presumptive barosensory axon (A) encountered in subepicardial connective tissue in the sino-atrial region of a rat heart. Besides a Schwann-like cell (Sch), it is accompanied by a number of small axons some of which contain aggregations of small vesicles, indicative of their efferent nature (*). M: myocardial cell. The scale represents 1 μm.

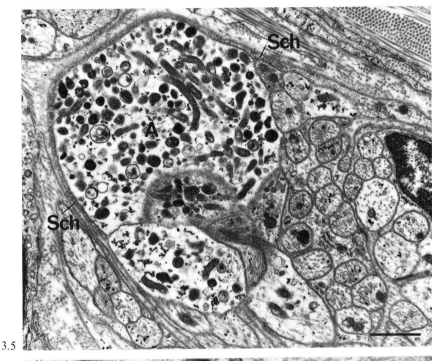

3.5

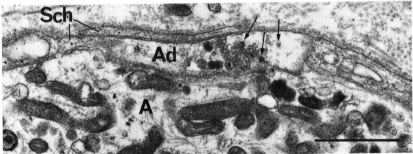

3.6

Fig. 3.5. From the atrium of a dog. A presumptive barosensory axon (A) shares a common Schwann sheath (Sch) with a number of thin, unmyelinated axons during its course in the subepicardial connective tissue. The scale represents 1 μm.

Fig. 3.6. From the atrial epicardium of another dog. A large axon (A) loaded with mitochondria and dense bodies is in direct contact with an adrenergic axon (Ad) which contains at least three small granular vesicles (arrows). Sch: Schwann sheath. The scale represents 1 μm.

tor axons being closely associated with efferent axons, those within the heart are nearly always accompanied by a number of small unmyelinated axons, many of which are probably efferents in view of their size and their occasional inclusion of masses of small vesicles. It has been recognised for some time, indeed, that the terminal portions of the adrenergic, cholinergic and sensory axons in the heart are often together enclosed by a common Schwann sheath in connective tissue (see Yamauchi, 1973). Furthermore, there are many cases in the heart where a presumptive sensory axon terminal is in direct contact with the accompanying small axons (Figs 3.5 and 3.6), and the case shown in Fig. 3.6 is of particular interest since it suggests the occurrence of a direct contact between a sensory axon and an adrenergic efferent axon that contains small granular vesicles.

The presumptive barosensory axons in the heart also appear to be more intensively covered by thin attenuations of Schwann-like cells, and so to possess a much smaller proportion of naked axolemma than the barosensory axons present in the aorta. What may be related to this difference is a tendency for the intra-axonal small vesicles of uniform size to be less frequently encountered in the former than the latter axons.

Ultrastructure of chemoreceptor end-organs in the arch of the aorta (group 3 aortic bodies) of the dog

A puppy, 3 days of neonatal age, was used. A 50 μm tissue distance of the dog aortic body located at the outer margin of the adventitia of the arch of aorta (Fig. 3.7), was studied by means of serial sampling at 0.5 ± 0.4 μm intervals of 0.1 μm sections for electron microscopic examination. In another series of 300 serial sections, each group of two consecutive sections was mounted on successive, Formvar-coated, one-hole grids and all of the sections were examined for relevant profiles of one particular paraganglionic cell that was lying in close proximity to the perikaryon of microganglia in the aortic region. Since the paranganglionic cells invariably possess intracytoplasmic membrane-bound granules, they have been often referred to as 'granule-containing (GC) cells' by electron microscopists; the latter term is also used in the following.

There seem to be at least two types of axonal endings on the GC cells in the dog aortic body. One, the chemosensory-type endings, is much more frequently encountered than the other, the efferent-type ending: all of the 44 GC cells studied had a direct contact at least with the former type endings whereas only 6 out of the 44 GC cells had contact with the latter.

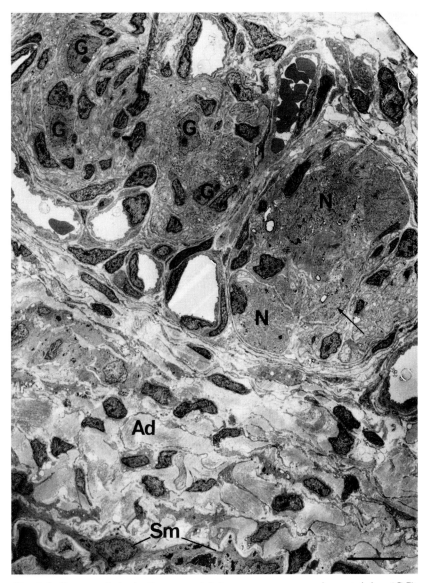

Fig. 3.7. Dog aortic body tissue, containing numerous granule-containing (GC) cells (G) closely associated with blood capillaries and nervous elements, is shown to be located at the outer margin of the adventitia (Ad) of the arch of the aorta near its junction with the ductus arteriosus. A GC cell (arrow) is lying near the nerve cell bodies (N) of a microganglion. Sm: smooth muscle cell. Scale: 10 μm.

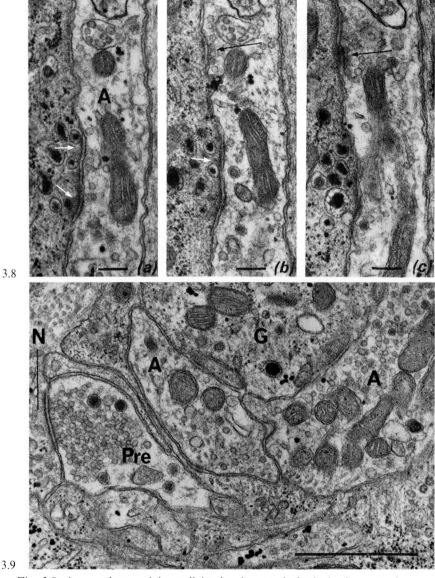

3.8

3.9

Fig. 3.8. A granule-containing cell in the dog aortic body is shown to form reciprocal synapses (at white and black arrows) with a chemosensory-type axon (A). Levels *(a)* and *(b)* are from a group of five thin sections mounted on a grid, and the level *(c)* is from the next group of five sections. Scale: 0.2 μm.

Fig. 3.9. This shows typical appearances of chemosensory-type axons (A) related to

Additionally, the chemosensory-type endings are larger in size and contain greater numbers of mitochondria (2.9 ± 1.6 per μm^2 axoplasmic area: means $\pm$ standard deviation of 20 endings) and lower concentrations of small clear vesicles (30.8 ± 15.6), as compared with the efferent-type endings (concentrations of mitochondria and small clear vesicles being 0.6 ± 1.1 and 119.1 ± 23.4, respectively). The chemosensory-type axon in the aortic bodies, like that in the carotid body, connects through reciprocal synapses to the GC cells in the chemoreceptor end-organ (Fig. 3.8*a–c*). As seen in the present observations, 14 out of 55 GC cells examined were in reciprocal synaptic junction with the chemosensory-type axon, while 3 had only a synapse to a chemosensory-type axon, 12 had only a synapse from a chemosensory-type axon, and 3 had synapses with either of the two kinds of polarities, but in which reciprocal relationships between pairs of GC cell and axon were not confirmed.

The efferent-type axon formed axons to GC cell synapses only. Of the four cells that were postsynaptic to efferent-type axons, however, two were shown to be also in reciprocal synapse with chemosensory-type axons. The small clear vesicles within efferent-type axons, like those contained within the efferent preganglionic axons terminating on the nearby nerve cell bodies (Fig. 3.9), were densely packed in more or less localised areas of the axoplasm, showing a contrast with the vesicles that were rather evenly distributed within the chemosensory-type axons. The efferent-type axons terminating on the GC cells in the dog aortic body contained vesicles that were 13–15% smaller in size than the vesicles within the chemosensory-type axon (Fig. 3.10*a–b*).

With respect to the connection between GC cells and the postganglionic neurons of microganglia in the aortic bodies no instances were encountered where the two elements made a direct contact or a synapse. Even in the closest proximity of a GC cell process to a postganglionic neuron the former was separated from the perikaryon of the latter by a gap of about 80 nm in width which was filled by a thin layer of satellite cell cytoplasm.

(caption continued)

a granule-containing cell (G), as well as of a preganglionic axon terminal (Pre) in synapse with a neuronal perikaryon (N), in the dog aortic body tissue. Note a content of numerous mitochondria and scattered vesicles within the chemosensory-type axon, and a rather dense packing of vesicles within the latter axon. The scale represents 1 μm.

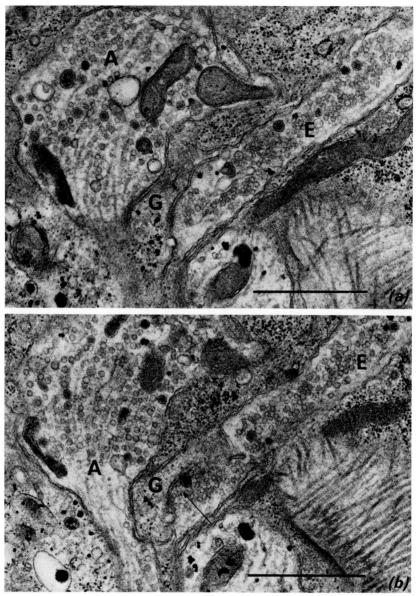

Fig. 3.10. Two sectional levels are used to show that an efferent-type axon (E) in the aortic body of a dog, in level *(b)*, is presynpatic (arrow) to a granule-containing cell (G) which becomes identifiable as such because of its content of a membrane-bound granule in level *(a)*. Note, also, a difference in size between the small vesicles contained within chemosensory-type axons (A) and those within efferent-type axons. The scale represents 1 μm.

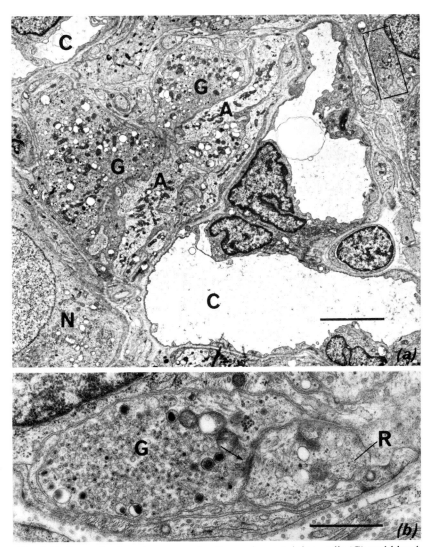

Fig. 3.11. Part of a tiny glomus, made of granule-containing -cells (G) and blood capillaries (C) and located in the sino-atrial region of a dog heart, is shown. The G cells are supplied by chemosensory-type axons (A). Boxed area in *(a)* is enlarged in *(b)*, where a GC cell process (G) is shown to be presynaptic to a neuronal process that contains ribosomes (R) and is therefore likely to be the dendrite of one of the nearby postganglionic neurons. Scale in *(a)* represents 5 μm and in *(b)* 1 μm.

**Comparisons between the chemoreceptor end-organs in the heart and
those in the aorta**

It seems reasonable to assume that the tiny glomera distributed within the
heart may be the end-organs of physiologically detectable cardiac chemoreceptors (Armour, 1973) in the cases where the intracardiac glomera are
composed of GC cell groups intimately associated with blood capillaries
and are innervated by chemosensory-type axons, thus showing similarities
in morphological features to the glomera constituting the carotid and
aortic bodies. Fig. 3.11a shows that this situation exists in the dog atrium.
The chemosensory-type axons in it resemble the ones found in the carotid
and aortic bodies of the same animal species in having large profiles and in
containing numerous mitochondria, together with a number of small
vesicles, distributed rather evenly within the axoplasm. It is to be noted also
that the presumptive chemoreceptor end-organ in the dog heart is located
in close proximity not only to the blood capillaries but also to the
postganglionic neurons of the cardiac ganglia.

Some of the GC cells contained in the cardiac ganglia of the rat and
turtle, on the other hand, have been demonstrated to be in reciprocal
synaptic junction with the axon terminals, which contain small clear
vesicles (Yamauchi, Fujimaki & Yokota, 1975; Yamauchi, Yokota &
Fujimaki, 1975). In the turtle, most of these terminals survive the bilateral
removal of the ganglion trunci vagi (homologous to the nodose ganglion in
mammals) as well as the removal of the thoracic spinal cord, which strongly
suggests that they are efferent and vagal postganglionic. In this respect,
however, there may also be a possibility of an intracardiac location of the
perikaryon of sensory neurons. The axon terminals that form reciprocal
synapses with the intracardiac GC cells in the turtle are as heavily (the
concentration of vesicles being 122 ± 49 per μm^2 axoplasmic area; mean $\pm$ -
standard deviation of 20 samples) vesiculated as the preganglionic axon
terminal to the cardiac ganglia ($159 \pm 44 \ \mu m^{-2}$) and show a lower concentration of intra-axonal mitochondria ($2.15 \pm 1.07 \ \mu m^{-2}$) than the latter
($3.34 \pm 1.57 \ \mu m^{-2}$), thus exhibiting features different from the typical
chemosensory-type of axonal ending.

Concerning the case in the rat, the concentrations per unit axoplasm of
mitochondria and small clear vesicles in the axonal endings which were in
reciprocal synapses with intracardiac GC cells were 1.67 ± 0.87 and
37.8 ± 17.23, respectively (mean $\pm$ standard deviation of 20 samples).
Because these concentrations are somewhat similar to those in the

chemosensory-type axons seen in the aortic bodies, a possibility must be considered that GC cells in the rat cardiac ganglion may be innervated by chemosensory axons rather than cholinergic efferent axons as assumed previously (Yamauchi, Yokota & Fujimaki, 1975) and thus may constitute a chemoreceptor end-organ in the heart. Nevertheless, some GC cells in the heart of the dog (Fig. 3.11*b*), rat and turtle (Yamauchi, 1976) are presynaptic to the postganglionic neuron, suggesting a role played by these cells in modulating the autonomic outflow toward the heart. Whether a single GC cell has synapses with both the chemosensory and the postganglionic neurons or whether the two different kinds of neurons are connected to different GC cells in the heart remains open to question. It is noted here again that the occurrence of two such GC cells, synaptically connected to both chemosensory-type and efferent-type axons, was shown in the present observations of the chemoreceptor end-organ in the dog aorta.

References

Abbott, C.P. & Howe, A. (1972). Ultrastructure of aortic body tissue in the cat. *Acta anat.*, **81**, 609–19.

Ábrahám, A. (1969). *Microscopic innervation of the heart and blood vessels in vertebrates including man.* Pergamon Press: Oxford.

Armour, J.A. (1973). Physiological behavior of thoracic cardiovascular receptors. *Am. J. Physiol.*, **225**, 177–85.

Butler, P.J. & Osborne, M.P. (1975). The effects of cervical vagotomy (decentralization) on the ultrastructure of the carotid body of the duck, *Anas platyrhynchos. Cell Tiss. Res.*, **163**, 491–502.

Chiba, T. (1972). Fine structure of the baroreceptor nerve terminals in the carotid sinus of the dog. *J. Electron Microsc.*, **21**, 139–48.

Chiba, T. & Yamauchi, A. (1970). On the fine structure of the nerve terminals in the human myocardium. *Z. Zellforsch.*, **108**, 324–38.

Coleridge, H.M., Coleridge, J.C.G. & Howe, A. (1970). Thoracic chemoreceptors in the dog. *Circulation Res.*, **26**, 235–47.

Coleridge, J.C.G., Hemingway, A., Holmes, R.L. & Linden, R.J. (1957). The location of atrial receptors in the dog: a physiological and histological study. *J. Physiol.*, **136**, 174–97.

Comroe, J.H., Jr (1964). The peripheral chemoreceptors. In *Handbook of Physiology*, sect. 3, vol. 1, ed. W. O. Fenn & H. Rahn, pp. 557–83. Am. Physiol. Soc.: Washington, D.C.

Dogiel, A.S. (1898). Die Sensiblen Nervenendigungen im Herzen und in dem Blutgefassen der Saugetiere. *Arch. mikrosk. Anat.*, **52**, 44–68.

Ehinger, B., Falck, B., Persson, H. & Sporrong, B. (1968). Adrenergic and cholinesterase-containing neurons of the heart. *Histochemie*, **16**, 197–205.

Ellison, J.P. (1974). The adrenergic cardiac nerves of the cat. *Am. J. Anat.*, **139**, 209–26.

Ellison, J.P. & Hibbs, R.G. (1974). Catecholamine-containing cells of the guinea pig heart: an ultrastructural study. *J. mol. cell. Cardiol.*, **6**, 17–26.

Hansen, J.T. & Yates, R.D. (1975). Light, fluorescence and electron microscopic studies of rabbit subclavian glomera. *Am. J. Anat.*, **144**, 477–90.

Hess, A. & Zapata, P. (1972). Innervation of the cat carotid body: normal and experimental studies. *Fedn Am. Socs exp. Biol.*, **31**, 1365–82.

Heymans, C. & Neil, E. (1958). *Reflexogenic areas of the cardiovascular system.* J. & A. Churchill: London.

Hollinshead, W.H. (1939). The origin of the nerve fibers to the glomus aorticum of the cat. *J. comp., Neurol.*, **71**, 417–26.

Hollinshead, W.H. (1940). The innervation of the supracardial bodies in the cat. *J. comp. Neurol.*, **73**, 37–47.

Howe, A. (1956). The vasculature of the aortic bodies in the cat. *J. Physiol.*, **134**, 311–18.

Jacobowitz, D. (1967). Histochemical studies of the relationship of chromaffin cells and adrenergic nerve fibres to the cardiac ganglia of several species. *J. Pharmac. exp. Ther.*, **158**, 227–40.

Kennedy, W.R., Webster, H. de F. & Yoon, K.S. (1975). Human muscle spindles: fine structure of the primary sensory ending. *J. Neurocytol.*, **4**, 675–95.

King, A.S., King, D.Z., Hodges, R.D. & Henry, J. (1975). Synaptic morphology of the carotid body of the domestic fowl. *Cell Tiss. Res*, **162**, 459–73.

Knoche, H. & Schmitt, G. (1963). Über Chemo- und Pressoreceptorenfelder am Coronarkreislauf. *Z. Zellforsch.*, **61**, 524–60.

Knoche, J. & Schmitt, G. (1964). Beitrag zur Kenntnis des Nervengewebes in der Wand des Sinus caroticus. *Z, Zellforsch.*, **63**, 22–36.

McDonald, D.M. & Mitchell, R.A. (1975). The innervation of glomus cells, ganglion cells and blood vessels in the rat carotid body: a quantitative ultrastructural analysis. *J. Neurocytol.*, **4**, 177–230.

Miller, M.R. & Kasahara, M. (1964). Studies on the nerve endings in the heart. *Am. J. Anat.*, **115**, 217–34.

Mitchell, G.A.G. (1956). *Cardiovascular innervation.* E. & S. Livingstone: Edinburgh.

Muratori, G. (1962). Histological observations on the cervico-thoracic paraganglia of amniotes. *Archs int. Pharmacodyn. Thér.*, **140**, 217–26.

Nettleship, W.A. (1936). Experimental studies on the afferent innervation of the cat's heart. *J. comp. Neurol.*, **64**, 115–33.

Nonidez, J.F. (1935). The aortic (depressor) nerve and its associated epithelioid body, the glomus aorticum. *Am. J. Anat.*, **57**, 259–301.

Nonidez, J.F. (1937). Distribution of the aortic nerve fibers and the epithelioid bodies (supracardial 'paraganglia') in the dog. *Anat. Rec.*, **69**, 299–317.

Nonidez, J.F. (1941). Studies on the innervation of the heart. II. Afferent nerve endings in the large arteries and veins. *Am. J. Anat.*, **68**, 151–89.

Papka, R.E. (1974). A study of catecholamine-containing cells in the hearts of fetal and postnatal rabbits by fluorescence and electron microscopy. *Cell Tiss. Res.*, **154**, 471–84.

Purves, M.J. (1975). *The peripheral arterial chemoreceptors.* Cambridge University Press: London.

Rees, P.M. (1967). Observations on the fine structure and distribution of presumptive baroreceptor nerves at the carotid sinus. *J. comp. Neurol.*, **131**, 517–48.

Seto, H. (1935). Ueber zwischen Aorta und Arteria pulmonalis gelegene Herzparaganglien. *Z. Zellforsch.*, **22**, 213–31.

Smirnow, A. (1895). Ueber die sensiblen Nervenendigungen im Herzen bei Amphibien und Säugetieren. *Anat. Anz.*, **10**, 737–94.

Tranum-Jensen, J. (1975). The ultrastructure of the sensory end-organs (baroreceptors) in the atrial wall of young mini-pigs. *J. Anat.*, **119**, 255–75.

Woollard, H.H. (1926). The innervation of the heart. *J. Anat.*, **60**, 345–73.

Yamauchi, A. (1973). Ultrastructure of the innervation of the mammalian heart. In *Ultrastructure of the mammalian heart*, ed. C. E. Challice & S. Virágh, pp. 127–78. Academic Press: New York.

Yamauchi, A. (1976). Ultrastructure of chromaffin-like adrenergic interneurons in the autonomic ganglia. In *Functional morphology of chromaffin, enterochromaffin and related cells*, ed. R. E. Coupland & T. Fujita, pp. 128–30. Elsevier: Amsterdam.

Yamauchi, A., Fujimaki, Y. & Yokota, R. (1975). Reciprocal synapses between cholinergic postganglionic axon and adrenergic interneuron in the cardiac ganglion of the turtle. *J. Ultrastruct. Res.*, **50**, 47–57.

Yamauchi, A., Yokota, R. & Fujimaki, Y. (1975). Reciprocal synapses between cholinergic axons and small granule-containing cells in the rat cardiac ganglion. *Anat. Rec.*, **181**, 195–210.

Zypen, E. van der (1974). On catecholamine-containing cells in the rat interatrial septum. *Cell Tiss. Res.*, **151**, 201–18.

Discussion

(a) The speaker was asked how the region of atria containing the granule-containing cells was identified, and whether these cells were invariably associated with vagal ganglia. He replied that he usually looked at the sinus region and there was nearly always a relationship with the vagal ganglia. Further experiments on other regions of the heart, e.g. the atrial septum, were in progress.

(b) A point of view was expressed that while it is a common assumption that synaptic vesicles are always associated with a functional chemical synaptic linkage, there are examples of cells in electric fishes where this is not so. There are cells in which there is evidence of electrical coupling and which are associated with tight junctions; synaptic vesicles were described on one side of the membrane, presumably the presynaptic membrane, but evidence for chemical transmission across this synapse did not exist. The speaker agreed that he made an interpretation of chemical transmission from the presence of vesicles.

(c) It was suggested that the different type of axon described in the atrial appendage could be either sympathetic or parasympathetic. The speaker

said that he believed that it was a parasympathetic postganglionic axon which had been observed. He speculated that the granules in the granule-containing cells of the atrium contained dopamine and other catecholamines.

SECTION TWO

Electrophysiology of cardiac receptors

4. Electrophysiology of atrial receptors

A. S. PAINTAL

Types of atrial receptors

At present, four groups of functionally different sensory receptors are known to exist in the right and left atria of the heart. These are the type A and type B receptors (both with medullated fibres), the receptors with non-medullated fibres running in the vagi (Thorén, 1976; Kappagoda, Linden & Sivananthan, 1977) and the atrial endings with fibres running along with the sympathetic nerves (Uchida & Murao, 1974; Lombardi, Malliani & Pagani, 1976). In addition one should also keep in mind the intermediate type atrial receptors because although they can be regarded as extreme variations of type A and type B atrial receptors (Paintal, 1963a), it may turn out that the information they convey might be significantly different from that conveyed by the type A and type B atrial receptors.

Differences between right-sided and left-sided endings

It is essential to note that endings of one side convey information about haemodynamic changes that is different from that conveyed by the endings of the other side at any particular moment. A good example of this difference is provided by the respiratory fluctuations in activity caused by variations in inflow of blood into the two sides. As a result, the increase in activity in the left atrial type B endings during inspiration lags behind that occurring in the endings of the right side by one to three cycles. This amounts to a lag of about 0.5–1 s which is considerable as far as central processing information is concerned. Another example of an important temporal difference is provided by the responses of the type B atrial receptors during premature ventricular contractions – the right-sided ones respond less during premature ventricular contraction while the opposite occurs in the case of the left-sided endings (Fig. 4.1). There is a clear

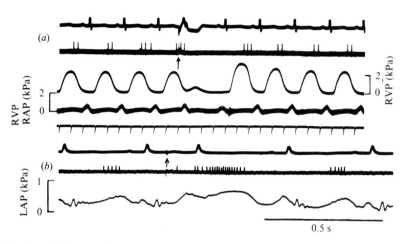

Fig. 4.1. Differences between the responses of right *(a)* and left *(b)* atrial type B receptors during premature ventricular contraction produced by application of ectopic stimuli to the right atrium at arrows. Note that the discharge is decreased during premature ventricular contraction in the right atrial receptor *(a)* and increased in the left atrial one *(b)*. The uppermost trace in each record is the ECG followed by record of impulses. The lowermost trace is that of atrial pressure. There is also a record of right ventricular pressure (RVP) in *(a)*. RAP is right atrial pressure and LAP is left atrial pressure. (From Paintal, 1963*b*.)

relation between the amount of activity and the time of occurrence of the premature contraction during the cardiac cycle. In the case of the right-sided endings (Fig. 4.2*b*) the relation is the opposite of that occurring in the left-sided endings (Fig. 4.2*a*). This interesting difference first seen in cats (Paintal, 1963*a,b*) has been confirmed by Fahim (1977) in dogs in several receptors. Indeed so consistent is the difference that one can use it to establish the location of an ending to one or the other side in dogs with intact chests (Fahim, 1977).

Possible reflex effects

Keeping in mind the differences between the right-sided and left-sided endings the aim should now be to try to study the reflex effects of each of the four groups of each side separately. This cannot be overemphasised not only because the reflex effects of those of one side will have important temporal differences from those of the other but also because the reflex effects of each type of ending of any one side may be qualitatively and

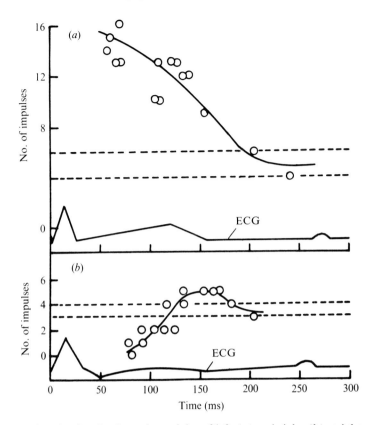

Fig. 4.2. Graphs showing how the activity of left *(a)* and right *(b)* atrial type B receptors during premature ventricular contaction varies with the position of the ectopic stimulus in the cardiac cycle. Abscissa indicates the interval between the QRS complex of the ECG and the stimulus to the right atrium. The ordinates indicate the number of impulses during premature ventricular contraction. The interrupted lines represent the range of normal activity (Paintal, 1963*b*).

quantitatively different from the effects produced by the others. It is unwise to assume that all the endings in one atrium will have identical reflex effects (e.g. Hakumäki, 1970; Arndt, Brambring, Hindorf & Röhnelt, 1971) just as it would be unwise to assume that the reflex effects of all the receptors of one muscle will be the same. Indeed it is important to learn from the fact that even though the natural stimulus for the primary and secondary endings of the muscle spindle is the same and the differences in the responses of the two are so small (as compared with the obvious and marked differences between the type A and type B atrial receptors), the reflex effects of the two are quite different from each other (see Matthews, 1972). Unfortunately, at present it is not easy to stimulate only one type of sensory receptor without affecting the others in some way. For example, although increase in the volume of the atrium is the best way to stimulate the type B atrial receptors, this stimulus can also excite to varying degrees the other three groups of receptors, i.e. the type A receptors, the receptors with non-medullated fibres (Thorén, 1976) and the receptors with fibres running along the sympathetic nerves. It is therefore necessary to design experiments which will ensure that the effects of only one type of receptor can be studied.

Differences between type A and type B receptors – gaps in knowledge

In spite of valuable new information obtained as a result of recent work, some of which has been reported in this symposium, there are still important gaps in our knowledge about certain aspects of the behaviour of atrial receptors. Notable among these is our knowledge about the natural stimulus for the type A atrial receptors and the structural basis underlying the different patterns of discharge in the type A and type B receptors. Indeed the only certain knowledge relates to the natural stimulus of the type B atrial receptors for which it is established that its natural stimulus is the pulsatile increase in the volume of the atrium (Paintal, 1953; 1963*a, b*). In the case of the type A receptors it is also certain that contraction *per se* of the atrium leading to mechanical deformation of the ending, possibly by increasing the tension on it, is an important factor in producing the initial one or more impulses in the type A receptors, because after making an allowance for total conduction time from the ending to the recording electrodes it has been shown that these impulses are produced before the *a* wave starts to rise (Paintal, 1963*a*) (Fig. 4.3). The importance of this (i.e. contraction) is highlighted by the fact that the receptors can be stimulated effectively in an atrium that is slit widely so that blood flows out freely (Paintal, 1963*a*). Finally, Recordati, Lombardi, Bishop & Malliani (1976) have shown that the discharge in type A receptors is a function of the active tension developed by atrial muscle during contraction. It thus appears that the type A receptors might signal the strength of atrial contraction. It should be noted that related factors, e.g. the direction in which contraction develops, and whether the atrium contracts against open or closed A–V valves, influence the activity of the receptor. For example, in Fig. 4.3*c* it can be seen that the contractions caused by ectopic stimuli produced a higher

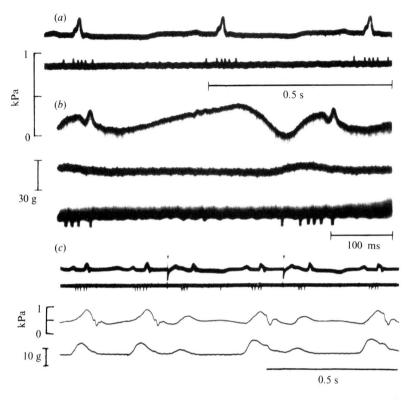

Fig. 4.3. Impulses in two fibres from atrial type A receptors: *(a)* is a section of a continuous record with ECG and impulses in a fibre from a left atrial receptor; total conduction time 12 ms; *(b)* is a sweep consisting of (from top downwards) left atrial pressure, left atrial myogram (indicating contraction of the left atrium) and impulses in the fibre. The first burst of impulses in the sweep, i.e. *(b)*, corresponds to the first burst in *(a)*. Note the first impulse of each burst is attributable to atrial contraction and not the *a* wave because it appears before the *a* wave (Paintal, 1963*a*). *(c)* is a record of activity in a right atrial type A receptor during normal and ectopic beats produced by stimulating the right atrium at the stimulus artifacts. The lower two traces represent right atrial pressure and the myogram. Total conduction time in the fibre was 8 ms. Note increased frequency of discharge during ectopic contractions in spite of smaller tension and *a* wave. Such responses suggest that these contractions occurred against partly closed A–V valves as revealed by the pressure tracing (Paintal, 1972).

frequency of discharge even though the rise in atrial pressure and strength of atrial contraction was smaller than that developed during normal contraction.

Another important gap in our knowledge relates to the mechanism

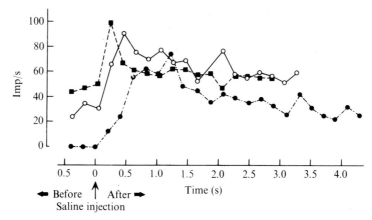

Fig. 4.4. Responses from a type A (■ – – – ■), type B (o———o) and intermediate type
(● – · – ●) atrial receptors in three cats to injection of saline at arrow into the isolated
in situ atrium containing the ending. The ordinate indicates the average frequency
of discharge (Gupta, 1977*b*).

underlying the different patterns of discharge characteristic of type A and
type B atrial receptors. At one time it was thought that differences in
adaptation rates may be an important factor (see Paintal, 1963*a*). However,
it was shown recently by Arndt, Brambring, Hindorf & Röhnelt (1974) that
stretching strips of the atrium sinusoidally produced similar responses in
both type A and type B receptors. It was felt that the identical responses
observed by Arndt *et al.* may have been caused by the fact that they used
strips of the atrium. Gupta therefore studied the effect of distending the
isolated atrium *in vivo* on both type A and type B atrial receptors and he
found that, under these conditions also, the type A receptors behaved very
much like the type B atrial receptors (Fig. 4.4) (Gupta, 1977*b*). There was
also the possibility that the pattern of discharge may be different because of
differences in the location of the two types of receptors in different parts of
the atrium. However, in a systematic study involving the punctate location
of 131 endings, Gupta did not find any decisive differences in the location
of the endings; both seemed to be concentrated at the junctions of the veins
with the atrium on both sides (Gupta, 1977*a*; this volume).

Correlation between structure and pattern of discharge

From Gupta's observations (see above) one is forced to conclude that the
differences between the type A and type B receptors must lie in certain

differences in the structural relationship of the endings to the tissues, notably muscular elements, in which the endings lie. As suggested by Whitteridge (1953) the structural relationship must be such that the type A receptors are effectively in series and the type B in parallel with the contractile elements with which the endings are associated. One should keep in mind that it is not necessary for an ending to have a strict anatomically in-series relationship to the contractile elements in order that it is stimulated during atrial contraction.

Ideally, it would be desirable to study histologically an atrial ending from which impulses had been recorded earlier. Some attempts have been made in the past in this direction. However, this approach would seem to be beset with pitfalls because even if it is assumed that exact localisation of the receptor by punctate stimulation can be made infallible (which would seem difficult in the case of the atrium), it is not possible to be certain that the ending identified histologically is the same as that from which the responses were recorded because of the density of the receptors in certain parts of the atrium such as the junction of the veins with the atrium on both sides (Nonidez, 1937). Indeed the position is even more complex because both the type A and type B receptors are located in the same region (Gupta, 1977*a*; this volume). One can appreciate the kind of difficulties involved if one considers the carotid baroreceptors. Here, with the present technical limitations it would be practically impossible to identify a receptor histologically as one from which impulses had been recorded earlier. However, with further refinements in both electrophysiological and histological techniques it may become possible in the future to study the structure of the very atrial receptor from which impulses were recorded.

Changes in pattern of discharge

Considerable attention has been paid to studying the changes in pattern of discharge in these receptors, notably from the type A pattern to the type B pattern and vice versa under unusual experimental conditions. It is not surprising that such changes can be made to occur although there are some type A receptors which, in spite of several experimental manoeuvres, continue to retain their characteristic pattern of discharge (Arndt *et al.*, 1971; Kappagoda, Linden & Mary, 1976). However, it would be more meaningful to study the changes that take place under the type of haemodynamic conditions likely to occur in animals, e.g. of the magnitude occurring during severe muscular exercise. At the same time it is important not to confuse the issue by keeping in mind that the type B receptors are functionally *quite* different from the type A receptors, as the natural stimulus for the type B receptors is the pulsatile increase in atrial volume which is definitely not the case with type A receptors. As pointed out earlier (Paintal, 1963*a*) this difference is obvious when impulses are recorded simultaneously in type A and type B receptors located in the same chamber – the type B will show typical fluctuations in activity corresponding with the respiratory variations in the inflow of blood but there will be hardly any change in the activity of the type A receptors. Arndt *et al.* (1971) and Recordati *et al.* (1976) have confirmed that changes in atrial volume have hardly any effect on type A receptors.

Role of paranatural stimuli

While considering alterations in the pattern of discharge one should keep in mind that although sensory receptors are ordinarily stimulated by their natural stimulus, they are also stimulated under certain conditions by stimuli that occur under physiological conditions but which are not the natural stimulus, i.e. paranatural stimuli (Paintal, 1976). (A good example is provided by the Pacinian corpuscle whose natural stimulus is vibration but which also responds to the pulsatile stimulus provided by the pulsations of the vessels in their vicinity (see Paintal, 1976).) This aspect should be kept in mind when considering the appearance of impulses at unusual times in the cardiac cycle in connection with the discharge patterns of the type A and type B atrial receptors. One important and useful criterion for identifying that a particular stimulus is a paranatural stimulus is that the impulses produced by it do not provide information of the kind that the receptor ordinarily supplies by virtue of the relation between the natural stimulus and the response of the ending (Paintal, 1976). For example, after haemorrhage, occasional type B receptors acquire a couple of impulses during the *a* wave of the atrial pressure curve at a time when the impulses during the *v* wave have almost disappeared (Paintal, 1963*a*). These impulses during the *a* wave are practically constant and therefore cannot be of much significance since they do not provide any information about changes in atrial volume.

Role of antidromic depression and the 'off effect'

While seeking possible explanations underlying the characteristic patterns of activity in type A and type B receptors one could look for clues provided in situations where the discharge of one type of receptor is converted into the pattern of the other. For example, if the atrium is grossly distended then some type A receptors acquire impulses during the *v* wave and when the number of these impulses becomes large then the impulses during the *a* wave may disappear, giving the impression that the receptor has become a type B receptor. The reverse can happen during severe haemorrhage in the case of some type B receptors (Paintal, 1963*a*) but this does not occur in the majority of them. It is, therefore, possible that the *v* burst of impulses are themselves responsible for depressing the ending so that *a* impulses are not generated (in the succeeding period, i.e. during the *a* wave) as it is known that orthodromic (and antidromic) impulses depress the ending. The duration of depression of the ending depends on the number and frequency of the burst (Paintal, 1959*b*). The effects of antidromic stimulation were therefore examined on arterial baroreceptors and type A and type B atrial receptors in order to estimate the possible role the *v* impulses may play in determining the generation of *a* impulses.

The vagus was stimulated at the root of the neck with stimuli that were clearly suprathreshold for the afferent fibre being examined. It was not necessary to atropinise the cat as the stimuli used were usually not strong enough to slow the heart. Moreover, the effects of the brief (100 ms) train of stimuli manifested their effect only in the subsequent cycle. The train of antidromic stimuli were applied during those cycles where the discharge of impulses remained reasonably uniform from cycle to cycle for at least three or four cycles as often obtained during the pause following expiration. In this way one could compare the effects of a train of stimuli on a particular burst with the control burst occurring in the preceding cycle as in Figs 4.5 and 4.6.

As expected (see Paintal, 1959*b*) the depression, indicated by the reduction in the number and frequency of impulses following antidromic stimulation, increased with the number and frequency of stimuli; the effect varied in different endings. Depression was more marked in type B atrial receptors because they have a lower frequency of discharge. In fact, as shown in the case of the two fibres in Fig. 4.6, the discharge could be abolished or reduced considerably by relatively high frequency repetitive stimulation.

As expected, the amount of depression also varied with the interval between the train of antidromic stimuli and the normal rhythmic burst of impulses, the smaller the interval the greater the depression. It was also seen that if the frequency of train of antidromic stimuli was much higher than the frequency of discharge of orthodromic impulses, and this train was applied during the initial part of the normally occurring burst of impulses, then the number and frequency of the remaining impulses was considerably reduced (Fig. 4.6). The reduction depended on the ratio of the frequency of repetitive stimulation to the frequency of discharge of impulses – the greater the ratio the greater the reduction (compare Fig. 4.5 with Fig. 4.6). These observations indicate that since the after effects of antidromic and orthodromic impulses are identical (Eyzaguirre & Kuffler, 1955; Diamond, Gray & Inman, 1958; Paintal, 1959a) then the discharge of impulses in the latter part of a burst is not only determined by the physical magnitude and the natural stimulus, but also by the orthodromic impulses that preceded them.

The effects of a variable train of antidromic stimuli applied at varying periods before the appearance of the *a* burst in type A receptors with no *v* impulses was noted. Those with a high-frequency *a* discharge were little, if at all, affected by a high-frequency train of stimuli applied just before the initiation of the *a* burst. In others, tetanisation at a time when *v* impulses usually appear delayed the first impulse of the *a* burst by a few ms only. Even this small effect was present only with high-frequency stimulation. Stimulation at a frequency usually encountered in the *v* burst, i.e. less than 100/s had no visible effect whatever on the *a* burst. This therefore indicates that the absence of the *a* burst under conditions when a *v* burst appears is not caused by the depressant effects of the train of *v* impulses. One has, therefore, to consider the possibility that the absence of the *a* impulses in type A receptors when *v* impulses appear during the enhanced *v* wave (e.g. during occlusion of the pulmonary artery in right atrial receptors) may be caused by the dynamic 'off effect' (Katz, 1950) occasioned by the downslope of the enhanced *v* wave. The role of this factor needs to be assessed.

Can two functionally different endings be connected to the same sensory fibre?

Finally, one has to consider the possibility suggested by Dr P. S. Rao (personal communication) that one branch of an atrial fibre might be connected to an ending responding as a type B receptor and another branch connected to an ending behaving as a type A receptor. Such a possibility has been entertained before in the case of ventricular pressure receptors (see

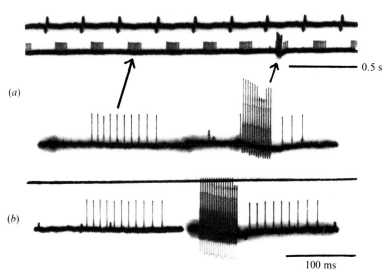

Fig. 4.5. Effect of repetitive antidromic stimulation at 320 Hz for 51 ms on an arterial baroreceptor before *(b)* and during *(a)* the initial part of the normal burst of impulses. Depression of activity in *(a)* is less than that in the small-spiked fibre in Fig. 4.6 in spite of longer duration of stimulation in *(a)*. The sections of the continuous record above to which the two sweeps in *(a)* correspond are indicated by arrows. The corresponding continuous record for *(b)* has been omitted.

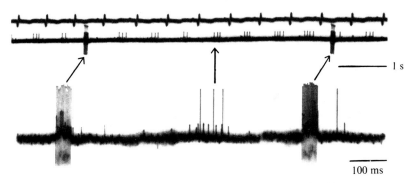

Fig. 4.6. Effect of repetitive antidromic stimulation at 300 Hz for 40 ms on two type B atrial receptors. Activity in the small-spiked fibre can be seen clearly only in the three sweeps below recorded along with the continuous record of ECG and impulses above. The sweeps correspond to the sections of the continuous record indicated by arrows. In the first sweep the train of stimuli were applied before the first impulse in the fibre with large spikes was initiated. The frequency of stimulation is about eight to nine times the frequency of discharge in the large-spiked fibre. Note the depression after stimulation.

Paintal, 1972). However, one must keep in mind that it contradicts the concepts associated with Müller's law of specific energies. Therefore, unless exceptions to the law are established beyond doubt in some simple system that leads to unequivocal conclusions, one should be cautious about entertaining such a possibility for the atrial receptors. At present it is known that although an afferent fibre may have multiple endings, e.g. the tough corpuscle of the skin (Iggo & Muir, 1969) it is generally accepted that each of the several endings are activated in the same way by the same stimulus.

Conclusions

(1) At present at least four types of sensory receptors are known to exist in the right and left atria, the type A and type B receptors with medullated fibres, the endings with vagal non-medullated fibres, and the endings with fibres running along the sympathetic nerves.

(2) There are no reasons for assuming that the reflex effects of these four types of endings are the same. The aim should therefore be to study the eight reflex mechanisms (four from each side) separately.

(3) The type A and type B atrial receptors are quite different from each other. The natural stimulus for the type B receptors is the pulsatile increase in atrial volume. The natural stimulus for the type A receptors has not been established but it is not increase in atrial volume. It is possible that the type A receptors signal strength of atrial contraction.

(4) The rate of adaptation of type A receptors to maintained distension of the atrium is the same as that of type B receptors.

(5) The type A and type B receptors are located in approximately the same regions of the atrium, i.e. mainly at the junction of the veins with the atrium.

(6) In view of (4) and (5) above it follows that the differences between the type A and type B receptors must lie in the structural relationship of the endings to the contractile elements associated with them.

(7) A change in the pattern of discharge from A-type to B-type (and vice versa) can be produced in some receptors by applying suitable experimental manoeuvres.

(8) The loss of *a* impulses in type A receptors on acquisition of *v* impulses is not caused by antidromic depression of the endings by the train of *v* impulses themselves. It could be caused by the 'off effect' engendered by the enhanced *v* wave.

References

Arndt, J.O., Brambring, P., Hindorf, K. & Röhnelt, M. (1971). The afferent impulse traffic from atrial A-type receptors in cats. Does the A-type receptor signal heart rate? *Pflügers Arch. ges. Physiol.*, **326**, 300–15.

Arndt, J.O., Brambring, P., Hindorf, K. & Röhnelt, M. (1974). The afferent discharge pattern of atrial mechanoreceptors in the cat during sinusoidal stretch of atrial strips *in situ*. *J. Physiol.*, **240**, 33–52.

Diamond, J., Gray, J.A.B. & Inman, D.R. (1958). The depression of the receptor potential in Pacinian corpuscles. *J. Physiol.*, **141**, 117–31.

Eyzaguiree, C. & Kuffler, S.W. (1955). Further study of soma, dendrite, and axon excitation in single neurons. *J. gen. Physiol.*, **39**, 121–53.

Fahim, M. (1977). A method for localizing atrial type B receptors in the dog. *Clin. exp. Pharmac. Physiol.*, **4**, 295–302.

Gupta, B.N. (1977*a*). The location and distribution of type A and type B atrial endings in cats. *Pflügers Arch. ges. Physiol.*, **367**, 271–5.

Gupta, B.N. (1977*b*). Studies on the adaptation rate and frequency distribution of type A and type B atrial endings in cats. *Pflügers Arch. ges. Physiol.*, **367**, 277–81.

Hakumäki, M.O.K. (1970). Function of the left atrial receptors. *Acta physiol. scand.*, **79** (supp. 344), 1–54.

Iggo, A. & Muir, A.R. (1969). The structure and function of a slowly adapting touch corpuscle in hairy skin. *J. Physiol.*, **200**, 763–96.

Kappagoda, C.T., Linden, R.J. & Mary, D.A.S.G. (1976). Relative occurrence of type A and type B atrial receptors in the cat. *J. Physiol.*, **256**, 26–7P.

Kappagoda, C.T., Linden, R.J. & Sivananthan, N. (1977). The receptors which mediate a reflex increase in heart rate. *J. Physiol.*, **266**, 89–90P.

Katz, B. (1950). Depolarization of sensory terminals and the initiation of impulses in the muscle spindle. *J. Physiol.*, **111**, 261–82.

Lombardi, F., Malliani, A. & Pagani, M. (1976). Nervous activity of afferent sympathetic fibers innervating the pulmonary veins. *Brain Res.*, **113**, 197–200.

Matthews, P.B.C. (1972). *Mammalian muscle receptors and their central actions*. Edward Arnold: London.

Nonidez, J.F. (1937). Identification of the receptor areas in the venae cavae and pulmonary veins which initiate reflex cardiac acceleration (Bainbridge's reflex). *Am. J. Anat.*, **61**, 203–31.

Paintal, A.S. (1953). A study of right and left atrial receptors. *J. Physiol.*, **120**, 596–610.

Paintal, A.S. (1959*a*). Intramuscular propagation of sensory impulses. *J. Physiol.*, **148**, 240–51.

Paintal, A.S. (1959*b*). Facilitation and depression of muscle stretch receptors by repetitive antidromic stimulation, adrenaline and asphyxia. *J. Physiol.*, **148**, 252–66.

Paintal, A.S. (1963*a*). Vagal afferent fibres. *Ergebn. Physiol.*, **52**, 74–156.

Paintal, A.S. (1963*b*). Natural stimulation of type B atrial receptors. *J. Physiol.*, **169**, 116–36.

Paintal, A.S. (1972). Cardiovascular receptors. In *Handbook of sensory physiology*, vol. III/I., ed. E. Neil, pp. 1–45. Springer-Verlag: Berlin.

Paintal, A.S. (1976). Natural and paranatural stimulation of sensory receptors. In *Sensory functions of the skin*, ed. Y. Zotterman, pp. 3–12. Pergamon Press: London.

Recordati, G., Lombardi, F., Bishop, V.S. & Malliani, A. (1976). Mechanical stimuli exciting type A atrial vagal receptors in the cat. *J. appl. Physiol.*, **38**, 397–403.

Thorén, P. (1976). Atrial receptors with nonmedullated vagal afferents in the cat. *Circulation Res.*, **38**, 357–62.

Uchida, U. & Murao, S. (1974). Afferent sympathetic nerve fibres originating in left atrial wall. *Am. J. Physiol.*, **227**, 753–58.

Whitteridge, D. (1953). Electrophysiology of afferent cardiac and pulmonary fibres. *Abstr. XIX int. Physiol. Congr.*, 66–72.

Discussion

(a) In answer to a question about the frequencies of antidromic stimulation applied to the nerve fibres from the atrial receptors Paintal replied that he had used stimuli varying from very low to very high frequencies (up to 300 Hz) and each frequency had been investigated at the same phase of the respiratory cycle, that is, during expiration. There appeared to be little evidence of any antidromic suppression; there was only a slight delay in the appearance of the first impulse.

(b) Recordati made a comment on the differences between type A and type B atrial receptors. He felt it important that the mechanical components of the stimulus which activated the receptors should be defined very clearly in the beating heart. These mechanical components should be characterised in terms of changes in the length of the muscle during atrial filling and contraction, and the differences in the dynamic components of the mechanical stimulus under these two circumstances. For example, the dynamic component during the atrial filling phase is relatively low compared with that during the atrial contraction phase. Possible changes in threshold should also be considered. He felt that the responses of type A and type B receptors to these aspects should be investigated. Paintal replied that while, in his view, electrophysiological properties of the receptors may be the same, he felt that what was being discussed in this session was the natural stimulus to the receptors. But he agreed that other receptor properties must also be studied very carefully.

(c) In reply to a question about the existence of atrial receptors in various species Paintal said that experiments from his laboratories indicated that the monkey did not have any type A receptors. However, this species did have a high proportion of type B receptors. Studies in his laboratory had been unable to find any conventional type A or type B receptors in the rabbit (see Discussion, K. Floyd, p. 24). Gilmore indicated that he had been able to record type A atrial receptors in the monkey. He

took an opposite view to that of Paintal on the natural stimulus of type B receptors. The argument that the natural stimulus to atrial receptors is the pulse wave of atrial pressure is based on the observation that hysteresis can be shown when impulse discharge is plotted against v wave pressure but cannot be shown when the impulse discharge is plotted against pulse pressure. Gilmore said that he had been able to demonstrate hysteresis when impulse frequency was plotted against pulse pressure as well as peak v wave pressure. He argued, therefore, that it was likely that the natural stimulus to the receptors was stretch not pulsatile pressure in the atrium. Paintal agreed with the conclusion that stretch of the atrial muscle was responsible for stimulating the type B receptors and that atrial filling was represented by the amplitude of the v wave. He was unable to account for the differences in the results obtained in the two laboratories.

(d) Ledsome made the point that the differences in the ratio of A type to B type atrial receptors in cats and dogs might be a reflection of differences in the vascular conditions in the two animals. If it were possible to set the vascular volumes and/or the heart rates to the same levels in these two animals then the ratios might well be the same.

5. Neurophysiological properties of atrial mechanoreceptors*

J. O. ARNDT

This presentation will focus mainly on the observations made in our laboratory on mechanoreceptors of the veno-atrial regions with myelinated fibres of vagal origin. Our observations convinced us, first, that the atrial afferents irrespective of their discharge type originate from identical mechanoreceptors like muscle spindles, tendon organs, and arterial baroreceptors in particular. Secondly, our experiments made us think that these receptors convey two specific types of information to the central nervous system: (1) about the fullness of the thoracic circulation and (2) of the heart rate. Obviously such a dual afferent function fits well with a large body of evidence which suggests that heart rate and also fluid balance can reflexly be influenced by receptors present in the thorax (Bainbridge, 1915; Gauer, Henry & Sieker, 1961; Ledsome & Linden, 1964; Gauer & Henry, 1976).

Admittedly, it is still a matter of debate how the functional differentiation of these receptors is exactly brought about. Nevertheless, certain differences between the two most prominent atrial bursts – the type A and type B burst as defined and classified by Paintal (1953, 1963b) – can be understood when the general principles of receptor physiology in relation to the receptors' working conditions *in vivo* are considered.

Classification of the atrial afferents

The pure A- and B-bursts originate from receptors in the atria and their adjacent vessels (Nonidez, 1937; Johnston, 1968; Coleridge, Hemingway,

* The support of the Deutsche Forschungsgemeinschaft through the Schwerpunktprogramm 'Receptorphysiologie' (Ar 64/4) and the Sonderforschungsbereich Kardiologie Düsseldorf is gratefully acknowledged.

Holmes & Linden, 1957; Gupta, 1977a), but they differ from each other in: (1) their temporal occurrence in the cardiac cycle; (2) their discharge patterns; and (3) their functional behaviour.

(1) *Difference in the temporal occurrence.* The A-burst coincides with the *a* wave of the atrial pressure curve, i.e. it occurs during the atrial contraction when atrial pressure increases in face of decreasing atrial dimensions. The earlier spikes in the burst occur prior to atrial pressure changes, and the peak burst coincides with the slope of the pressure curve rather than the peak (Paintal, 1963a, b). It is also seen that these endings generate impulses in the empty and widely slit open atria which proves that the pressure is not indispensable for their activation (Jarisch & Zottermann, 1948; Paintal, 1963a, b). In contrast the B-type burst coincides with the *v* wave of the atrial pressure curve, i.e. during the atrial filling phase when the dimension and pressure change in the same direction. Further, it is well documented that the activity is sensitive to passive stretch (Paintal, 1972).

(2) *Differences in the discharge pattern.* The differences in the discharge patterns are also noteworthy. The A-burst is characterised by a short burst of constant duration and peak frequencies around 200 Hz in the intact animal. In contrast, the B-burst is of low frequency of around 30 to 40 Hz and the duration as well as the frequency of the burst vary markedly with changing atrial mechanics (Paintal, 1972; Arndt, Brambring, Hindorf & Röhnelt, 1971).

(3) *Differences in the functional behaviour.* Both fibre categories also seem to differ functionally. This particular aspect will be discussed in more detail below, but in brief there seems to be general agreement that the B-type activity follows closely atrial stretch (Paintal, 1953, 1972) whereas the A-type burst is rather insensitive to varying atrial mechanics (Arndt *et al.*, 1971; Recordati *et al.*, 1976; Rao & Fahim, 1976).

It must be added for completion that one sometimes encounters atrial afferents which fire during both atrial systole as well as atrial diastole (Langrehr, 1960a,b; Neil & Joels, 1961). These were put by Paintal (1963a,b) into a separate category and called intermediate-type endings. They are equally distributed in the veno-atrial regions like the pure type A and type B endings (Gupta, 1977a). Further, in a study on the frequency distribution of atrial receptors they were found to be relatively small in number amounting to about 10% (Gupta, 1977*b*). It appears unlikely that they form a separate functional group. Rather they might be viewed as extreme variations of the pure type A and type B endings. Curiously, Kappagoda, Linden & Mary (1976) were able to 'convert' most of the pure bursts into intermediate types. Yet it remains to be shown whether the 'converted' fibres retain the functional characteristics of the parent burst or not, as might appear likely at least in the case of the type A-endings studied by Recordati *et al.* (1976).

Functional characteristics of the type A and type B bursts

Historically, the pioneer work on the cardiac afferent innervation was primarily undertaken in the hope of identifying the afferent arc of certain heart rate reflexes (Amann & Schäfer; 1943, Jarisch & Zottermann, 1948; Whitteridge, 1948; Neil & Zottermann, 1950; Schäfer, 1950). The complexity and also the interconvertibility of the various bursts made Schäfer (1950) in his review doubt that specific information could be gained from the cardiac receptors. Later the idea of the specificity of atrial receptors, in particular, was brought again into discussion with respect to the B-type activity. This issue was stimulated by the work of Gauer, Henry and colleagues who brought a completely new aspect into debate by recognising the particular relationship between the stretch conditions of the thoracic circulation and the excretory function of the kidney, since increasing thoracic blood volume caused a diuresis while antidiuresis occurred when the thoracic blood volume was reduced. This and the observation that left atrial balloon distension caused diuresis which was blocked by vagal cooling at 9 °C led them to think that this reflex was initiated by atrial mechanoreceptors with myelinated fibres and thus they hypothesised that this reflex plays a part in the regulation of the blood volume (see Gauer & Henry, 1976). Whatever the efferent mechanism of this reflex may be, the functional behaviour of the B-type endings is suitable for conveying information about the fullness of the thoracic circulation. This notion is supported by subsequent observations.

B-type endings

Generally the type B discharge follows closely atrial filling in the isolated atrium (Paintal, 1953; Chapman, 1958; Gupta, 1977*b*). Thus it is not too surprising to find that the type B discharge correlates well with changes in total blood volume (Henry & Pearce, 1956; Gupta, Henry, Sinclair & von

Baumgarten, 1966), with central blood volume (Langrehr & Kramer, 1960) and also with central venous pressure (Langrehr, 1960*a,b*) which by and large is a measure of central blood volume.

The quantitative relationship between blood volume changes and the activity of the B-type receptors came out clearly in experiments on closed-chest dogs (Gupta *et al.*, 1966). The responses of six B-type atrial receptors in six dogs are shown in Fig. 5.1. According to this, a change in blood volume of only 10% reduces the number of spikes per burst to 50% and a change of 20% in the blood volume produces an 80% drop in the spike count. A similar relationship holds also for the average discharge rate, which is a running average of the number of spikes generated per unit time and which in all likelihood is the information-carrying parameter (Douglas, Ritchie & Schaumann, 1956; Gupta *et al.*, 1966; Eysel & Grüsser, 1970; McKean, Poppele, Rosenthal & Terzuolo, 1970). Thus there are good reasons for the idea that the fullness of thoracic circulation can be signalled through the type B afferents.

A-type endings

Naturally, the question arises whether the other atrial activities subserve the same function. Or specifically: does the most prominent A-burst follow the same principles? Dickinson's (1950) work is often quoted as showing that the A-type activity also follows atrial stretch. This conclusion rests on one single example in which the correlation between the burst frequency and the atrial pressure is indeed strong if one considers the extreme pressure levels of 13.7 kPa. However, in the physiological range of right atrial pressures which may be 2.6 kPa at the most, the correlation does not exist. And, furthermore, possibly because of the experimental conditions in this case, i.e. chest opened with extensive surgery with removal of the lungs on one side, the discharge frequency given as 10 Hz is far lower than that which is typical in the intact animal where the average A-burst frequency lies at around 150 Hz (Arndt *et al.*, 1971). The problem involved here will be discussed in more detail later, but it should be mentioned in passing that the receptors' working range is quite important for their response. Generally, all mechanoreceptors are characterised by a non-linear stimulus–response curve with a threshold on the one end, a saturation range at the other and a more or less linear part between these two non-linearities. If *in vivo* a receptor operates in the saturation range as indicated in the case of the A fibres by the extreme frequency discharge for this fibre category, the varying stimulus condition will not have much effect. However, the same

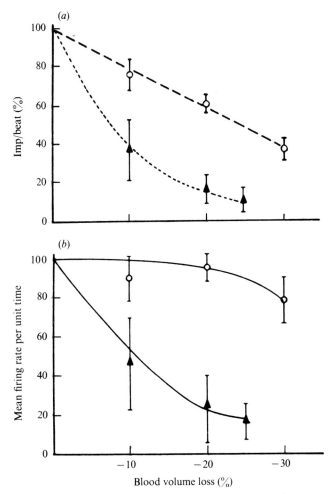

Fig. 5.1. Comparison of the mean firing rate per second of six aortic (o) and six atrial (▲) fibre preparations during stepwise reduction of blood volume. *(a)* Presents data as mean impulses per beat and *(b)* data in terms of firing rate per second, i.e. with correction for the change in heart rate. (Vertical lines indicate the s.d.) (From Gupta *et al.,* 1966.)

receiver will be rendered sensitive to the stimulus conditions when for whatever reason its operating point is shifted towards the threshold range.

This last problem applies also to the results of Homma & Suzuki (1966), who found a close correlation between A-type fibre activity and the pressure in the isolated atrium, but the maximal discharge frequency of 40 Hz at atrial pressure of 3.3 kPa is also extremely low.

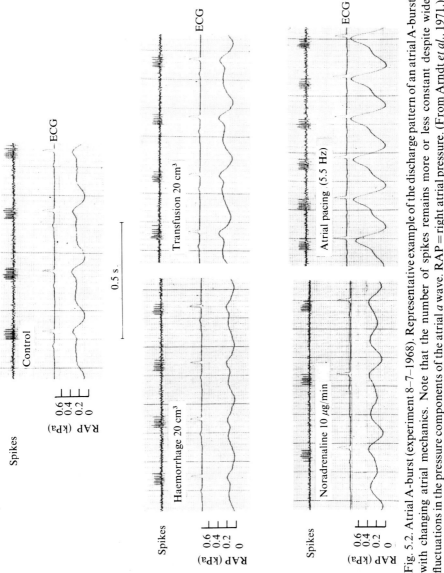

Fig. 5.2. Atrial A-burst (experiment 8–7–1968). Representative example of the discharge pattern of an atrial A-burst with changing atrial mechanics. Note that the number of spikes remains more or less constant despite wide fluctuations in the pressure components of the atrial *a* wave. RAP = right atrial pressure. (From Arndt *et al.*, 1971.)

So far the conclusions rest on experiments which distorted the *in vitro* conditions. It is interesting to note that the A-type burst behaviour in the intact animal has attracted less attention than the B-type, possibly because those who work with dogs seldom encounter A-type activity. This lack of information prompted our analysis of the A-burst response in intact, anaesthetised cats (Arndt *et al.*, 1971). Stimuli like blood volume changes, noradrenaline infusions and atrial pacing were used to provoke extreme changes in the atrial dynamics, and surprisingly the A-burst was found to be amazingly insensitive to changing atrial stimulus conditions.

A representative experiment in Fig. 5.2 exemplifies this point. Note the wide variations in the pressure components, i.e. the foot of the *a* wave as an indicator of the background stretch, the amplitude and also the slope of the atrial pressure curve which, according to several authors, has been assumed to be the most critical stimulus parameter for these endings (Struppler & Struppler, 1955, Langrehr, 1960*a, b*). Yet, neither the number of spikes per cardiac cycle, nor the spike-to-spike intervals (i.e. the frequency of the

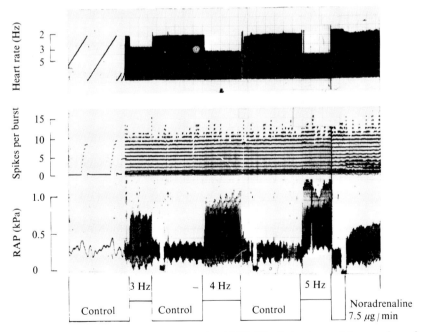

Fig. 5.3. Atrial A-burst (experiment 8–25–1969). The behaviour of the number of spikes per burst during several minutes (25 mm on paper equals 1 min). The nerve discharge (centre trace) is unaffected by the changes in atrial pressure (atrial pacing, noradrenaline infusion). RAP = right atrial pressure. (From Arndt *et al.*, 1971.)

burst) vary accordingly. This lack of responsiveness is even more impressive when the receptor response is followed over time spans of minutes as in Fig. 5.3. Again one recognises the appreciable changes in the atrial pressure conditions during atrial pacing and on infusion of noradrenaline, but there is no corresponding effect in the spike count per

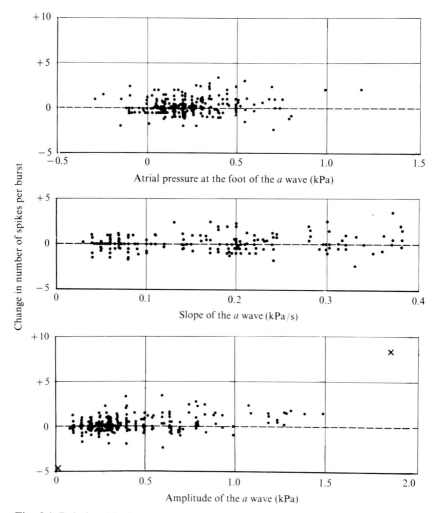

Fig. 5.4. Relationship between the change in number of spikes per burst and the three parameters of the atrial pressure: the foot, the slope and the amplitude of the *a* wave. Data from 16 fibres in 16 experiments. Note the parallel distribution of the data along the zero lines. There is no correlation. (From Arndt *et al.*, 1971.)

heart beat. Certainly, the test procedures are not mirrored in the receptor response.

Similar results were obtained from a total of 16 fibres from 16 cats being summarised in Fig. 5.4. The changes in the number of spikes per cardiac cycle were plotted against several of the possible stimulus parameters: the pressure at the foot of the *a* wave, its amplitude and also its slope. The data points are clearly distributed parallel to the x-axis around the zero line, i.e. there is no correlation between the receptor response and any of the stimulus parameters. Surely at the extreme ends the receptor discharge may change more markedly. The two crosses in the lower diagram are cases in point. On the one end the receptor discharge decreased by about five spikes when the ending was excited by atrial contraction in the isolated atrium under isotonic conditions. On the other end the spike count increased by eight when the atrium contracted against the closed A–V valves during a period of A–V dissociation.

Obviously, the number of spikes per cardiac cycle is only one receptor parameter which could convey information. We therefore also looked into the behaviour of the spike frequency parameters, i.e. peak frequency and the burst frequency which is computed from the average of the spike-to-spike intervals of a burst. Five out of 16 fibres were analysed more rigorously. Their data – the atrial pressure and the spikes – were stored on magnetic tape and later analysed on a computer. Spike-to-spike interval histograms for ten successive heart beats were derived for any of the above experimental test procedures. These histograms were then related to the respective atrial pressures as is shown in Fig. 5.5. The unresponsiveness of these parameters is obvious again. Neither the peak frequency nor the duration of the A-burst is affected by the drastic variations in the pressure components. Essentially similar results were obtained in all five fibres. Fig. 5.6 shows the averages with a standard error for two frequency parameters: that for the maximum instantaneous frequency and that for the average burst frequency for controls and for the pacing experiments chosen because they caused the largest variation in atrial mechanics. Two points should be stressed: first, *in vivo* the peak frequency of these endings amounts to about 200 Hz and even the burst frequency, around 150 Hz, is remarkably high. Secondly, these two parameters do not increase in spite of massive changes in atrial mechanics.

Similar observations by other investigators on the behaviour of A-type endings in closed-chest animals are lacking. However, it appears from the work by Hakumäki (1970) in the open-chest cat that the A-type endings

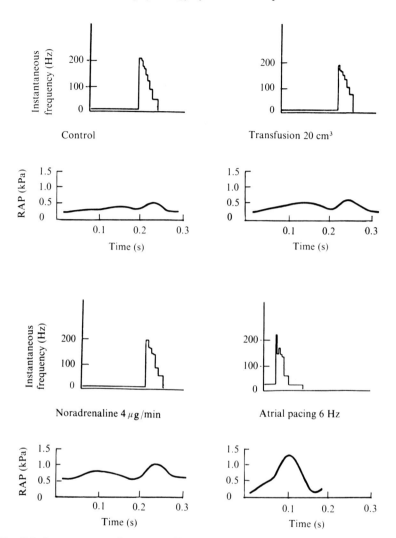

Fig. 5.5. Instantaneous frequency. Representative example. Averages from ten successive heart cycles for either condition. The discharge pattern is not sensitive to changes in the atrial *a* wave. (From Arndt *et al.*, 1971.)

respond to volume changes to a lesser extent than B-type endings. Curiously enough, the author found increases in A-type activity with increasing as well as with decreasing blood volume. The unresponsiveness of the type A endings to varying blood volume is supported by recent observations by Recordati *et al.* (1976) and also by Rao & Fahim (1976).

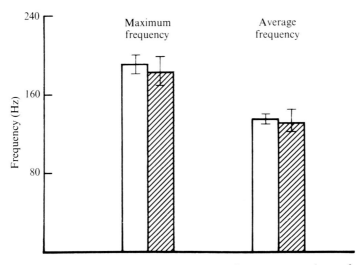

Fig. 5.6. Maximal instantaneous and average frequency per burst for two conditions: controls (open columns) and atrial tachycardia (atrial pacing, hatched columns). Mean values and deviations from the mean for five experiments. Atrial pacing changes atrial mechanics in a most pronounced way without significantly changing the discharge pattern. (From Arndt *et al.*, 1971.)

Table 5.1. *Receptor properties in the closed-chest versus open-chest animals*

Reference			Spikes per cardiac cycle	Average discharge rate (spikes/s)	Maximum instantaneous frequency (Hz)	Burst frequency (Hz)
Arndt *et al.* (1971)	Control (closed chest)	$\bar{x}$	7.2	24.2	204	135
		S.D.	±2	±7.6	±10.3	±6
		n	16	16	5	5
Recordati *et al.* (1976)	Control (open chest)	$\bar{x}$	3.8	9.2	93.6	73.6
		S.D.	±0.6	±0.6	±14.4	±10.4
		n	16	16	16	16
	Sympathetic stimulation	$\bar{x}$	5.0	14.4	123.5	101.2
		S.D.	±0.5	±0.8	±6.4	±6.3
		n	14	14	14	14
	Parasympathetic stimulation	$\bar{x}$	1.4	3.1	23.2	24.4
		S.D.	±0.5	±0.3	±12.3	±11.0
		n	9	9	9	9

S.D. = standard deviation.

Yet with respect to the effect of inotropic interventions our own observation seem to conflict with those of Recordati and his colleagues, who clearly documented opposite effects of sympathetic and parasympathetic interventions on the A-type fibre discharge in their experiments. However, a comparison of the discharge properties of their fibres and ours in Table 5.1 reveals that all receptor parameters during the controls in the open-chest animals are about half of those in the intact animals, indicating an appreciable shift in the working range along their stimulus–response curves.

The above point is convincingly brought out from the results of these workers in Fig. 5.7. The peak frequency of an A-type ending was plotted against certain possible stimuli, i.e. dimensional parameters in the upper diagrams and pressure parameters in the lower. The control curves and also

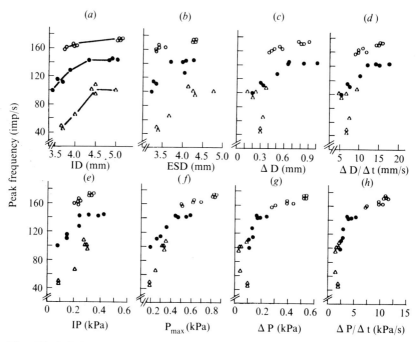

Fig. 5.7. Relationship between peak frequency of discharge in the burst, atrial dimension and pressure for one receptor studied during changes in atrial volume performed under control conditions (o) and sympathetic (•) and vagal (▵) stimulation. ID = initial dimension; ESD = end-systolic dimension; ΔD = amount of systolic shortening; $\Delta D/\Delta t$ = mean rate of shortening; IP = pressure at the foot of the *a* wave; P_{max} = peak of the A wave; ΔP = amplitude of the *a* wave; $\Delta P/\Delta t$ = slope of the upstroke of the *a* wave. (From Recordati *et al.*, 1976.)

the lower ones under parasympathetic stimulation have the well-known appearance of a stimulus–response curve of a mechanoreceptor with a threshold, a saturation range and a linear part between the two. But in this A-type ending this relationship obviously holds only when there is a low state of activity. By contrast, at the higher state of activity – here under sympathetic influence – the receptor hardly responds to the parameter changes. Since in the closed-chest animal the receptors fire at far higher frequencies than even under sympathetic stimulation in the above experiments, the variations in the sympathetic or parasympathetic tone appear unlikely to have an effect on the A-type ending *in vivo*. Hence it is not surprising that in our experiments no changes in the receptor response during the infusion of noradrenaline were seen. It seems therefore safe to conclude that under *in vivo* conditions the A-type endings are neither sensitive to stretch nor are they affected by the autonomic nervous system.

Do A-type receptors signal heart rate?

A discussion of the possible significance of the impulse traffic in the A-type afferents for the control of the circulation must then take into account the fact that obviously these endings do not transmit information about the atrial pressure components. Interestingly, the constancy of the number of spikes per cardiac cycle must yield an absolutely linear relationship between heart rate and the average discharge rate, i.e. the number of spikes per unit time as is demonstrated in Fig. 5.8. From this, one is tempted to speculate whether the type A endings convey information about the heart rate.

Of course, the last hypothesis needs experimental confirmation. But obviously the question of the actual heart rate level is not trivial if one thinks of its importance for the economy of cardiac work, for the optimal coupling of the heart to the arterial tree and, in particular, for the adaptation of the circulation to stress (Brömser, 1935; Taylor, 1964). The rather peculiar A-type activity, with its constant discharge over wide working ranges being present even in the widely slit-open atrium thus having no stretch-related threshold, appears well suited to convey information about the heart rate. In any event, other mechanoreceptors of the circulation cannot do so under all conditions because they might be silent when the stimulus parameters are below their threshold but fire continuously when the stimulus parameters work above threshold at all times (Arndt *et al.*, 1974, 1975).

It might be argued whether the average discharge rate could be decoded

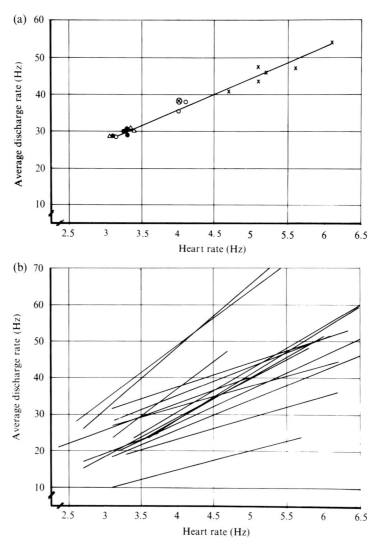

Fig. 5.8. Relationship between average discharge rate (spikes per second) and heart rate. *(a)* Representative example; *(b)* calculated regression lines for all 16 fibres. ●, control; ▵, change in volume ± 20 cm³; ○, noradrenaline, 5 ug/min; ⊗, noradrenaline + atrial pacing; ×, atrial pacing (Arndt *et al.*, 1971).

by the central neurons. This requires neurons which can store and compute the afferent inflow over time periods longer than one cardiac cycle. There is indeed support for this. Koepchen *et al.* (1967) observed cardiovascular neurons in the reticular formation whose discharge was maintained for as long as 30 s after the afferent input which had caused it had been withdrawn. Finally Kidd (this volume) found a category of reticular activity which increased slowly within 10–20 s after sudden balloon distentions of the veno-atrial junctions.

Basic neurophysiological properties of atrial mechanoreceptors

The differences between the bursts with respect to their functional behaviour described above, in the time of their occurrence in the cardiac cycle and in the discharge patterns, suggested the existence of two different receptor types (Struppler & Struppler, 1955; Paintal, 1963*a,b*). This, however, is not the case because all atrial afferents show the characteristics of slowly adapting mechanoreceptors when studied under comparable conditions. This came out when atrial receptor zones containing the ending previously identified in the intact animal were exposed to sinusoidal length changes of varying frequency (1–10 Hz) and amplitudes (1.5–5.0 mm) at given pre-stretch. The sinusoidal stimulation was preferred to other stimulus modes because it is convenient to quantify the dynamic sensitivity of the receptors and to test for the linearity of the system. The technical details were described previously (Arndt *et al.*, 1974), but briefly, single-fibre activity was recorded from the vagi in the neck of cats. The chest was opened for receptor localisation while the nerve recording continued. Then the ventricles were removed and a strip of the atrium containing the ending was cut free towards the pulmonary vessels. This resulted in atrial strips usually 1.5 cm wide and 2–3 cm long with one free end and the opposite end of the strip remained naturally connected to the mediastinum. The free end of the atrial strips was attached to a mechanical sine-wave generator. To increase the survival time of the nerve endings the preparations were superfused within the thorax with oxygenated tyrode solution.

A-type and B-type receptor response to identical stimulation

All receptors responded in a similar fashion when subjected to passive stretch as in Fig. 5.9. The differences between the A-type and the B-type discharge with respect to the time of occurrence and their frequency characteristics vanished, and bursts occurred with similar patterns and spike counts when both endings were stretched sinusoidally. No differences in the responses between any of the bursts were noticed when the stimulus parameters were varied. For example, the number of spikes generated per

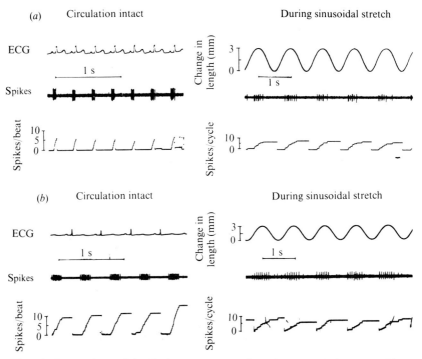

Fig. 5.9. Comparison of the discharge patterns of *(a)* A-type and *(b)* B-type fibres with intact circulation and during sinusoidal stimulation of the receptor area *in situ*. Note the similarities in discharge with sinusoidal stimulation (right-hand part of the figure). (From Arndt *et al.*, 1974.)

stimulus cycle decreased hyperbolically when the stimulus frequency was increased from 1 to 10 Hz. The regularity of the response is demonstrated for a B-type ending and for all fibres (Fig. 5.10). The differences in the individual response at 1 Hz is related to differences in the control conditions with respect to pre-stretch and amplitude of stretch which, however, remained constant during the frequency test run. The hyperbolic function is plausible because the time during which the stimulus works at the receptor site must decrease with increasing stimulus frequency, i.e. with decreasing stimulus periods.

Dynamic response characteristics of atrial receptors

Curiously, the product of the number of spikes per stimulus cycle and heart rate, i.e. the average discharge rate, is independent of the stimulus frequency. At first sight this is somewhat surprising. One would expect that

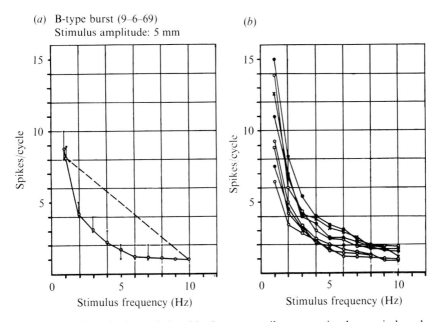

Fig. 5.10. Quantitative relationship between spikes per stimulus period and stimulus frequency (initial stretch and stimulus amplitude constant for each run). *(a)* shows a representative example. Note the hyperbolic relationship between the two parameters, the small spontaneous fluctuations in receptor discharge by only one spike, and the agreement in discharge at one cycle for the start and the end of this run. o, averages; bars are ±fluctuations. *(b)* shows results from eight fibres in eight experiments. Note the hyperbolic relationship irrespective of the fibre type. ●——●, A-type burst; o——o, B-type burst; ×——×, intermediate-type burst. (From Arndt *et al.,* 1974.)

the dynamic receptor component should add to the overall receptor response because the slope of the forcing function must necessarily increase with increasing stimulus frequencies. In fact, the velocity sensitivity was found to be relatively small in all endings. The sinusoidal stimulation allowed quantification of the dynamic receptor response from the phase relationship between the forcing function and the output function of the receptor. How this was tested is demonstrated in Fig. 5.11 for a type A fibre. The traces represent from top to bottom: the imposed change of length, i.e. static or sinusoidal, the spikes and the interspike intervals. The shortest trace in the *y*-direction represents the maximum instantaneous frequency. The phase relationship can be determined from the time relationship between peak stretch and peak frequency. For easy comparison of the phase relationship the time sweep was expanded for the stimulus frequencies of 1 and 2 Hz (recording at the bottom) such that one stimulus period covers the same distance. When sinusoidal stimulation is substituted for the static stimulation above threshold as indicated by the continuous discharge (see recordings on top) then the static discharge is modulated continuously according to the forcing function and there is no phase shift between peak stretch and peak frequency. A small phase shift, the peak frequency leading the peak stretch, occurs when the stimulus frequency is increased from 1 to 2 Hz at given amplitude and pre-stretch (recordings on bottom). Note that the fibre discharges intermittently with the positive-going part of the sine wave because the ending was silent before dynamic stimulation, i.e. the static stimulus was below threshold. Accordingly only the above-threshold path of the forcing function was effective.

Sometimes it was rather difficult to determine the phase angles because of spontaneous fluctuations in the instantaneous frequency as is visualised in Fig. 5.11 by the recording at low static stretch. However, reliable results

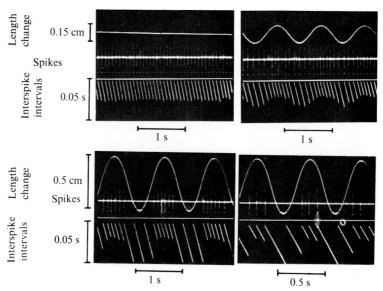

Fig. 5.11. Response of an A-type ending to static and dynamic stimulation in the linear part of the response curve (on top) and near the threshold (at the bottom). Note the perfect sinusoidal modulations of the interspike intervals without phase shift between the forcing function and the receptor response in the top recording and the intermittent discharge with a small time lapse between maximum length and the shortest spike-to-spike interval in the recording at the bottom.

were obtained in five fibres for different stimulus conditions with respect to stimulus frequency and amplitude. The results are summarised in Table 5.2. The phase angles tend to increase with both increasing stimulus frequency as well as with increasing stimulus amplitude. They are relatively small in any event thus showing that the velocity sensitivity of the receptors does not add much to their overall response, and most important: there is, contrary to the assumption of Struppler & Struppler (1955) and of Langrehr (1960a,b), obviously no difference between the two receptor types. Certainly, none of the fibres had ever a phase angle of $-90°$ which would indicate a pure velocity sensitivity.

Table 5.2. *Phase angles between sinusoidal length changes and instantaneous frequency in A-type and B-type endings at different stimulus frequencies and amplitudes*

Receptor type	Stimulus amplitude (mm)	Phase angle at different stimulus frequencies (degrees)			Static discharge (spikes/s)
		1 Hz	2 Hz	4 Hz	
A	1.5	0	—	—	15
	5.0	-14	—	—	15
A	1.5	0	0	—	10
	5.0	-9	-14	—	10
B	3.0	-13	-18	-36	4/10
	5.0	-12	-18	-54	19/20
B	1.5	0	-9	—	20/29
	5	0	-18	-28	18/22
B	3	0	-18	—	sporadic
	5	-9	-24	-36	sporadic

Static response characteristics of atrial receptors

Furthermore all atrial endings respond similarly to static stimulation in that all can be made to fire continuously at a certain pre-stretch. This was known for type B endings (Paintal, 1953, Chapman, 1958) but according to our own observations this holds also for the type A ending as in Fig. 5.11 (Arndt *et al.*, 1974) in agreement with recent experiments by Gupta (1977b). Curiously, Chapman & Pankhurst (1976), who did not detect differences in the adaptation properties of atrial mechanoreceptors not selected for different types, maintain that none of their receptors showed true static discharge. According to their own data, however, this statement does not seem to apply to high states of stretch, such as those in Fig. 4 of their paper when static discharge is clearly present. In fact, extrapolation of the adaptation curves to zero frequency suggests an adaptation time of several hundred seconds (see Fig. 5 of their paper).

Interaction between static and dynamic stimuli and the physiological working range

Finally it is of some interest to look closer into the interaction between static and dynamic stimuli because this tests the linearity of the system and gives a clue about the physiological working range of the receptors. This aspect can be appreciated more completely by the analysis of Fig. 5.12: this B-ending was silent at some static stretch below threshold, as indicated by the absence of any spike activity during this condition. With increasing stretch it fired continuously at a frequency around 20 Hz and finally at even higher pre-stretch the fibre fired continuously with about 60 Hz. The lower record shows the effect of sinusoidal stimulation on the ending. This results in a burst during the above-threshold path of the forcing function such that the fibre fires intermittently during short time periods. If the dynamic stimuli are applied when there is a higher pre-stretch as in the centre of the record, the fibre operates during static stimulation above threshold and accordingly discharges continuously. In this situation, sinusoidal length changes produce bursts with longer duration and higher frequencies because the stimulus now works for a longer time above the threshold. Furthermore, the frequency increases because the receptor is being pushed up on its response curve, certainly not because of an additionally dynamic receptor effect which is clearly shown by the absence of a phase lag between stimulus and response function. At the highest static activity the ending operates in the saturation range. The sinusoidal length changes modulate the instantaneous frequency perfectly only during the negative-going parts of the sine wave whereas the instantaneous frequency cannot follow peak stretch as indicated by the plateau of the interspike intervals.

Since B-type fibres discharge intermittently *in vivo* with bursts of variable duration and of low frequency one must conclude that these endings work physiologically near their thresholds. This explains also why these endings are sensitive to varying stretch and also the amplitude of the *v* wave (Paintal, 1963*a, b*).

Curiously, the same basic properties apply to A-endings also as is shown

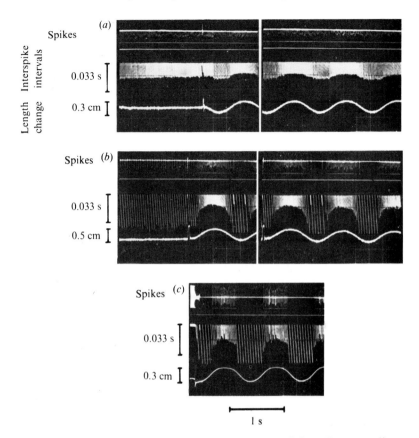

Fig. 5.12. Interplay between static and dynamic stimuli in a B-type ending at different pre-stretches, i.e. *(a)* in the saturation on top, *(b)* above threshold in the centre, and *(c)* below threshold at bottom. Note the continuous discharge with a plateau in the interspike intervals in the saturation and the decrease in the burst duration from the above- and below-threshold condition. Note, furthermore, the absence of appreciable phase angles between the forcing function and the receptor response.

in Fig. 5.11. In this case one recognises again the perfect modulation of the static discharge frequency by the sinusoidal length changes. And as in the B-fibre, dynamic stimulation near the threshold generates bursts only during the above-threshold path of the forcing function.

Thus, the atrial receptors belong basically to a homogeneous population as far as their excitatory properties are concerned. Yet *in vivo* the A-type endings fire bursts with frequencies in the saturation range, i.e. with extremely high frequencies for fibres with conduction velocities at about 20 m/s (Paintal, 1963*a, b*). By contrast, the B-type endings although also firing bursts maintain a relatively low frequency at around their thresholds.

Why are atrial receptors different in vivo?

One is then left with the question, what makes these basically identical atrial endings different *in vivo?* Whitteridge (1953) suggested that somehow the A-endings must be excited directly by the contraction of the atrial musculature and there are indeed several arguments in favour of this idea, but exactly how this is brought about still remains puzzling. That a direct, active excitation of the A-endings might be responsible for their high state of activity is corroborated by observations in the tendon organ which were shown to be about 50 times more sensitive to active than to passive excitation (Houk & Simon, 1967). We originally speculated that A-type endings might be located in the atria where they are apt to be excited by atrial contraction. By contrast, the B-endings were expected to be located in the adjacent vessels, thus being more under the influence of the passive stretch conditions. But this hypothesis has been made untenable by Gupta (1977*a*). Most recently, interesting differences between atrial receptors in the isolated atrium were reported by Chapman & Pankhurst (1976). They identified two receptor populations with respect to the subthreshold excitability. For one group the subthreshold excitability decreased with time (HPL = High-Pass Latency receptors in their terminology); for the other group, subthreshold excitability increased with time (LPL = Low-Pass Latency receptors). The authors suggested that these differences are dominated by different tissue mechanisms in the sense that the HPL receptors are coupled in series but the LPL receptors in parallel with visco-elastic elements. It remains to be shown whether the former correspond to the A-type and the latter to the B-type fibres.

Concluding remarks

In conclusion, the atrial afferent nerves of the myelinated fibre category originate from identical mechanoreceptors of the slowly adapting type sharing their basic properties with other mechanoreceptor-like muscle spindles, tendon organs and the baroreceptor of arterial system in particular (Arndt *et al.,* 1974, 1975). Nevertheless, *in vivo* these endings are differentiated in such a way that the B-type endings operate in the threshold range of their function curves where they are apt to respond most sensitively to varying stretch conditions. By contrast, the A-type endings are pushed into saturation by a hitherto unknown mechanism thus being rendered insensitive to varying stretch.

Certainly, according to neurophysiological evidence the information transmitted by the two fibre groups is not the same. Admittedly it has not been possible to show experimentally as yet that the two endings subserve different reflexes, but at least the B-type endings appear well suited to measure and signal the fullness of the thoracic circulation which, as the integrative part of the low pressure system, is closely linked with blood volume regulation (Gauer & Henry, 1976). From this viewpoint it is of some interest to note that with respect to its pressure–volume relationship the pulmonary circulation, including the four chambers of the heart, constitutes a functional unit in that the pressures throughout this system are intimately related with the blood volume. Except for the A-type, all afferent activities from this vasculature, like that from atrial B-type endings, that from the pulmonary arteries (Whitteridge & Pearce, 1951; Coleridge & Kidd, 1960) and even that from the non-myelinated C fibre category (Thorén, Donald & Shepherd, 1976) correlate closely with the pressures which by and large are functions of blood volume. One wonders, therefore, whether all of these receptors subserve the same function, namely the regulation of blood volume, with the ultimate aim of guaranteeing the filling of the heart. Clearly, the A-type discharge does not follow the same functional principle but could well signal heart rate. One is even tempted to speculate whether the receptors from the low pressure side of the circulation transmit information about the cardiac output.

References

Amann, A. & Schäfer, H. (1943). Über sensible Impulse im Herznerven. *Pflügers Arch. ges. Physiol.*, **246**, 757–89.

Arndt, J.O., Brambring, P., Hindorf, K. & Röhnelt, M. (1971). The afferent impulse traffic from atrial A-type receptors in cats. Does the A-type receptor signal heart rate? *Pflügers Arch. ges. Physiol.*, **326**, 300–15.

Arndt, J.O., Brambring, P., Hindorf, K. & Röhnelt, M. (1974). The afferent discharge pattern of atrial mechanoreceptors in the cat during sinusoidal stretch of atrial strips *in situ*. *J. Physiol.*, **240**, 33–52.

Arndt, J.O., Dörrenhaus, A. & Wiecken, H. (1975). The aortic arch baroreceptor response to static and dynamic stretches in an isolated aorta-depressor nerve preparation of cats *in vitro*. *J. Physiol.*, **252**, 59–78.

Bainbridge, F.A. (1915). The influence of venous filling upon the rate of the heart. *J. Physiol.*, **50**, 65–84.

112   *Electrophysiology of cardiac receptors*

Brömser, Ph. (1935). Über die optimale Beziehung zwischen Herztätig-keit und physika-lischen Konstanten des Gefäßsystems. *Z. Biol.*, **96**, 1–10.

Chapman, K.M. (1958). Transient response studies of the blood volume receptor system in dogs. *WADC tech. Rep.*, 58–101.

Chapman, K.M. & Pankhurst, J.H. (1976). Strain sensitivity and directionality in cat atrial mechanoreceptors *in vitro*. *J. Physiol.*, **259**, 405–26.

Coleridge, J.C.G., Hemingway, A., Holmes, R.L. & Linden, R.J. (1957). The location of atrial receptors in the dog: a physiological and histological study. *J. Physiol.*, **136**, 174–97.

Coleridge, J.C.G. & Kidd, C. (1960). Electrophysiological evidence of baroreceptors in the pulmonary artery of the dog. *J. Physiol.*, **150**, 319–31.

Dickinson, C.J. (1950). Afferent nerves from the heart region. *J. Physiol.*, **111**, 399–407.

Douglas, W.W., Ritchie, J.M. & Schaumann, W. (1956). A study of effect of the pattern of electrical stimulation of the aortic nerve of the reflex depressor responses. *J. Physiol.*, **133**, 232–42.

Eysel, U.Th. & Grüsser, O.J. (1970). The impulse pattern of muscle spindle afferents. *Pflügers Arch. ges. Physiol.*, **315**, 1–26.

Gauer, O.H. & Henry, J.P. (1976). Neurohormonal control of plasma volume. *International Review of Physiology II*, **9**, 145–90.

Gauer, O.H., Henry, J.P. & Sieker, H.O. (1961). Cardiac receptors and fluid volume control. *Prog. cardiovasc. Dis.*, **4**, 1–26.

Gupta, B.N. (1977a). The location and distribution of type A and type B atrial endings in cats. *Pflügers Arch. ges. Physiol.*, **367**, 271–5.

Gupta, B.N. (1977b). Studies on the adaptation rate and frequency distribution of type A and type B atrial endings in cats. *Pflügers Arch. ges. Physiol.*, **367**, 277–81.

Gupta, P.D., Henry, J.P., Sinclair, R. & von Baumgarten, R. (1966). Responses of atrial and aortic baroreceptors to non-hypotensive hemorrhage and to transfusion. *Am. J. Physiol.*, **211**, 1429–37.

Hakumäki, M.O.K. (1970). Function of the left atrial receptors. *Acta physiol. scand.*, **344**, 1–54.

Henry, J.P. & Pearce, J.W. (1956). The possible role of cardiac atrial stretch receptors in the induction of changes in urine flow. *J. Physiol.*, **131**, 572–85.

Homma, S. & Suzuki, S. (1966). Phasic properties of aortic and atrial receptors observed from their afferent discharge. *Jap. J. Physiol.*, **16**, 31–41.

Houk, J. & Simon, W. (1967). Responses of Golgi-tendon-organs to forces applied to muscle tendon. *J. Neurophysiol.*, **30**, 1466–77.

Jarisch, A. & Zotterman, Y. (1948). Depressor reflexes from the heart. *Acta physiol. scand.*, **16**, 31.

Johnston, B.D. (1968). Nerve endings in the human endocardium. *Am. J. Anat.*, **122**, 621–30.

Kappagoda, C.T., Linden, R.J. & Mary, D.A.S.G (1976). Atrial receptors in the cat. *J. Physiol.*, **262**, 431–46.

Koepchen, H.P., Langhorst, P., Seller, H., Polster, J. & Wagner, P.H. (1967). Neuronale Aktivität im unteren Hirnstamm mit Beziehung zum Kreislauf. *Pflügers Arch. ges. Physiol.*, **294**, 40–64.

Langrehr, D. (1960a). Entladungsmuster und allgemeine Reizbedingunger von Vorhofsrecep-toren bei Hund und Katze. *Pflügers Arch. ges. Physiol.*, **271**, 257–69.

Langrehr, D. (1960*b*). Beziehungen zwischen Vorhofsreceptoraktivitäter und Herzmechanik von Hund und Katze bei verschiedenen Kreislaufzuständen. *Pflügers Arch. ges. Physiol.*, **271**, 270–82.

Langrehr, D. & Kramer, K. (1960). Beziehungen der mittleren Impulsfrequenz von Vorhofsreceptoren zum thorakalen Blutvolumen. *Pflügers Arch. ges. Physiol.*, **271**, 797–807.

Ledsome, J.R. & Linden, R.J. (1964). A reflex increase in heart rate from distension of the pulmonary vein–atrial junctions. *J. Physiol.*, **170**, 456–73.

McKean, T.A., Poppele, R.E., Rosenthal, N.P. & Terzuolo, C.A. (1970). The biologically relevant parameter in nerve impulse trains. *Kybernetika*, **6**, 168–70.

Neil, E. & Joels, N. (1961). The impulse activity in cardiac afferent vagal fibres. *Naunyn-Schmiedebergs Arch. exp. Path. Pharmak.*, **240**, 453–60.

Neil, E. & Zotterman, Y. (1950). Cardiac vagal afferent fibres in the cat and the frog. *Acta physiol. scand.*, **20**, 160–5.

Nonidez, J.F. (1937). Identification of the receptor areas in the venae cavae and pulmonary veins which initiate reflex cardiac acceleration (Bainbridge's reflex). *Am. J. Anat.*, **61**, 203–31.

Paintal, A.S. (1953). A study of right and left atrial receptors. *J. Physiol.*, **120**, 596–610.

Paintal, A.S. (1963*a*). Vagal afferent fibres. *Ergebn. Physiol.*, **52**, 74–131 and 148–56.

Paintal, A.S. (1963*b*). Natural stimulation of type B atrial receptors. *J. Physiol.*, **169**, 116–36.

Paintal, A.S. (1972). Cardiovascular receptors. In *Handbook of sensory physiology*, vol. III/I, *Enteroceptors*, ed. E. Neil, pp. 1–45. Springer-Verlag: Berlin, Heidelberg and New York.

Rao, P.S. & Fahim, M. (1976). Effect of propranolol in the relationship between atrial systolic pressure and type A atrial receptor discharge in cats. *Arch. int. Pharmacodyn. Thér.*, **222**, 137–47.

Recordati, G., Lombardi, F., Bishop, V.S. & Malliani, A. (1975). Response of type B atrial vagal receptors to changes in wall tension during atrial filling. *Circulation Res.*, **36**, 682–91.

Recordati, G., Lombardi, F., Bishop, V.S. & Malliani, A. (1976). Mechanical stimuli exciting type A atrial vagal receptors in the cat. *Circulation Res.* **37**, 397–403.

Schäfer, H. (1950). Elektrophysiologie der Herznerven. *Ergebn. Physiol.*, **46**, 71–125.

Struppler, A. (1955). Afferente vagale Herznervenimpulse und ihre Beziehung zur Hamodynämik. *Z. Biol.*, **107**, 416–28.

Struppler, E. & Struppler, A. (1955). Über spezielle Charakteristika afferenter vagaler Herznervenimpulse und ihre Beziehungen zur Herzdynamik. *Acta physiol. scand.*, **33**, 219–31.

Taylor, M.G. (1964). Wave travel in arteries and the design of the cardiovascular system. In *Pulsatile blood flow*, ed. E. O. Attinger, pp. 343–67. McGraw-Hill: New York.

Thorén, P., Donald, D.E. & Shepherd, J.T. (1976). Role of heart and lung receptors with nonmedullated vagal afferents in circulatory control. *Circulation Res.*, **38**, 2–9.

Whitteridge, D. (1948). Afferent nerve fibres from the heart and lungs in the cervical vagus. *J. Physiol.*, **107**, 496–512.

Whitteridge, D. (1953). Electrophysiology of afferent cardiac and pulmonary fibres. *Abstr. XIX, Int. Physiol. Congr.*, 66–72.

Whitteridge, D. & Pearce, J.W. (1951). The relation of pulmonary arterial pressure variations to the activity of afferent pulmonary vascular fibres. *Q. J. exp. Physiol.*, **37**, 177–88.

Discussion

(a) It was asked whether there was any systematic difference in the initial lengths from which the analyses of the receptor responses to stretch was started. Arndt indicated that the pre-stretch condition could not be defined since one end of the strip had to be left attached to the heart in order to protect the innervation. Thus, only changes in stretch could be considered.

(b) The point was made that during the analysis it was assumed that the system was linear. However, it is probable that the type A atrial receptors were operating in a saturation range and the question was raised whether phase changes could be adequately described since in the saturated condition the response would be non-linear. Arndt replied that the fibres had been investigated over a linear section of the operating range since this was indicated by the perfect modulation of the instantaneous frequency-forcing function. He indicated that while there may well be a theoretical argument that frequency was undefined during burst firing, for the continuous discharge from the atrial receptor the phase relationships were no different from those in the linear range.

(c) At this point in the discussion several questions arose about the concept that the type A atrial receptors only work in the saturated range and that they signal changes in heart rate. Recordati briefly described some experiments in the anaesthetised cat in which instantaneous changes in right atrial dimensions were measured with a sonomicrometer technique and the heart rate was altered over a wide range by pacing. When heart rate was changed from 140 to 190 beats/min the peak frequencies of discharge from the type A receptors were unchanged; when heart rate was increased up to 200–210 beats/min the peak frequency increased. However, at these heart rates there was frequent A–V dissociation and, when the atrium was contracting against a closed A–V valve there were large increases in peak frequency from the receptor. He concluded that under these circumstances it was likely that the 'strength' of atrial contraction was the major factor causing the increases in frequency from the receptor rather than changes in heart rate. Arndt replied that these were not spontaneous changes and the heart was paced. He felt it important that the atrium must be working in the 'saturation range' for the receptor and so there was little opportunity for them to respond to alterations in the natural stimulus. Arndt added that, so far as he was aware, the atrial receptor was the only one which functioned in such a way so as to work within its saturation range. There were no other receptors which could sense heart rate in this way. The point was made

from the floor that very frequently when working with anaesthetised cats there were very large amounts of catecholamines circulating and the sympathetic system was very active. Under these circumstances it may be that the saturation is an artifact of the preparation. In reply Arndt indicated that even in the conscious cat there was a wide range of heart rates, e.g. when sleeping the heart rate went down to 70 beats/min while during exercise the heart rate could be well above 180 beats/min.

6. Vagal afferent C fibres from the ventricle*

D. G. BAKER, H. M. COLERIDGE &
J. C. G. COLERIDGE

Some 30 years ago Amann & Schäfer (1943) and Jarisch & Zotterman (1948) recorded afferent vagal impulses arising from endings in the cardiac ventricles of the cat. The potentials, which were small, were thought to travel in thin, slowly conducting nerve fibres. Coleridge, Coleridge & Kidd (1964) and Sleight & Widdicombe (1965) recorded activity from single fibres of this type in dogs and, by measuring conduction velocity, showed that they were C fibres. Observations on the stimulation of these endings by chemicals capable of evoking depressor effects, as well as by various physiological procedures, were confirmed and extended in later studies (Muers & Sleight, 1972; Öberg & Thorén, 1972a, b; Thorén, 1972, 1977; Thames, Donald & Shepherd, 1977).

Ventricular C fibres are part of a widespread system of unmyelinated afferent vagal fibres from the heart and great vessels (Sleight & Widdicombe, 1965; Coleridge et al., 1973; Thorén, 1976; Thorén, Saum & Brown, 1977). Vagal C fibres are of particular importance for the ventricles because they constitute the major part of the afferent innervation of these chambers. Myelinated vagal fibres with mechanoreceptive endings are richly supplied to the atrio-venous junctions on the input side of the pumps, and to the outflow regions of the two great arteries, but there are surprisingly few in the ventricular pumps themselves (Paintal, 1955; Coleridge et al., 1964).

Results of previous electrophysiological studies may be summarised briefly, as follows. Ventricular C fibres travel in both vagus nerves and most originate in the wall of the left ventricle. The few that arise from the right ventricle appear to have generally similar properties to those from the left.

* Supported by US Public Health Service Program Project Grant HL-06285 from the National Heart, Lung and Blood Institute.

Firing under control conditions is typically sparse and irregular, but in some fibres one or two spikes may discharge with each heart beat. Many endings become more active and develop a rhythmic pattern of discharge when ventricular pressure is increased by venous infusion or aortic occlusion and, under these conditions, the effective stimulus appears to be related to an increase in ventricular volume (Thorén, 1977).

The mechanism of the response of these endings to mechanical distortion is not wholly understood, however, for their discharge can also be increased by injection of adrenaline (Sleight & Widdicombe, 1965; Muers & Sleight, 1972), by severe haemorrhage (Öberg & Thorén, 1972b) (when the heart becomes smaller, not larger) and by occlusion of the coronary sinus (Muers & Sleight, 1972). Left ventricular C fibre endings are also stimulated by asphyxia or by coronary arterial occlusion (Thorén, 1972), an effect believed to be caused by mechanical changes in the heart, rather than by a specific chemosensitivity of the ending to changes in pO_2 or pCO_2.

Some endings appear to be located superficially in or near the epicardium, while others are situated more deeply in the myocardium; hence the names 'epicardial', 'epimyocardial' and 'myocardial', which have been used by various investigators. Epicardial and myocardial endings may differ in their response to various experimental procedures, but such differences are believed to be related to the position of the endings in the ventricular wall, rather than to the existence of functionally different types of ending. In brief, the evidence obtained over the past 12 years is largely in favour of a single, functionally homogeneous group of ventricular C fibres whose endings are stimulated by a variety of exotic chemicals but which are also capable of responding to mechanical changes occurring in the ventricular wall. Our observations suggest, however, that the ventricular wall may be supplied by afferent vagal C fibres of more than one type.

Methods

Experiments were carried out in dogs anaesthetised with sodium pentobarbitone (35–40 mg·kg^{-1} i.v.) or chloralose (80 mg·kg^{-1} i.v.). The chest was opened, and the lungs were ventilated by a Harvard pump. Airway carbon dioxide was monitored by a Beckman LB-1 analyser. Afferent impulses were recorded from 'single-' or few-fibre slips of the right or left cervical vagus nerve. We measured conduction velocities of afferent fibres and selected for investigation fibres with velocities of less than 2.0 m/s. We determined the approximate location of the ending whose impulses we were recording by comparing the effects of occluding the pulmonary artery and aorta. At the end of the experiment, we determined the precise location of the ending by exploring the heart and great vessels with a fine probe or bristle (Coleridge, Hemingway, Holmes & Linden, 1957). In some experiments we stimulated the receptor site electrically and measured the conduction time to the recording electrodes. Tracheal pressure, atrial pressure and arterial blood pressure (in the femoral artery, descending thoracic aorta or aortic arch) were recorded with Statham P23 Gb strain gauges. Left ventricular pressure was recorded by a catheter pressure transducer (Millar Instruments Inc., Model PC 350) inserted through a carotid artery or pulmonary vein.

Location of endings

A crucial step in the experiments we shall describe was to determine the location of the afferent ending. It has sometimes been the practice for an investigator to designate an ending as 'ventricular' on the basis of results of somewhat indirect procedures. For example, if impulse activity is reduced by occluding the left atrio-venous junction and increased by occluding the ascending aorta and by pressing on the left ventricle, the ending is assumed to be in the left ventricle. Attempts to determine the location of receptors in the living animal, however, may be misleading, as illustrated by the records in Fig. 6.1. The ending in Fig. 6.1*a* was in the wall of the left ventricle immediately upstream to the aortic valves, that in Fig. 6.1*b* was in the ascending aorta just downstream to the valves. Both endings were stimulated by occluding the descending thoracic aorta and both were stimulated by probing the base of the left ventricle – but only one was in the ventricular wall. Without killing the dog and opening the heart, we would have been unable to determine the precise location of either ending. If we had relied upon the results of our provisional estimate of location we would have included within our 'tally' of ventricular C fibre endings one possessing different characteristics and patterns of response.

By carefully exploring the non-beating heart one may sometimes be able to determine the depth of the ending in the cardiac wall. In Fig. 6.2*a* the sparse, irregular discharge characteristic of many ventricular C fibres is shown. The ending was vigorously stimulated by stroking a localised area on the surface of the right ventricle with a cotton wool-tipped probe (Fig. 6.2*b*). Finally, we opened the heart and, by comparing the effects of stroking the epicardial and endocardial surfaces, we located the ending

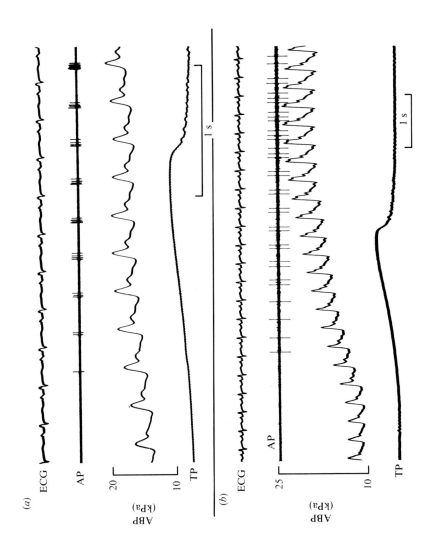

Fig. 6.1. Effect on two cardiovascular C fibres of partially occluding the descending thoracic aorta (experiments in *(a)* and *(b)* were performed in different dogs). *(a)* Activity recorded from a left vagal fibre whose afferent ending was subsequently found to be located in the left ventricular wall about 1.0 cm from the aortic valves. The fibre was silent during the control period; aortic occlusion began 3 s before the beginning of *(a)*. *(b)* Impulses recorded from left vagal C fibre arising from an ending in the wall of the ascending aorta, 3–4 mm beyond the aortic valves. The fibre was completely silent in the control period; aortic occlusion is signalled by the increase in blood pressure. In both *(a)* and *(b)* the afferent ending was vigorously stimulated in the beating heart by gently probing the left ventricular wall. ECG, electrocardiogram; AP, action potentials recorded from afferent vagal slip; ABP, arterial blood pressure (in this case measured in the aortic arch); TP, tracheal pressure (upstroke representing inflation).

Fig. 6.2. Method for determining the location of a ventricular C fibre ending. *(a)* Control; note sparse and irregular spontaneous discharge. *(b)* Bursts of impulses produced by lightly stroking the wall of the right ventricle with a probe tipped with cotton wool. The dog was then anaesthetised deeply and was bled to death. The right ventricle was opened. *(c)* Bursts of impulses produced by stroking a small area (approximately 5 mm in diameter) of the right ventricular epicardium with a bristle. Stroking the endocardial surface of the ventricle had no effect. Note different oscillograph recording speed in *(c)*. ECG, electrocardiogram; AP, action potentials recorded from afferent vagal slip; ABP, arterial blood pressure; LVP, left ventricular pressure; TP, tracheal pressure (upstroke representing inflation).

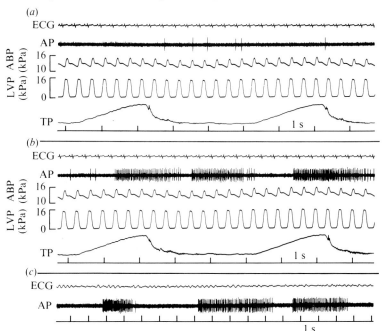

precisely to a small area of the right ventricular epicardium (Fig. 6.2c), and were able to stimulate the ending from the epicardial surface after cutting away the endocardium and much of the myocardium. Using similar techniques, one may show that some ventricular endings are in the endocardium while others are in the myocardium.

Of a total of 53 ventricular C fibres whose impulses we recorded, 20 were in the right vagus and 33 in the left. Eight endings were in the wall of the right ventricle and 45 in the left. As far as we were able to determine by carefully exploring the non-beating heart, 16 were in or near the epicardium, 9 were in the myocardium and 4 were in or very near the endocardium. We were unable to determine the depth of the remaining 24 endings, but since they were not stimulated by lightly stroking the epicardium with a fine bristle – but were stimulated by pressure with a fine glass rod – they were probably in the deeper layers of the ventricular wall.

Mechanical stimulation

In dogs, adrenaline has been found to be a more effective stimulus to ventricular C fibre activity than aortic occlusion (Sleight & Widdicombe, 1965; Muers & Sleight, 1972). Moreover, ventricular receptors in cats become strikingly active after a sudden, severe reduction in cardiac filling (Öberg & Thorén, 1972b), the mechanism of stimulation being thought to be the vigorous squeezing of the ventricle at a reduced cardiac volume. Nevertheless, recent studies of Thorén (1977) and Thames *et al.* (1977), in which the relationship of C fibre discharge to ventricular pressure was examined while left ventricular volume was increased, showed convincingly that ventricular C fibre activity increases with diastolic pressure (hence with diastolic volume) in the ventricle, although the relationship may be influenced by the inotropic state of the ventricle (Thorén, 1977) or by changes in afterload (Thames *et al.*, 1977). To take an example from Thorén's paper, graded aortic occlusion caused no change in ventricular C fibre discharge as systolic pressure increased from 16.0 to 29.3 kPa. Thereafter, however, end-diastolic pressure began to rise and C fibre discharge increased proportionately, even though there was little further increase in systolic pressure.

Many ventricular C fibres in our experiments on dogs responded to aortic occlusion, in much the same way as Thorén (1972) and Thames *et al.* (1977) have described in cats. We examined the effect of aortic occlusion on 38 left ventricular C fibres: 27 fibres were stimulated, 18 firing with an

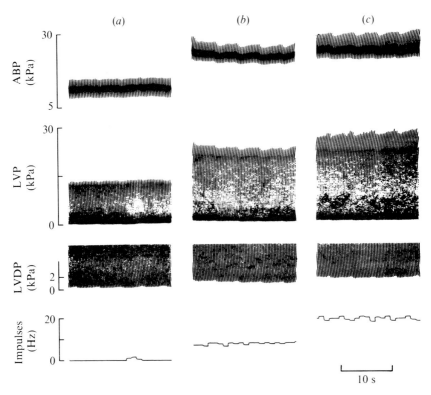

Fig. 6.3. Stimulation of a left ventricular C fibre ending by graded occlusion of the descending thoracic aorta (arterial blood pressure recorded in the aortic arch). Impulse frequency is recorded by a rate-meter; LVDP, left ventricular diastolic pressure. ABP, arterial blood pressure; LVP, left ventricular pressure.

obvious cardiac rhythm (Fig. 6.1*a*) and 9 with a generally irregular discharge; there was little or no response in the remaining 11 fibres. Fig. 6.3 shows three stages of graded aortic occlusion in an experiment on one of the fibres responding with a cardiac rhythm of discharge. Activity appeared to be related more linearly to changes in diastolic pressure than to changes in systolic pressure. The relationship between left ventricular end-diastolic pressure and firing in four sensitive ventricular C fibres of this type is shown in Fig. 6.4. At a mean left ventricular end-diastolic pressure of 2.0 kPa, activity in the fibres had risen from a mean control value of 0.6 imp/s to a mean of 9.5 imp/s, a sensitivity that compared well with that observed by Thorén (1977) in cats.

In view of the evidence linking C fibre activity with ventricular volume, it

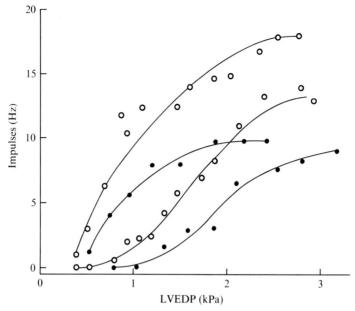

Fig. 6.4. Effect of increasing left ventricular end-diastolic pressure (LVEDP) on the impulse frequency of four left ventricular C fibre endings. Left ventricular pressure was gradually increased by progressive aortic occlusion and intravenous infusion of dextran.

is perhaps surprising that Muers & Sleight (1972) found that in dogs, whether the endings were stimulated by aortic occlusion or by injection of adrenaline, they were activated in systole, 100–200 ms after the Q wave of the ECG. Thorén (1977), after making the necessary phase correction to allow for the time taken for the impulse to travel to the recording electrode, agrees that in the cat also systolic activation is the rule. A careful examination of the precise point in the cardiac cycle at which the impulse is generated may tell us something about the mechanism of stimulation, so that it seems worthwhile to pursue this point a little further.

We measured total conduction time from the afferent ending to the recording electrode in 10 of 18 left ventricular C fibres with a cardiac rhythm. When stimulated by aortic occlusion, seven of the endings appeared to be regularly activated during systole. (When stimulated chemically, e.g. by veratridine, the evoked discharge frequently had no consistent relationship to the cardiac cycle.) Activity from one of these endings is shown in Fig. 6.1a. Silent during the control period, the ending was stimulated by aortic occlusion. When allowance is made for conduc-

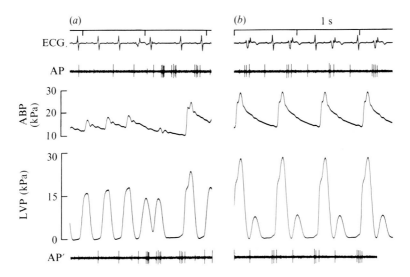

Fig. 6.5. Relationship between the activity of a left ventricular C fibre and the phase of the cardiac cycle. Arterial blood pressure was recorded from the aortic arch. AP', action potential trace displaced to the left to allow for the conduction time between the afferent ending and the recording electrodes. A few seconds before record *(a)*, the descending thoracic aorta was partially occluded with a snare. Interval of 4–5 s between *(a)* and *(b)*. ECG, electrocardiogram; AP, action potentials recorded from afferent vagal slip; ABP, arterial blood pressure; LVP, left ventricular pressure.

tion time (275 ms), each burst of impulses is seen to be generated at the peak of systolic ejection in the *preceding* cycle. There were notable exceptions, however, and some endings were more variable in the timing of their discharge. Fig. 6.5 shows activity in a ventricular C fibre during occlusion of the thoracic aorta. Conduction time from receptor to electrodes was estimated, and in mounting a duplicate action potential trace (AP') at the bottom, we have realigned it to indicate the points in the cycle at which impulses were generated. In the early part of occlusion (Fig. 6.5*a*), the fibre failed to fire with every cycle but when it did, the impulse was clearly generated in diastole. Later in the occlusion period (Fig. 6.5*b*), alternate extra systoles developed, and the burst of impulses occurring with each augmented beat was essentially late diastolic – certainly each burst appeared to be completed before the onset of the ejection phase of systole. And earlier in the record (Figs. 6.5*a*), when two ectopic beats failed to open the aortic valves, there was a burst of activity with each. These observations fit well with the idea that activity in this fibre was related to the diastolic size of the ventricle, and that impulses were generated either during passive stretch of the ventricular wall at end diastole, or during the development of active tension at end-diastolic volume during the isovolumetric phase, as in the case of the two beats that failed to open the valves (Fig. 6.5*a*).

If some ventricular C fibres are activated in systole and some in diastole, one need not be too surprised. Indeed one might argue that such variability of behaviour is to be expected of distorsion receptors embedded at different depths in the walls of hollow organs that undergo a regular but complex sequence of filling and emptying. It would be remarkable if an ending located in the epicardium at the apex of the left ventricle had a pattern and timing of discharge identical to that of an ending in the left ventricular muscle deep to the attachment of a papillary muscle. Atrial receptors, for example, show two well-defined patterns of activity in relation to the cardiac cycle, according to whether they are stimulated during atrial con-

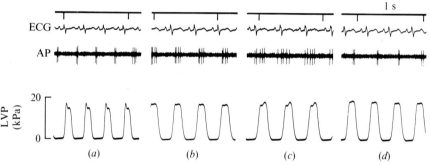

Fig. 6.6. Spontaneous change in timing of the discharge of a left ventricular C fibre, observed over a period of 90 min. The ending was located in the left ventricular myocardium about 1 cm from the aortic valves. There was an interval of 30 min between successive records. The conduction velocity of the afferent fibre was measured in each interval and was unchanged (1.1 m/s). ECG, electrocardiogram; AP, action potentials recorded from afferent vagal slip; LVP, left ventricular pressure.

traction or atrial filling (Paintal, 1953). It is possible that there is one basic type of atrial receptor whose firing pattern is determined by the location of the ending in the veno-atrial wall. Kappadoga, Linden & Mary (1976), however, found in many instances that the patterns of discharge could be converted, one into the other, by a number of physiological procedures.

In the case of two of our ventricular C fibres, the timing of the discharge was considerably more labile, as is well illustrated by the records in Fig. 6.6.

Activity in this fibre was recorded at 30-min intervals, with no experimental interventions between. The conduction velocity was measured at each interval and remained constant, but the cardiac timing of the discharge changed considerably over the total recording period of 90 min, varying from mid-diastole to systolic ejection. Clearly, such lability of firing pattern does little to further our understanding of mechanical events at the afferent terminal.

Stimulation by chemicals

So far we have dealt with C fibres that are sensitive to mechanical changes in the ventricles. In searching for these afferent vagal fibres, we found aortic occlusion to be the most convenient method for increasing ventricular pressure and stimulating the often silent endings. But these are not the only ventricular C fibres in the vagus nerve. There are others that are not stimulated at all by ventricular distension. Endings in the latter category may be provisionally identified by their response to chemicals such as capsaicin, phenyl diguanide and veratridine injected into the blood stream. In effect, ventricular C fibres display a spectrum of responses to mechanical and chemical stimuli. At the one extreme are endings that often fire with a cardiac discharge, are readily stimulated by ventricular distension and are not stimulated by capsaicin or phenyl diguanide. At the other extreme are endings that are not stimulated by ventricular distension, never fire in rhythm with the heart beat and are vigorously stimulated by capsaicin or phenyl diguanide or both. Between these two extremes are endings that can be stimulated by ventricular distension and also by one or both chemicals, and that fire with an irregular discharge whatever the stimulus. The method of selection largely determines the character of the sample, and this explains why the general behaviour of the ventricular C fibres originally described by Coleridge *et al.* (1964) is so different from that of the C fibres described later by Thorén (1972). In the former instance, endings were selected largely on the basis of their response to drugs; in the latter, selection was based on the effect of aortic occlusion.

We show two examples to illustrate this range of responses. Fig. 6.7 depicts the record of activity from a C fibre ending that was extremely sensitive to phenyl diguanide but was unresponsive to aortic occlusion or hypoxia. (Phenyl diguanide has already been shown by Paintal (1973) to stimulate ventricular C fibres in cats, but its action on these endings in dogs has not previously been demonstrated.) The pronounced cardiac slowing and increase in arterial blood pressure (Fig. 6.7) were due to a vagal reflex,

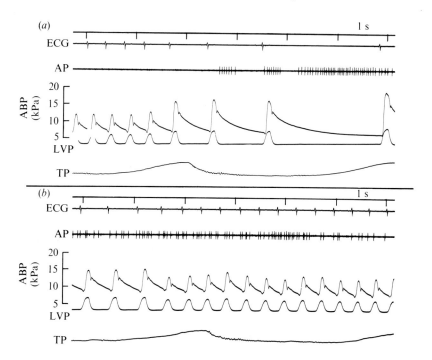

Fig. 6.7. Left ventricular C fibre ending stimulated by phenyl diguanide (10 $\mu g \cdot kg^{-1}$) which was injected into the right atrium 5 s before the beginning of record *(a)*. Interval of about 4 s between *(a)* and *(b)*. The ending was not stimulated by occluding the aorta or by ventilating the lungs with 5% O_2 in N_2 for 50 s. ECG, electrocardiogram; AP, action potentials recorded from afferent vagal slip; ABP, arterial blood pressure; LVP, left ventricular pressure; TP, tracheal pressure (upstroke representing inflation).

usually attributed to stimulation of the arterial chemoreceptors. Fig. 6.8 shows the responses of another left ventricular C fibre ending that was stimulated by ventricular distension *(a)* as well as by three chemicals *(b, c and d)*. The ending fired irregularly in response to both mechanical and chemical stimulation, and it is noteworthy that its responses to the three aortic occlusions were not proportionate to the changes in aortic pressure (Fig. 6.8*a*). It is also interesting that this fibre, whose activity was recorded in experiments made in collaboration with Dr M. Lipton, was stimulated vigorously by injecting meglumine diatrizoate (Renografin – 76) into the left ventricle (Fig. 6.8*d*). Renografin is a radio-opaque substance used in diagnostic coronary arteriography in man.

The following observations appear to be generally applicable: ventricu-

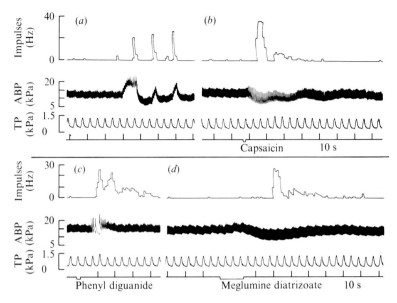

Fig. 6.8. Left ventricular C fibre stimulated by distending the left ventricle and by injecting chemicals into the blood-stream; impulses recorded by a rate-meter. *(a)* Effect of briefly occluding the descending thoracic aorta three times; blood pressure was recorded from the aortic arch. *(b)* Capsaicin (10 μg·kg^{-1}) injected into the right atrium. *(c)* Phenyl diguanide (10 μg·kg^{-1}) injected into the right atrium. *(d)* Meglamine diatrizoate (20 cm^3) injected into the left ventricle through a catheter inserted via a pulmonary vein. A control injection of saline had only a trivial effect. At no time did this ending fire with a cardiac rhythm of discharge. ABP, arterial blood pressure; TP, tracheal pressure (upstroke representing inflation).

lar C fibre endings that fire with a cardiac rhythm are rarely stimulated by capsaicin (2/11 tested) or phenyl diguanide (0/10 tested); endings with an irregular discharge throughout are frequently stimulated by capsaicin (18/21 tested) and phenyl diguanide (7/16 tested). Veratridine, by contrast, stimulates most ventricular C fibres (23/28 tested), whether they have an irregular or a cardiac rhythm of discharge. These results provide further evidence that ventricular C fibres are not a homogeneous group.

Hypoxia

It is generally agreed that ventricular C fibre endings, although susceptible to a variety of chemical agents, do not function as physiological chemoreceptors: that is to say, they are not directly stimulated by a decrease in pO_2 or an increase in pCO_2 (Coleridge *et al.*, 1964; Sleight & Widdicombe, 1965; Thorén, 1972). Thus, although Thorén (1972) described an increase of ventricular C fibre activity in cats after approximately 40 s of severe asphyxia or coronary occlusion, the augmented discharge appeared to be caused by progressive distension of the ventricular wall resulting from impairment of myocardial metabolism rather than by a direct chemical effect upon the ending. The discharge had a pulsatile rhythm in the earlier stages of the response, although ultimately it often became continuous and irregular, and its appearance usually coincided with an increase in left ventricular diastolic pressure.

Armour (1973) has suggested that chemoreceptors proper are situated in the ventricular wall. He recorded activity in myelinated fibres in the thoracic cardiac nerves of dogs, and found that asphyxia, coronary occlusion or injection of cyanide induced an irregular discharge that appeared to originate in endings in the interventricular septum. If Armour's observations have not obtained wider recognition it is because a reasonable doubt exists about the reliability of his methods for determining the exact location of the endings. Even so, results obtained recently in our laboratory go some way to support Armour's claim that chemosensitive endings are present in the ventricular wall.

We found that mechanosensitive ventricular C fibres in dogs are stimulated by hypoxia, in much the same way as Thorén (1972) has described in cats. Thus the impulse frequency of seven ventricular C fibre endings (all of which were stimulated to fire with a cardiac rhythm when the aorta was occluded) increased three- to ten-fold when the lungs were ventilated with 5% O_2 in N_2. In the example shown in Fig. 6.9, impulse activity increased from an average of 1.2 imp/s *(a)* to 4.7 imp/s after 50 s of hypoxia *(b)*. Although firing frequency varied from beat to beat during hypoxia (Fig. 6.9*b*), the discharge had a clear cardiac rhythm and there seems no reason to doubt that stimulation resulted from mechanical changes in the ventricle.

We also examined the effect of hypoxia on three left ventricular C fibre endings of a different type from those described above. The endings were stimulated when the gas ventilating the lungs was changed from O_2 to 5% O_2 in N_2, impulse frequency increasing from 0.1–0.7 imp/s to 3.8–8.2 imp/s (Fig. 6.10). The discharge was quite irregular and had no obvious relation to the cardiac cycle. A striking feature of the response of all three endings was that periodically the discharge was concentrated into groups of two to five intense bursts of activity, the groups being separated by periods of

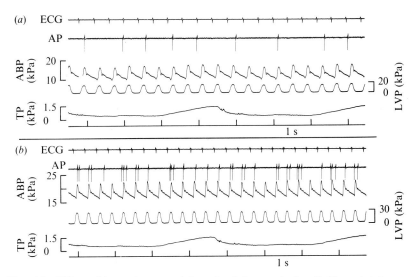

Fig. 6.9. Effect of hypoxia on activity of a left ventricular C fibre. *(a)* Lungs ventilated with air. *(b)* After the lungs had been ventilated with 5% O_2 in N_2 for 50 s; note the development of a cardiac rhythm of discharge. This ending was also stimulated by occluding the descending thoracic aorta; it was not stimulated when capsaicin or phenyl diguanide (10 $\mu g \cdot kg^{-1}$) was injected into the left atrium. ECG, electrocardiogram; AP, action potentials recorded from afferent vagal slip; ABP, arterial blood pressure; TP, tracheal pressure (upstroke representing inflation).

silence lasting several seconds (Fig. 6.10*c*). Aortic chemoreceptors in dogs often acquire a 'bursting' pattern of discharge when stimulated by hypoxia, but we have never seen them display such an intensely sporadic discharge as that encountered in these experiments. The three endings were not stimulated when the descending aorta was occluded to produce a larger increase in aortic blood pressure than occurred during hypoxia. Like the aortic chemoreceptors (Paintal, 1967), these endings were stimulated by phenyl diguanide (10 $\mu g \cdot kg^{-1}$, injected into the left atrium). Although not mechanosensitive in the conventional sense, the three endings could, like the chemoreceptors of the aortic and carotid bodies, be stimulated in a somewhat irregular and uncontrolled way by firm pressure with a blunt probe – indeed this was how we obtained our first clue to their approximate location. They were located more accurately by exploring the heart with a stimulating electrode after the dog had been killed. Each of the three endings was in the dorsal wall of the left ventricle within 1 cm of the A–V ring. Activity was abolished by localised infiltration of the ventricular wall with 1% lidocaine, or by localised incision of the ventricular wall. We do not know whether the endings were situated in aberrant groups of aortic body glomus cells, or whether they represented some other type of chemo-sensitive ending. In any event, they provided additional support for the belief that afferent C fibres arising from the ventricle are a varied group.

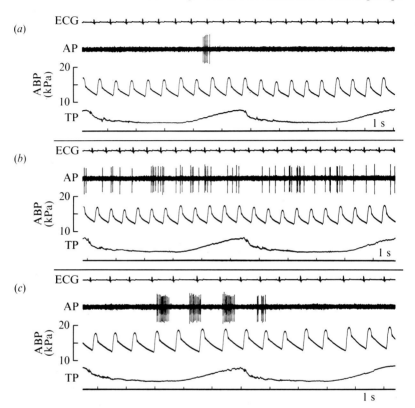

Fig. 6.10. Effect of hypoxia on impulse activity recorded from a left vagal C fibre whose ending was located in the dorsal wall of the left ventricle. *(a)* The lungs were ventilated with O_2; *(b)* the lungs had been ventilated with 5% O_2 in N_2 for 45 s; *(c)* 30 s later; the lungs were still ventilated with 5% O_2 in N_2. This ending was not stimulated by aortic occlusion; it was stimulated by left atrial injection of phenyl diguanide 10 μg·kg^{-1}. ECG, electrocardiogram; AP, action potentials recorded from afferent vagal slip; ABP, arterial blood pressure; TP, tracheal pressure (upstroke representing inflation).

Stimulation by prostaglandins

Finally, we shall briefly describe some observations, to be published in full elsewhere, that may shed light on one of the mechanisms by which ventricular C fibres are stimulated by hypoxia, for they have to do with the possibility that the endings may be activated by chemical agents formed within the heart itself. Among the latter substances, the prostaglandins (PGs) have aroused great interest in recent years and have been shown to be released by the myocardium during hypoxia and myocardial ischaemia (Berger *et al.*, 1976; Needleman, 1976). When injected into the blood-stream prostaglandins produce changes in the tone of vascular smooth muscle and in cardiac contractility and rate, which vary with the animal species and type of prostaglandin. Some of the cardiovascular effects of prostaglandins may be brought about reflexly by stimulation of afferent vagal endings in the heart. Thus Koss & Nakano (1976) found that injection of $PGF_{2\alpha}$ in cats caused cardiac slowing and a fall in arterial blood pressure. The effects, which were most pronounced when $PGF_{2\alpha}$ was injected at the coronary openings, were abolished by vagotomy. Little is known about the effect of prostaglandins on impulse activity in cardiovascular afferent fibres, apart from the observation that $PGF_{2\alpha}$ and PGE_2 have no direct effect on baroreceptor and chemoreceptor firing in the carotid sinus nerve (McQueen & Belmonte, 1974). Recently, however, Coleridge *et al.* (1976) found that PGE_2 injected into the blood-stream vigorously stimulated afferent C fibre endings in the lungs and airways.

We therefore examined the effects on ventricular C fibres in dogs of injecting PGE_2 or $PGF_{2\alpha}$ into the left atrium, in doses that produce reflex cardiovascular effects. In brief, stimulation of unmyelinated vagal afferents in the ventricles was selective, in that fibres having a pronounced cardiovascular rhythm of discharge under control conditions or in response to aortic occlusion were unaffected by prostaglandins in the doses used (PGE_2, 10 $\mu g \cdot kg^{-1}$; $PGF_{2\alpha}$, 5 $\mu g \cdot kg^{-1}$). On the other hand, ventricular fibres that fired irregularly and were relatively insensitive to an increase in ventricular pressure, but were readily stimulated by other chemicals, were strongly stimulated by PGE_2. The effect was prolonged, sometimes persisting for 5 min or more. Stimulation did not appear to be secondary to changes in ventricular pressure or volume, for some ventricular C fibre endings were stimulated by both PGE_2 and $PGF_{2\alpha}$, which have opposite effects on ventricular pressure. The 'chemoreceptor-like' ventricular C fibre endings that were stimulated by hypoxia were also stimulated by PGE_2 and fired in characteristic bursts. Neither PGE_2 nor $PGF_{2\alpha}$ appeared to act directly on the cardiovascular endings of myelinated vagal fibres (e.g. aortic baroreceptors or atrial receptors), and any changes in firing that did occur could be explained by changes in blood pressure or cardiac volume.

Summary

In conclusion, the afferent C fibres that comprise almost the entire vagal input from the cardiac ventricles are functionally diverse. Far from responding only to noxious stimuli, some are mechanoreceptors capable of signalling changes in the inotropic state of the ventricles as well as changes in its volume. Others are chemosensitive, and their importance may lie in their stimulation by substances produced in the myocardium when metabolic demands upon it are high. A few may even be chemoreceptors in the conventional sense of being stimulated by hypoxia, hypercapnia and a fall in pH. Activity in all types may increase in hypoxia or myocardial ischaemia, although the mechanism of stimulation is different in each. They may be involved in reflex effects as diverse as the human 'fainting' reaction (Öberg & Thorén, 1972a, b) and the vomiting that often accompanies coronary thrombosis in man (Abrahamsson & Thorén, 1973). It seems likely that their functional importance is concerned with 'protective' rather than with 'regulatory' reflexes, reducing metabolic demands on the myocardium by slowing heart rate and diminishing afterload.

References

Abrahamsson, H. & Thorén, P. (1973). Vomiting and reflex vagal relaxation of the stomach elicited from heart receptors in the cat. *Acta physiol. scand.*, **88**, 433–9.

Amann, A. & Schäfer, H. (1943). Über sensible Impulse im Herznerven. *Pflügers Arch. ges. Physiol.*, **246**, 757–89.

Armour, J.A. (1973). Physiological behaviour of thoracic cardiovascular receptors. *Am. J. Physiol.*, **225**, 177–85.

Berger, H.J., Zaret, B.L., Speroff, L., Cohen, L.S. & Wolfson, S. (1976). Regional cardiac prostaglandin release during myocardial ischemia in anesthetized dogs. *Circulation Res.*, **38**, 566–71.

Coleridge, H.M., Coleridge, J.C.G., Dangel, A., Kidd, C., Luck, J.C. & Sleight, P. (1973). Impulses in slowly conducting vagal fibers from afferent endings in the veins, atria and arteries of dogs and cats. *Circulation Res.*, **33**, 87–97.

Coleridge, H.M., Coleridge, J.C.G., Ginzel, K.H., Baker, D.G., Banzett, R.B. & Morrison, M.A. (1976). Stimulation of 'irrritant' receptors and afferent C-fibres in the lungs by prostaglandins. *Nature (Lond.)*, **264**, 451–3.

Coleridge, H.M., Coleridge, J.C.G. & Kidd, C. (1964). Cardiac receptors in the dog, with particular reference to two types of afferent ending in the ventricular wall. *J. Physiol.*, **174**, 323–39.

Coleridge, J.C.G., Hemingway, A., Holmes, R.L. & Linden, R.J. (1957). The location of atrial receptors in the dog: a physiological and histological study. *J. Physiol.* **136**, 174–97.

Jarisch, A. & Zotterman, Y. (1948). Depressor reflexes from the heart. *Acta physiol. scand.*, **16**, 31–51.

Kappagoda, C.T., Linden, R.J. & Mary, D.A.S.G. (1976). Atrial receptors in the cat. *J. Physiol.*, **262**, 431–46.

Koss, M.C. & Nakano, J. (1976). Reflex bradycardia and hypotension produced by prostaglandin $F_{2\alpha}$ in the cat. *Br. J. Pharmac. Chemother.*, **56**, 245–53.

McQueen, D.S. & Belmonte, C. (1974). The effects of prostaglandins E_2, A_2 and $F_{2\alpha}$ on carotid baroreceptors and chemoreceptors. *Q. Jl exp. Physiol.*, **59**, 63–71.

Muers, M.F. & Sleight, P. (1972). Action potentials from ventricular mechanoreceptors stimulated by occlusion of the coronary sinus in the dog. *J. Physiol.*, **221**, 283–309.

Needleman, P. (1976). The synthesis and function of prostaglandins in the heart. *Fedn Proc. Fedn Am. Socs exp. Biol.*, **35**, 2376–81.

Öberg, B. & Thorén, P. (1972a). Studies on left ventricular receptors, signalling in non-medullated vagal afferents. *Acta physiol. scand.*, **85**, 145–63.

Öberg, B. & Thorén, P. (1972b). Increased activity in left ventricular receptors during hemorrhage or occlusion of caval veins in the cat. A possible cause of the vaso-vagal reaction. *Acta physiol. scand.*, **85**, 164–73.

Paintal, A.S. (1953). A study of right and left atrial receptors. *J. Physiol.*, **120**, 596–610.

Paintal, A.S. (1955). A study of ventricular pressure receptors and their role in the Bezold reflex. *Q. Jl. exp. Physiol.*, **40**, 348–63.

Paintal, A.S. (1967). Mechanism of stimulation of aortic chemoreceptors by natural stimuli and chemical substances. *J. Physiol.*, **189**, 63–84.

Paintal, A.S. (1973). Sensory mechanisms involved in the Bezold–Jarisch effect. *Aust. J. exp. Biol. med. Sci.*, **51**, 3–15.

Sleight, P. & Widdicombe, J.G. (1965). Action potentials in fibres from receptors in the epicardium and myocardium of the dog's left ventricle. *J. Physiol.*, **181**, 235–58.

Thames, M.D., Donald, D.E. & Shepherd, J.T. (1977). Behavior of cardiac receptors with vagal nonmyelinated afferents during spontaneous respiration in cats. *Circulation Res.*, **41**, 694–701.

Thorén, P. (1972). Left ventricular receptors activated by severe asphyxia and by coronary artery occlusion. *Acta physiol. Scand.*, **85**, 455–63.

Thorén, P. (1976). Atrial receptors with nonmedullated vagal afferents in the cat: discharge frequency and pattern in relation to atrial pressure. *Circulation Res.*, **38**, 357–62.

Thorén, P. (1977). Characteristics of left ventricular receptors with nonmedullated vagal afferents in cats. *Circulation Res.*, **40**, 415–21.

Thorén, P., Saum, W.R. & Brown, A.M. (1977). Characteristics of rat aortic baroreceptors with nonmedullated afferent nerve fibers. *Circulation Res.*, **40**, 231–7.

Discussion

(a) When asked whether the variability of firing pattern observed from ventricular C fibres could be related to changes in the acid–base status of the blood, Coleridge indicated that he had not looked at this aspect.

(b) In relation to a possible chemoreceptor function for some C fibres, particularly those sensitive to changes in partial pressure of oxygen, the point was made that latency of the response (approximately 30–40 s) to 5% oxygen was rather long for a chemoreceptor. In reply, Coleridge indicated that recordings had been made from chemoreceptors under these circumstances and they did not take such a long time to become excited. The major difference between aortic chemoreceptors and the receptors attached to the C fibres was that aortic chemoreceptors produced a rather steady rate of firing which was maintained for some considerable time; the activity from ventricular C fibres, on the other hand, was much more irregular. In reply to a question about responses occurring over more physiological ranges during hypoxia and hypercapnia, Coleridge indicated that he was very well aware of the rather crude character of many of the changes; it was his intention in the future to make more sophisticated observations.

(c) Thorén confirmed the extreme variability of the responses of the receptors attached to vagal C fibres and pointed out that C fibres in the cat behaved in equally 'crazy' ways. He asked whether there was a difference between cat and dog ventricular C fibres in that many of those in the dog fired during systole while in the cat they fired mainly in diastole. Coleridge pointed out that although in his study the majority of fibres discharged during systole there were so many exceptions to this that he thought that it was likely that there was another variable, not yet defined. On the other hand, Thorén felt the major differences between the cat and the dog were concerned with the heart rate and the consequent cycle length. With a high heart rate the diastolic period was very short and this was probably the reason why it was difficult to observe diastolic firing in the cat.

(d) The problems of localisation procedures for receptors in the heart were discussed. Thorén indicated that he had never found receptors which, on the basis of rough location in the beating heart, were thought to be in the ventricle, but were eventually found to be located outside the heart. Coleridge indicated that in his experience, even in the very much bigger heart of the dog, it was frequently extremely difficult to be certain about the precise location of a receptor unless it had been located in the fully opened and exposed heart. He made the point that a small number of receptors accurately located were worth infinitely more than many less precisely localised.

(e) When asked why arterial blood pressure fell dramatically but heart rate did not change following infusions of prostaglandins, Coleridge said that the absence of a heart rate response might well be a result of damage to

vagal efferent fibres during the dissection. However, he would not wish to indicate that the effects of prostaglandins were the same as the Bezold–Jarisch effect since a wide variety of fibres was being stimulated. He made the point that cardiovascular changes following infusion of prostaglandins were quite dramatic and prolonged, the reflex effects could continue for 4–5 min and over this period the endings continued to fire. He indicated that the reflex changes were complex and likely to be the result of activation of a wide spectrum of afferent fibres; there was evidence that some effects may be caused by direct action on vascular smooth muscle. When asked whether prostaglandins influenced medullated fibres from the heart, Coleridge indicated that neither prostaglandins $F_{2\alpha}$ nor prostaglandins $E_{2\beta}$ had a direct effect on activity of baroreceptors or atrial receptors; any changes observed were the result of secondary changes in blood pressures.

(f) The suggestion was made that some of the variation in the activity of the C fibre afferents may be caused by alterations in either sympathetic or the vagal efferent activity which was then modulating the activity of receptors; small variations in the autonomic efferent output would not normally be controlled or observed. Coleridge said it was difficult to see why such alterations in activity should primarily effect C fibre afferents and not A fibre afferents.

7. Activity of atrial receptors under normal and pathological states

J. P. GILMORE & I. H. ZUCKER

Unencapsulated nerve endings which are situated in the endocardium of the atria and which discharge during the v wave of the atrial pulse have been designated as type B atrial receptors by Paintal (1953). These are believed to be the so-called atrial volume receptors which modulate the secretion of antidiuretic hormone. Although considerable work has been done describing the physiological responses to stimulation of these receptors, information is available concerning their natural stimulus, the extent to which they may be influenced by efferent neural activity and the extent to which they may be altered under pathological conditions. Therefore, we undertook an extensive investigation to characterise further type B atrial receptors in both the dog and non-human primate and to determine their possible physiological function.

The discharge of a typical type B left atrial receptor in the dog is shown in Fig. 7.1 (Gilmore & Zucker, 1974a). Panel (a) shows activity during the control period. During each cardiac cycle the discharge frequency paralleled the level of the v wave of the atrial pressure pulse and discharge stops abruptly immediately before atrial contraction at the time the A–V valve opens. Between panels (a) and (b) and (b) and (c) 150 cm³ of a saline blood mixture was infused intravenously. Discharge frequency increased with each increment in atrial pressure. These experiments demonstrate that the natural stimulus to type B receptors appears to be the level of the atrial v wave rather than atrial pulse pressure. This is consistent with the view that the natural stimulus for type B receptors is atrial distension. If type B receptors adapt they do not appear to be rapidly adapting. Left atrial pressure was increased by inflating a balloon that was previously placed in the left atrium of the dog while recording from a type B atrial receptor (Gilmore & Zucker, 1974a). Although spike discharge decreased

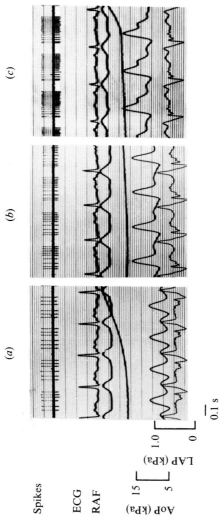

Fig. 7.1. Original recordings of a type B left atrial receptor. From above downwards the records are; the neurogram (spikes), electrocardiogram (ECG), right atrial force (RAF), a tracing of the phase of respiration (inspiration upward), aortic pressure (AoP) and left atrial pressure (LAP). (*a*) Control recording; (*b*) after a 150-cm^3 infusion of an isotonic saline-blood mixture; (*c*) after a 300-cm^3 infusion of a saline-blood mixture was infused. (Gilmore & Zucker (1974*a*), reproduced with permission.)

slightly after 5 min of inflation the decrease was not significant nor were significant changes found at the 10- and 15-min periods.

In view of the finding that the osmoreceptors which mediate the salt gland reflex in the goose appear to be located in the heart (Hanwell, Linzell & Peaker, 1972), it was of interest to examine the possibility that type B atrial receptors in the dog are osmosensitive. Fig. 7.2 shows a recording from a type B left atrial receptor before *(a)* and after 5 *(b)*, 10 *(c)* and 15 min *(d)* of the intravenous infusion of a 20% sodium chloride solution (Gilmore & Zucker, 1974*b*). It may be noted that over this wide range of infusion there was no alteration in discharge frequency. These results are representative of ten receptors studied for which plasma osmolality was increased by as much as 50 mOsmol·litre^{-1}. Thus, it appears that type B receptors are not osmosensitive.

In order to determine the extent to which type B atrial receptors might be sensitive to ischaemia, studies were done in which the left coronary artery was acutely occluded while recording was made from type B receptors. Although coronary occlusion increases type B atrial receptor discharge it could not provide information as to whether the increase in discharge was a result of the ischaemia or secondary to the accompanying elevation in left atrial pressure. In order to distinguish between the two possibilities, left atrial pressure was increased in a stepwise manner by an intravenous infusion of warm isotonic sodium chloride and the relationship between receptor discharge and atrial pressure determined. Atrial pressure was then reduced to the pre-infusion level and the main left coronary artery occluded so that atrial pressure was increased again over a wide range. As a result we were able to compare the relationship between left atrial pressure and receptor discharge when pressure was increased without ischaemia (volume expansion) and when left atrial pressure was increased as a result of ischaemia (coronary artery occlusion). The results from two of the six experiments are shown in Fig. 7.3 (Zucker & Gilmore, 1974*a*). For any given increase in left atrial pressure, the increase in receptor discharge was essentially the same whether the increase was a result of volume expansion or the result of coronary occlusion. From these experiments it may be concluded that the increase in type B atrial receptor discharge which is observed during acute coronary occlusion is a result of the associated increase in atrial pressure and not a direct effect of coronary ischaemia.

We next determined the extent to which autonomic efferent neural activity might influence the activity of type B atrial receptors in the dog. The basic question was whether or not efferent activity can set or modify

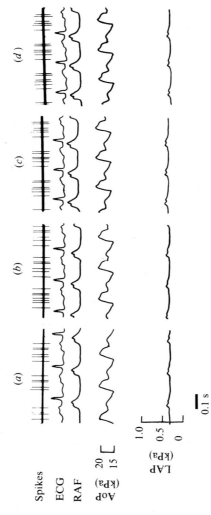

Fig. 7.2. Tracing of recordings of activity from a left atrial type B receptor. From above downward: electroneurogram (spikes), electrocardiogram (ECG), right atrial force (RAF), aortic pressure (AoP), left atrial pressure (LAP). All recordings were made during expiratory pause. *(a)* Control; *(b)* 5 min after infusing 20% sodium chloride; *(c)* 10 min after infusing 20% sodium chloride; *(d)* 15 min after infusing 20% sodium chloride. (Gilmore & Zucker (1974*b*), reproduced with permission.)

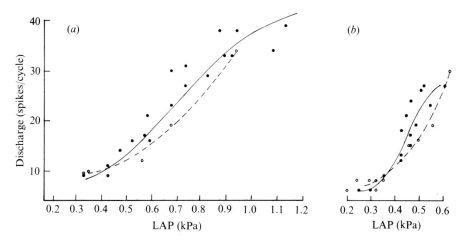

Fig. 7.3. Response of two left atrial receptors during coronary artery occlusion (●) and during volume expansion (○). Left atrial pressure (LAP) refers to pressure taken at peak of *v* wave of left atrial pulse.

receptor activity at any given level of atrial stretch. This question is of particular interest currently in view of the recent histological observation that efferent fibres may course with axons projecting from atrial receptors (Tranum-Jensen, Chapter 2 of this volume). In addition, efferent control of afferent activity has been claimed for arterial baroreceptors (Kedzi, 1954; Koizumi & Sato, 1969; Sampson & Mills, 1970; Bagshaw & Peterson, 1972). During sympathetic stimulation there was a large decrease in receptor discharge, an increase in atrial contractility as indicated by the right atrial force tracing, an increase in aortic pressure, and a substantial decrease in left atrial pressure. In order to determine if the decrease in receptor discharge was caused by a direct effect of sympathetic nerve stimulation or was secondary to the associated decline in atrial pressure, experiments were performed in which atrial pressure was increased by volume expansion and then subsequently decreased by bleeding in order to obtain the relationship between receptor discharge and atrial pressure while unloading the receptor. This procedure was necessary since the relationship between atrial pressure and type B receptor discharge showed hysteresis (Gilmore & Zucker, 1974a). Thus, it was necessary to establish an unloading curve in the absence of sympathetic stimulation and compare the results with those obtained during cardiac nerve stimulation. The two curves were not significantly different from each other (Zucker & Gilmore,

1974*b*). Similar results were obtained when noradrenaline was administered. During vagal nerve stimulation there was a substantial increase in
receptor discharge coincident with an elevation and prolongation of the *v*
wave of the atrial pulse, and atrial contractility was completely depressed
(Wahab, Zucker & Gilmore, 1975). In order to determine whether the
increase in receptor discharge was the direct effect of vagal nerve
stimulation or secondary to the associated haemodynamic changes the
following experiments were carried out. Atrio-ventricular pacing with a
delay of 150 ms was initiated and atrial pressure increased in a stepwise
manner by the intravenous infusion of isotonic sodium chloride. The
relationship between receptor discharge and atrial pressure was then
determined over a broad range, after which atrial pressure was reduced to
the pre-infusion level by bleeding and vagal nerve stimulation initiated at
an intensity which would produce a substantial bradycardia in the absence
of A–V pacing. Atrial pressure was once again increased over the same
range by the stepwise infusion of isotonic sodium chloride. When
compared with the control curve (pace) for any given level of atrial pressure,
vagal stimulation had no consistent influence on receptor discharge. These
studies clearly indicate that the affect of autonomic efferent nerve activity
on atrial receptor activity is an indirect one mediated through alterations in
atrial contractility or the increase in left atrial pressure, experiments were
Therefore, these experiments do not support the view that there is a direct
efferent neural control of type B receptor activity. Because of the potential
importance of atrial receptors in the regulation of fluid balance we next
directed our attention to the question concerning the extent to which type B
atrial receptor activity may be altered in pathological states, i.e. under
conditions of alterations in atrial contractility or chronic atrial distension.
Propranolol was used to depress acutely atrial contractility. Following the
injection of propranolol there was an increase in atrial receptor activity, a
decrease in right atrial force and arterial pressure and an increase in left
atrial pressure. In order to determine if the increase in receptor activity
following the injection of propranolol was the result of the depression of
atrial contractility or the increase in left atrial pressure, experiments were
carried out in which the relationship between left atrial pressure and
receptor discharge was determined over a broad range by the infusion of
isotonic sodium chloride before and after the administration of propranolol. Propranolol had no significant effect on the relationship between the
change in atrial pressure and change in type B receptor discharge.
Therefore, the increased receptor discharge observed after an acute

depression of atrial contractility appears to be a result of the associated increase in atrial filling pressure. Similar results were obtained using injections of pentobarbital sodium in doses which depress atrial contractility.

During clinical episodes of atrial tachyarrhythmias, such as atrial fibrillation and paroxysmal atrial tachycardia, a pronounced polyuria is often exhibited which lasts in some cases longer than the paroxysm itself (Wood, 1963). Although it has been shown that atrial pressure is often elevated during these episodes the contribution of increasing heart rate *per se* to atrial receptor discharge is not clear. Therefore, we undertook studies to characterise the discharge of type B atrial receptors in response to alterations in heart rate induced by electrical pacing of the atrium. Fig. 7.4 shows an experiment in which heart rate was altered over a broad range while recording from a type B atrial receptor in the dog (Zucker & Gilmore, 1976). Before the experiment, the animal was given a total of 400 cm^3 of isotonic sodium chloride to elevate receptor discharge so that the effects of increasing heart rate would be more easily discerned. With each increment in heart rate there was, in general, a decrease in the number of spikes per cardiac cycle. This decrease was associated with a progressive decrease in the peak *v* wave of the atrial pressure pulse. Fig. 7.5 shows the relationship between atrial receptor discharge in spikes per cardiac cycle and left atrial peak *v* wave pressure from one animal, heart rate ranging from 90 to 240 beats/min. Irrespective of the heart rate there is an excellent correlation between discharge per cardiac cycle and left atrial peak *v* wave pressure. Fig. 7.6 shows the relationship between discharge in spikes per minute (the product of spikes per cycle and heart rate) versus left atrial peak *v* wave pressure for the same receptor shown in Fig. 7.5. Again, irrespective of heart rate there is a good correlation between discharge in spikes per minute and left atrial peak *v* wave pressure. It was also found that there was no significant difference between the discharge at 90 beats/min and any of the higher heart rates. Therefore, it is apparent that during atrial tachycardia the input to the central nervous system in terms of the number of discharges per minute does not vary over a wide range of heart rates. From this we conclude that the polyuria observed during paroxysmal atrial tachycardia is not secondary to alterations in type B atrial receptor discharge. These results are not consistent with the suggestion of Boykin, Cadnapaphornchai, McDonald & Schrier (1975). They observed that the diuresis which accompanies atrial tachycardia is associated with an increase in atrial pressure and therefore suggested that atrial receptors might be involved in the response. However, these authors reported mean

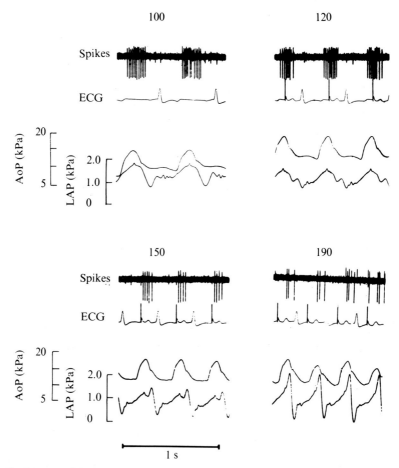

Fig. 7.4. An original recording of a type B atrial receptor at varying heart rates. The horizontal lines are time markers. The numbers above each panel denote the heart rate. At heart rates of 120–190 beats/min a pacing artifact is evident in the electrocardiogram (ECG) trace. Horizontal line at lower left indicated 1 s. AoP = aortic pressure; LAP = left atrial pressure. (Zucker & Gilmore, 1976, reproduced with permission.)

atrial pressure rather than peak v wave pressure and the latter correlates better with atrial receptor discharge. Mean atrial pressures increased with heart rate and this was undoubtedly caused by a progressive increase in the a wave of the atrial pulse (see Fig. 7.4). Although our experiments demonstrate clearly that the polyuria of atrial tachycardia cannot be the result of an increase in type B atrial receptor activity we have found that

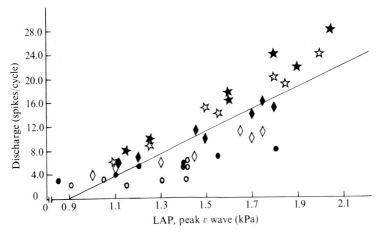

Fig. 7.5. The relationship between discharge in spikes per cardiac cycle and left atrial peak *v* wave pressure (Y 1.83X − 16.3; r = 0.85) at the following heart rates: ★ = 90 beats/min; ☆ = 120; ♦ = 150; ◊ = 180; ●, 210; ○, 240. LAP = left atrial pressure. (Zucker & Gilmore (1976), reproduced with permission.)

during atrial fibrillation or flutter there is a significant increase in type B and type A atrial receptor discharge when expressed as discharge per minute (Zucker & Gilmore, 1973).

In 1973 Greenberg and his associates reported that there was a significant impairment in the response of type B atrial receptors in dogs in which heart failure was induced by tricuspid insufficiency and pulmonary artery stenosis. They observed that at any given central venous pressure, type B atrial receptor discharge was significantly less in animals with heart failure than in normal control animals. They concluded that impairment of salt and water homeostasis in heart failure is caused, at least in part, by a decreased sensitivity of type B atrial receptors. Several years ago we had the opportunity to study a 16-year-old Chihuahua which had all the clinical signs of myocardial failure and the heart of this animal showed retraction and calcification of both mitral cusps. In addition the heart was greatly dilated and the left ventricle hypertrophied. A jetting lesion, indicative of aortic stenosis, was also noted. We successfully recorded from three type B atrial receptors in this animal. Fig. 7.7 compares the results from this animal with a group of normal animals. As Greenberg *et al.* (1973) found, there was for any given left atrial pressure a significantly diminished receptor discharge in the animal with heart failure when compared with the normal dogs. Since the availability of dogs which develop heart failure

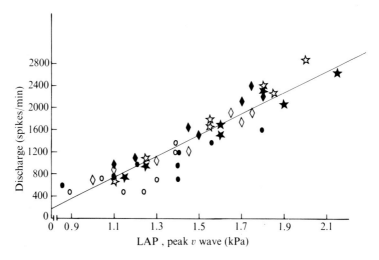

Fig. 7.6. The relationship between discharge in spikes per minute ((spikes per cycle) × heart rate) ($Y = 192.9X - 1380.4$; $r = 0.94$) versus left atrial peak v wave pressure for the same receptor illustrated in Fig. 7.5, at the following heart rates:★, 90 beats/min;☆, 120; ◆, 150; ◊, 180; ●, 210; ○, 240. LAP = left atrial pressure. (Zucker & Gilmore (1976), reproduced with permission.)

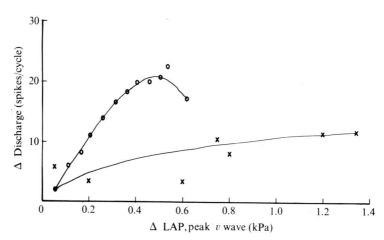

Fig. 7.7. Influence of heart failure on responsiveness of type B atrial receptors: ○——○, normal ($n = 24$); ×——×, heart failure ($n = 3$). LAP = left atrial pressure. See text for further description.

naturally was very limited we subsequently induced heart failure in a series of animals by establishing an arterio-venous anastomosis a few centimetres below the level of the renal arteries. We preferred this model to that used by Greenberg and his associates since it precluded entering the chest and thus the possibility of damaging cardiac neural pathways and thoracic adhesions. The electrophysiological results from these animals confirmed the finding that type B receptor sensitivity was significantly less than in normal dogs. To obtain further insight into the physiological consequences of this reduced type B atrial receptor sensitivity, experiments were undertaken in which we compared the renal response to atrial balloon distension of dogs in heart failure with a group of normal, sham-operated controls. In the sham-operated controls atrial distension was always associated with a significant increase in urine flow which was primarily caused by an increase in free water excretion. However, when atrial pressure was raised by balloon distension to a similar degree in dogs with congestive heart failure there was little or no change in urine flow. No consistent changes were observed in creatinine clearance or effective renal plasma flow in either group of animals. Preliminary data indicate that this decrease in sensitivity of type B atrial receptors in the dog in congestive failure as well as their failure to respond to physiological alterations in atrial pressure is secondary to structural changes in the receptor endings. We have observed that the receptor endings from dogs in congestive heart failure are grossly altered in that they tend not to show the typical unencapsulated endings. Those endings which are present do not show the classic arborisation seen with normal receptor endings. Therefore, we have tentatively concluded that these experiments support our hypothesis that the decrease in sensitivity of type B atrial receptors and the attenuation of the diuretic response to alterations of atrial pressure in heart failure is the result of structural alterations in the receptor endings that are presumed to modulate the secretion of antidiuretic hormone. At the same time it is possible that the reduced responsiveness of atrial receptors in chronic heart failure is the result of overdistension of the atrium rather than the result of structural changes in the receptor endings. This is suggested by Fig. 7.8 which shows the relation between type B atrial receptor discharge and left atrial pressure from three normal dogs. At high levels of atrial pressure and stretch there was a greatly reduced receptor discharge. In an attempt to reverse the depressed sensitivity of atrial stretch receptors in congestive heart failure we administered a cardiac glycoside to several of these animals. Fig. 7.9 indicates the response of a left atrial type B receptor in a

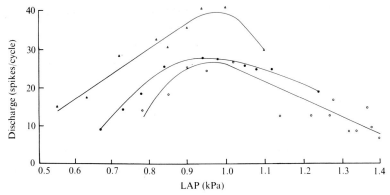

Fig. 7.8. The influence of acutely induced high levels of atrial pressure on type B atrial receptor discharge from three normal dogs. Atrial pressure was increased by the i.v. administration of isotonic sodium chloride. LAP = left atrial pressure.

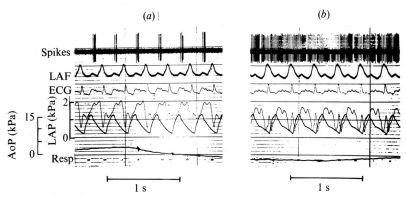

Fig. 7.9. Influence of ouabain on a type B receptor. See text for further description of figure. LAF = left atrial force; LAP = left atrial pressure; AoP = aortic pressure; ECG = electrocardiogram.

dog with an A–V fistula that resulted in chronic congestive heart failure. Panel *(a)* indicates the control discharge. Note the high left atrial pressure and wide aortic pulse pressure. Panel *(b)* was taken 2 min after intravenous administration of 1 mg of ouabain. Left atrial pressure is decreased while left atrial force is enhanced and receptor discharge is markedly potentiated. The results of this experiment clearly indicate that ouabain has a direct effect upon the receptor ending. The possibility therefore exists that part of the efficacious effect of cardiac glycosides in heart failure may be mediated through atrial receptor stimulation.

The experiments presented were carried out using the dog as the experimental preparation. However, little work has been done concerning the reflex effects of atrial stretch receptor stimulation in the non-human primate or in man himself. The contribution of atrial receptors to the diuresis which results in man during procedures such as negative pressure breathing, whole body water immersion, acceleration and weightlessness remains largely suggestive (Gauer & Henry, 1976). Although atrial stretch receptors have been demonstrated histologically in both the human (Johnston, 1968; Ábrahám, 1969) and monkey (Ábrahám, 1969) there has been only one study in which the activity of fibres originating from the atrium of the primate has been studied electrophysiologically (Chapman & Pearce, 1959). However, no attempt was made to determine the responsiveness of the receptors to alterations in atrial distension. Therefore we undertook a series of experiments in order to determine first the extent to which type B atrial receptors in the non-human primate are responsive to alterations in atrial pressure and, secondly, to determine the renal responses of the non-human primate to interventions which would presumably stimulate type B atrial receptors. Fig. 7.10 shows the discharge of a typical type B atrial receptor obtained from a monkey (Zucker & Gilmore, 1975). Panel *(a)* shows a control tracing, panel *(b)* after 20 cm³ of isotonic saline had been infused, and panel *(c)* after a total of 50 cm³ of saline were administered. As described for the dog and other species receptor discharge begins as the atrium fills and reaches its peak at the peak of the *v* wave of the left atrial pulse. There was a progressive increase in the

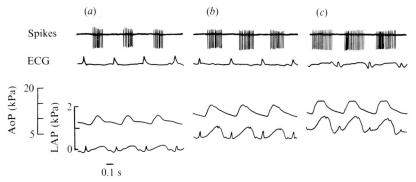

Fig. 7.10. A tracing of an original recording from a type B left atrial receptor in an open chest monkey. *(a)* Control; *(b)* following the i.v. infusion of 20 cm³ of isotonic saline; *(c)* following the i.v. infusion of a total of 50 cm³ of saline. Abbreviations from above downward: ECG = electrocardiogram; AoP = aortic pressure; and LAP = left atrial pressure. (Zucker & Gilmore (1975), reproduced with permission.)

number of spikes per cardiac cycle as atrial pressure was increased with each increase in volume. Although this experiment demonstrates that type B atrial receptors in the monkey are responsive to alterations in atrial pressure their sensitivity is substantially less than those in the dog. Fig. 7.11 compares the relationship between the change in discharge of type B atrial receptors and the change in the left atrial pressure or central venous

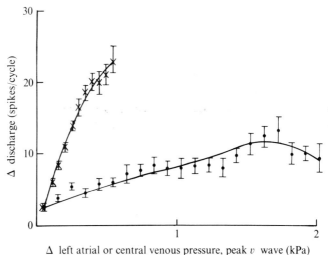

Δ left atrial or central venous pressure, peak v wave (kPa)

Fig. 7.11. The relationship between the change in discharge (spikes/cycle) and the change in left atrial or central venous pressure for dog and monkey. ×——×, dog ($n=24$); ●——●, monkey ($n=24$). Vertical bars are ± s.e.m. (Zucker & Gilmore (1975), reproduced with permission.)

pressure for the dog and monkey. Compared with the dog, for any given change in left atrial or central venous pressure, the discharge of type B atrial receptors is substantially diminished in the monkey. We are presently carrying out experiments in order to determine the renal responses of the monkey to increases in atrial pressure induced by inflation of an atrial balloon. Although preliminary, these experiments clearly show that compared with the normal dog significant increases in atrial pressure in the monkey, induced by atrial balloon inflation, have little or no effect on urine flow. We believe that the difference in type B receptor sensitivity between the dog and monkey is related to postural differences between the two species and raises the question as to the appropriateness of using the dog as a model for man in studying mechanisms involved in salt and water homeostasis and thus in the control of blood volume.

In recent years the concept has developed that whole body water immersion may be a good model for studying volume homeostasis in man (Gauer & Henry, 1976). When man is immersed to the neck, there is a redistribution of blood volume from the periphery into the thorax, resulting in an increase in central blood volume and cardiac stretch (Arborelius, Balldin, Lilja & Lundgren, 1972). This is associated with substantial increse in urine flow and sodium excretion (Behn, Gauer, Kirsch & Eckert, 1969; Epstein, Duncan & Fishman, 1972). It has been suggested that the diuresis is mediated by atrial receptors whose axons traverse the vagi (Arborelius *et al.,* 1972; Gauer & Henry, 1976). In order to determine the contribution of vagal pathways to the renal responses to immersion in the primate we carried out a series of experiments, one of which is shown in Fig. 7.12. The data were obtained from a vagotomised monkey which was immersed to the neck in a thermal neutral tank. During immersion there occurred a substantial increase in both urine flow and sodium excretion. This experiment as well as others clearly demonstrates that vagal pathways are not necessary for the renal responses to immersion on the non-human primate and thus presumably in man. In a second series of experiments which are in progress we have immersed vagotomised monkeys pretreated with both mineralocorticoids and antidiuretic hormone. In the four animals studied immersion was always accompanied by a substantial increase in urine flow and sodium excretion. These experiments therefore preclude the possibility that the renal responses to immersion, at least in the non-human primate, are dependent upon intact vagal pathways or alterations in the circulating levels of mineralocorticoids or antidiuretic hormone.

There is one interesting aspect of the haemodynamic responses to immersion that has been ignored by previous authors. This is the failure of the human (Arborelius *et al.,* 1972) and indeed our monkeys to respond to immersion with a substantial tachycardia despite a substantial elevation of atrial pressure. It would thus appear that although a Bainbridge reflex is clearly demonstrable in the dog by selective stimulation of atrial receptors (Linden, 1973) it plays little or no physiological role in the primate.

General conclusions

(1) The natural stimulus for type B atrial receptors is stretch, and over a broad physiological range the rate of change of atrial pressure or stretch has no significant effect on type B atrial receptor discharge.

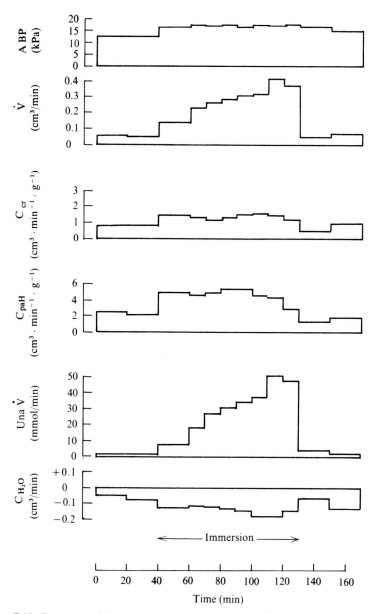

Fig. 7.12. Response of the bilaterally vagotomised monkey to water immersion. ABP = mean arterial blood pressure; $\dot{V}$ = urine volume; C_{cr} = exogenous creatinine clearance; C_{paH} = PAH clearance; Una $\dot{V}$ = sodium excretion; C_{H_2O} = free water clearance.

(2) Type B atrial receptors are not osmosensitive.

(3) Cardiac sympathetic nerve stimulation decreases and cardiac vagal nerve stimulation increases type B atrial receptor discharge. However, these effects are secondary to alterations in atrial stretch rather than caused by a direct effect on the receptor itself.

(4) The increase in type B atrial receptor discharge observed during coronary occlusion is the result of the associated increase in atrial stretch rather than ischaemia *per se*.

(5) There is an inverse relationship between heart rate and type B atrial receptor discharge per cardiac cycle. However, the discharge per minute is not significantly altered.

(6) The sensitivity of type B atrial receptor in the dog with chronic heart failure is significantly decreased. This reduced sensitivity is associated with a greatly reduced renal response to atrial pressure elevation and with structural changes in the receptor endings.

(7) The sensitivity of type B atrial receptors in the monkey is significantly less than that of the dog and we suggest that this difference is related to postural differences between the two species.

(8) The renal responses to immersion in the non-human primate, and presumably in man, are not dependent upon the presence of vagal pathways or alterations in the circulating levels of mineralocorticoids or antidiuretic hormone.

(9) In contrast to the dog, the Bainbridge reflex plays little, if any, role in the cardiovascular responses to elevation of central venous or atrial pressure in the monkey.

References

Ábrahám, A. (1969). *Microscopic innervation of the heart and blood vessels in vertebrates including man*. Pergamon Press: Oxford.

Arborelius, M. Jr, Balldin, U.I., Lilja, B. & Lundgren, D.E.G. (1972). Hemodynamic changes in man during immersion with the head above water. *Aerospace Med.*, **43**, 592.

Bagshaw, R.J. & Peterson, L.H. (1972). Sympathetic control of mechanical properties of the canine carotid sinus. *Am. J. Physiol.*, **22**, 1462–8.

Behn, C., Gauer, O.H., Kirsch, K. & Eckert, P. (1969). Effects of sustained intrathoracic vascular distension on body fluid distribution and renal excretion in man. *Pflügers Arch. ges. Physiol.*, **313**, 123–35.

Boykin, J., Cadnapaphornchai, P., McDonald, K.M. & Schrier, R.W. (1975). Mechanism of diuretic response associated with atrial tachycardia. *Am. J. Physiol.*, **229**, 1486–91.

Chapman, K.M. & Pearce, J.W. (1959). Vagal afferents in the monkey. *Nature (Lond.)*, **184**, 1237–8.

Epstein, M., Duncan, D.C. & Fishman, L.M. (1972). Characterization of the natriuresis caused in normal man by immersion in water. *Clin. Sci.*, **43**, 275–87.

Gauer, O.H. & Henry, J.P. (1976). Neurohormonal control of plasma volume. *International Review of Physiology II*, **9**, 145–90.

Gilmore, J.P. & Zucker, I.H. (1974a). Discharge of type B atrial receptors during changes in vascular volume and depression of atrial contractility. *J. Physiol.*, **239**, 207–23.

Gilmore, J.P. & Zucker, I.H. (1974b). Failure of the type-B atrial receptors to respond to increase in plasma osmolarity in the dog. *Am. J. Physiol.*, **27**, 1005–7.

Greenberg, T.T., Richmond, W.H., Stocking, R.A., Gupta, P.D., Meehan, J.P. & Henry, J.P. (1973). Impaired atrial receptor responses in dogs with heart failure due to tricuspid insufficiency and pulmonary artery stenosis. *Circulation Res.*, **32**, 424–33.

Hanwell, A., Linzell, J.L. & Peaker, M. (1972). Nature and location of the receptors of salt gland secretion in the goose. *J. Physiol.*, **226**, 453–72.

Johnston, B.D. (1968). Nerve endings in the human endocardium. *Am. J. Anat.*, **122**, 621–9.

Kedzi, P. (1954). Control by the superior cervical ganglion of the state of contraction and pulsatile expansion of the carotid sinus arterial wall. *Circulation Res.*, **2**, 367–71.

Koizumi, K. & Sato, A. (1969). Influence of sympathetic innervation on carotid sinus baroreceptor activity. *Am. J. Physiol.*, **216**, 321–9.

Linden, R.J. (1973). Function of cardiac receptors. *Circulation*, **48**, 463–80.

Paintal, A.S. (1953). A study of right and left atrial receptors. *J. Physiol.*, **120**, 596–710.

Sampson, S.R. & Mills, E. (1970). Effects of sympathetic stimulation on discharges of carotid sinus baroreceptors. *Am. J. Physiol.*, **218**, 1650–3.

Wahab, N.S., Zucker, I.H. & Gilmore, J.P. (1975). Lack of a direct effect of efferent cardiac vagal nerve activity on atrial receptor activity. *Am. J. Physiol.*, **229**, 314–17.

Wood, P. (1963). Polyuria in paroxysmal tachycardia and paroxysmal atrial flutter and fibrillation. *Br. Heart J.*, **25**, 273.

Zucker, I.H. & Gilmore, J.P. (1973). Left atrial receptor discharge during atrial arrhythmias in the dog. *Circulation Res.*, **33**, 672–7.

Zucker, I.H. & Gilmore, J.P. (1974a). Atrial receptor discharge during acute coronary occlusion in the dog. *Am. J. Physiol.*, **227**, 360–3.

Zucker, I.H. & Gilmore, J.P. (1974b). Evidence for an indirect sympathetic control of atrial stretch receptor discharge in the dog. *Circulation Res.*, **34**, 441–6.

Zucker, I.H. & Gilmore, J.P. (1975). Responsiveness of type B atrial receptors in the monkey. *Brain Res.*, **95**, 159–65.

Zucker, I.H. & Gilmore, J.P. (1976). The response of atrial stretch receptors to increases in heart rate in dogs. *Circulation Res.*, **38**, 15–19.

Responses of type A atrial vagal receptors to changes in atrial volume and contractility

G. RECORDATI

from Istituto Ricerche Cardiovascolari and CNR, University of Milano,
Milan, Italy

Using anaesthetised, immobilised cats with their chests open we studied the effects of changes in atrial contractility (induced either by propranolol infusion or by vagal stimulation) on the impulse activity of type A right atrial receptors at different atrial volumes and pressures (different static levels of stretch). The response of the receptors was assessed as instantaneous firing rate both during atrial contraction (active stretch) and filling (passive stretch).

For any given atrial pressure and volume (infusion of saline and withdrawal of blood) the response of the receptors to active and passive stretch was greater during control conditions than after propranolol infusion or vagal stimulation. During control conditions different receptors had a different sensitivity to changes in atrial pressure: these differences were decreased by propranolol and abolished by vagal stimulation, as under this latter condition all the receptors had similar firing rates at any given pressure. Moreover, during control conditions the response of the receptors was always greater to active than to passive stretch, while during vagal stimulation there was no difference between the two responses at any atrial pressure.

Our results indicate that for any given atrial pressure and volume (static level of stretch) the response of type A receptors to atrial contraction and filling is markedly influenced by the dynamic component of stretch, which is dependent on the inotropic state of atrial muscle.

The location and distribution of type A and type B atrial endings in cats

B. N. GUPTA

from Abteilung Experimentelle Anaesthesiologie, Institut für Anaesthesiologie,
Universität Düsseldorf, Düsseldorf, Germany

To explain the difference in discharge pattern of atrial endings a series of experiments was done. A total of 131 endings were localised by punctate stimulation, 44 being type A, 77 type B and 10 intermediate type; all were located on the dorsal wall of the atria with none on the ventral wall or in the appendage. On the right side, 74% of type A were located in the atria and 63% of type B in or near the veins. On the left side 67% of type A and 94% of type B were located in or near the veins. Thus, there appeared some difference in the location of type A and type B endings on the right side but on the left side both types of endings were mostly confined to the venous region. Further, on both right and left sides, these endings were present both in the central part of the atria and in or adjacent to veins. This suggests that the difference in discharge pattern is not because of a difference in the location but may be caused by some other reason such as a different arrangement in the atrial wall – as in muscle endings.

An hypothesis to explain the various patterns of discharge from atrial receptors

C. T. KAPPAGODA, R. J. LINDEN & D. A. S. G. MARY

*from the Department of Cardiovascular Studies, University of Leeds,
Leeds LS2 9JT, UK*

Atrial receptors have been classified into three types on the basis of their patterns of discharge; A, B and intermediate (Paintal, 1973). However, there have been reports of more than one pattern of discharge originating in a single receptor (e.g. Neil & Joels, 1961) – a finding which would argue against a rigid classification of these receptors. The present investigation was initially undertaken to determine the integrity of the pattern of discharge in a given receptor. The findings of the study were subsequently used to formulate an hypothesis to explain the various patterns of discharge observed.

Action potentials from single units were recorded from a cervical vagus in anaesthetised cats. Following identification of the units into type A, B or intermediate, an attempt was made to alter the pattern of discharge (i.e. convert) by: (i) infusion of dextran; (ii) haemorrhage; and (iii) administration of adrenaline. Finally each receptor was located by probing with a fine glass rod.

Thirty units, having atrial patterns of discharge, were obtained randomly in 30 cats. Of these, 25 units were located in the endocardium and consisted of 2 type A, 15 type B and 8 intermediate receptors. Conversion of the pattern of discharge was achieved in 16 out of 25 receptors. Of the remaining 5 units which were located outside the endocardium 3 possessed 'type A' and 2 'type B' patterns of discharge.

The study was then extended to selectively study type A units and it was found that all type A units which were successfully 'converted' were located in the endocardium, while those which could not be 'converted' were located at other sites.

From these results it is concluded that type A receptors in the cat are few and that conversion of the pattern of discharge in atrial receptors is a

159

relatively common phenomenon. The results of the experiments to convert the pattern of discharge in the receptors were then taken in conjunction with data relating to the location of these receptors in the endocardium to formulate an hypothesis to explain the various patterns of discharge in atrial receptors. It is suggested that there is probably one basic receptor whose pattern of discharge is dependent upon its location within the endocardium.

References

Neil, E. & Joels, N. (1961). The impulse activity in cardiac afferent vagal fibres. *Arch. exp. Path. Pharmak.*, **240**, 453–60.

Paintal, A.S. (1973). Vagal sensory receptors and their reflex effects. *Physiol. Rev.*, **53**, 159–227.

Behaviour of left atrial receptors with non-medullated vagal afferents in spontaneously breathing cats*

M. D. THAMES

from the Mayo Clinic and Mayo Foundation, Rochester, Minnesota 55901, USA

Activity from eight left atrial receptors with non-medullated vagal afferents (mean conduction velocity 1.2 m/s) was recorded in eight closed-chest, spontaneously breathing cats, anaesthetised with alpha chloralose. Receptors which responded to aortic occlusion with an increase in discharge were studied and their exact anatomical locations were determined at the conclusion of the experiment by probing the open heart. Under resting conditions, the mean discharge frequency for the group was 0.9 imp/s. Three of these receptors were normally silent and the remaining five fired at a peak rate of 3–5 imp/s during the end-inspiratory and early-expiratory phases of respiration and were silent during the remainder of the ventilatory cycle. When respiration was augmented by carbon dioxide inhalation or blood volume increased, the rate of discharge was a linear function of the atrial transmural pressure. A similar relationship was observed for left atrial B receptors. Of five left ventricular receptors with non-medullated afferents four were normally silent and none was influenced by respiration, even during periods of augmented receptor discharge induced by volume expansion. Thus, cardiac receptors with non-myelinated vagal afferents have a low resting discharge in spontaneously breathing cats. The behaviour of left atrial but not of left ventricular receptors is influenced by respiration.

* Supported by NIH Grants HL-5883, HL-6143, HL-01379-02 and by Council for Tobacco Research Grant 1020.

SECTION THREE

Reflex effects of atrial receptors

8. Atrial receptors and heart rate

R. J. LINDEN

In 1915 Bainbridge reported the results of infusing blood and saline into anaesthetised dogs. He then postulated that there was a reflex increase in heart rate from receptors in the right atrium, though there was no evidence in his paper to support this hypothesis. Nonidez (1937), in a purely histological study, demonstrated that there were nerve endings at the junctions between both venae cavae and the right atrium and suggested that these receptors 'initiate the stimulus resulting in reflex cardiac acceleration (Bainbridge reflex)'. Since that time numerous investigators, using infusion and perfusion techniques, have examined this problem, with varying results, mostly observing bradycardia as the response. Evidence that intravenous infusions do not always result in an increase in heart rate has been provided by Coleridge & Linden (1955), Pathak (1959), Ahmed & Nicoll (1963) and Hirsch, Boyd & Katz (1964). In particular, Coleridge & Linden (1955) pointed out that if the initial heart rate was slow then an infusion caused an increase in heart rate and conversely if the control rate was fast then the infusion tended to cause a decrease in heart rate.

Bainbridge (1915) considered this response to an intravenous infusion to be a discrete reflex with the efferent pathway in both the sympathetic and vagal nerves and the afferent pathway primarily in the vagi. He defined the stimulus as an increase in venous pressure even though there was no strict relationship between the increase in heart rate and the increase in venous pressure. He did not locate the receptors nor did he give evidence of the afferent pathway involved. Infusions administered during these experiments were disproportionately large compared with the size of the dogs and were sufficiently rapid to cause variations in arterial pressure. The increase in heart rate was considered to be related to both the rate and the volume of the infusion. However, infusions result in stimuli being applied over a wide

165

area of the circulation which inevitably precludes the conclusion that a change in heart rate is caused by excitation of any single reflex pathway.

Many attempts have been made to limit the stimulus to chambers of the right side of the heart and thereby stimulate the right atrial receptors. These experiments have led to conflicting results. Sassa & Miyazaki (1920) have obtained an increase in heart rate by distension of the junction between the superior vena cava and the right atrium. Aviado *et al.* (1951) found that increasing the pressure in the right atrium resulted in variable changes in heart rate and Goetz (1965) demonstrated an increase in heart rate which was not affected by sectioning the vagi. Donald & Shepherd (1963) reported an increase in heart rate following an intravenous infusion in the chronically denervated dog, and Pathak (1959) showed a similar response in anaesthetised dogs after sectioning the vagi. Experiments such as these, involving stimuli to the atrium with nerves cut, have resulted in small changes in heart rate which have been explained by the stretching of the sino-atrial node, though it is just possible that receptors other than the right or left atrial receptors which do not discharge into the vagi are responsible. Again, infusions have been made to anaesthetised animals with spinal section and increases in heart rate have been observed (Gupta & Singh, this volume). It is possible that an afferent mechanism such as the 'sympathetic afferents' described by Malliani (Chapter 16 of this volume) could explain these phenomena.

On the left side, Daly, Ludany, Todd & Verney (1937) reported that distension of the left atrium and pulmonary veins caused bradycardia and hypotension. Since then bradycardia has been assumed to be a characteristic response to distension of the left side of the heart (see reviews: Aviado & Schmidt, 1955; Heymans & Neil, 1958). Other investigations have failed to produce any changes in heart rate by distension alone and have attributed the bradycardia solely to distension of the left ventricle (e.g. Aviado & Schmidt, 1959).

However, it is not possible to explain fully the effects of intravenous infusions and perfusions and it is particularly difficult to attribute the cause of the responses to specific reflexes. Some of the effects observed must have involved responses from many other receptors besides the receptors in both the right and left atria. In addition, both the infusion experiments and the experiments in which attempts were made to stimulate only single chambers or parts of chambers in the right and left sides of the heart have not allowed the conclusion that any reflexes at all were attributable simply to the atrial receptors.

Investigations in Leeds

In view of this conclusion we, in Leeds, over the past few years, have developed techniques for stimulating only the atrial receptors. In this discussion the term 'atrial receptors' will be used to mean the unencapsulated endings described by Floyd in the first chapter of this volume. All the work I am reporting here has been completed in dogs lightly anaesthetised with chloralose. Care was taken to maintain a constant level of anaesthesia by infusions (1 $cm^3 \cdot kg^{-1}$) of chloralose (1 g·100 cm^3) every 10–15 min. During surgical procedures, which usually took about 2 h, the animals were given a slow infusion of dextran; reasons for the use of dextran have been discussed previously (Ledsome & Linden, 1964*b*). The pH, pCO_2 and pO_2 of arterial blood were frequently measured and kept within the normal range.

Knowing that the receptors were in the subendocardial tissue, mainly at the junction of veins and atria and in the appendages (Nonidez, 1937; Coleridge, Hemingway, Holmes & Linden, 1957; Coleridge, Coleridge & Kidd, 1964; Floyd, Linden & Saunders, 1972), techniques were developed for discretely stretching these areas. Small balloons about 2–3 mm long were made from fine rubber finger stalls tied on to a nylon tube (1 mm diameter). Each balloon was then inserted into a pulmonary vein of the left lung so that the tip lay at the junction of the pulmonary vein and left atrium. The left lung was tied off behind the entrance of the catheters into the veins so that air or blood could not enter or leave the left lung (see, e.g., Linden, 1973). The balloons could then be distended with 1 cm^3 of warm saline to stimulate discretely the atrial receptors without altering flow through or pressure in the atrium. A separate, slightly larger balloon which could accept 2–3 cm^3 of saline could be tied into the atrial appendage.

To investigate the effects of stimulating right atrial receptors Dr Kappagoda made a complex balloon catheter which we inserted through the right external jugular vein and distended the superior vena caval–right atrial junction (Kappagoda, Linden & Snow, 1972). One balloon occluded the superior vena cava and the blood from the head end of the animal was then pumped into a femoral vein; pressure in the superior vena cava above the occlusion was measured and kept constant throughout the experiment. Pressure in the right atrium was measured. The terminal balloon could then be distended so as to stretch the receptors at the vein–atrial junction without altering flow into, or pressure in, the right atrium.

Action potentials from atrial receptors

Distension of the balloons in the pulmonary vein–atrial junctions caused changes in trains of impulses in myelinated fibres in the vagi attached to atrial receptors. Action potentials were recorded in the neck from slips of vagal nerves (as described on pp. 118–19) until a receptor was obtained with a recognised atrial pattern of discharge and which also increased its discharge during distension of the balloon (Kidd, Ledsome & Linden, 1966): parts of a record from one of these experiments are shown in Fig. 8.1. There was a sparse (3–4 imp/beat) pulsatile discharge in a single fibre when the balloon was not distended with a large increase (up to about 20 imp/beat) during distension; at post mortem the receptor was located in the endocardium of the left upper pulmonary vein–atrial junction. It should be noted that during distension of the balloon the discharge is still pulsatile and has a pattern of discharge similar to that of an atrial receptor in an atrium following infusion, where the pressure might be 1–1.5 kPa. Thus, however 'unphysiological' the balloon stimulus is, the trains of impulses entering the medulla are similar in pattern to those observed within the normal range of high activity. It may also be noted is that there is no

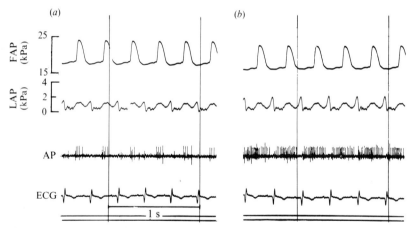

Fig. 8.1. Effect of distension of balloon on discharge of action potentials from left atrial receptor. From above: femoral arterial pressure (FAP) in kPa; left atrial pressure (LAP) in kPa; action potentials (AP) from slip of vagal nerve; ECG; two datum lines. *(a)* Before distension of the balloon; *(b)* during distension of the balloon with 1 cm³ of warm saline. Time between vertical lines in each panel is 1 s. Note (1) the sparse (3–4 imp/beat) pulsatile discharge in a single fibre with balloon not distended; (2) the great increase (about 20 imp/beat) in the pulsatile discharge during distension of the balloon. Location of receptor proved at post mortem to be in the endocardium of left upper pulmonary vein–atrial junction (Linden, 1973).

evidence of adaptation in the short strip which was taken after about 3 min of distension. Five other atrial receptors were examined in this investigation and all exhibited similar characteristics. Since that time we (Kappagoda, Linden & Sivananthan, 1977) have observed similar and consistent discharges in atrial receptors stimulated by distending balloons for the duration of the tests, in periods of up to 30 min. Examination of the performance of atrial receptors on the right side was completed in a similar manner; action potentials were recorded in vagal nerves in the neck and the terminal balloon was then distended with varying amounts of saline (Kappagoda, Linden & Snow, 1972). An example of one record is shown in Fig. 8.2, where an increased discharge of one receptor is shown in response to increase in distension of the terminal balloon on the complex catheter. It can be seen that, as the volume in the terminal balloon was increased in increments of 4 cm^3 (i.e. the superior vena caval–right atrial junction was stretched), the number of impulses per beat increased from about 6/beat to 12/beat. The impulse frequency returned to the control value after release

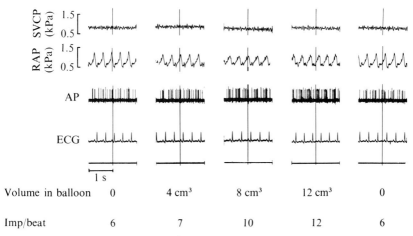

Fig 8.2. Effect of stretching the junction between the superior vena cava and the right atrium on a single unit in the cervical vagus. Each panel of experimental record shows from above downwards the superior vena caval pressure (SVCP), right atrial pressure (RAP), action potentials (AP) from a slip of the right cervical vagus and ECG. As the volume in the terminal balloon A was increased in increments of 4 cm^3 (i.e. as the superior vena caval–right atrial junction was stretched) the impulse frequency (imp/beat) increased from 6/beat to 12/beat. The impulse frequency returned to the control value after the release of the distension. This particular single unit showed a discharge during atrial filling and during atrial systole (Kappagoda, Linden & Snow, 1972).

of the distension. This particular single unit showed a discharge during atrial filling and during atrial systole and is typical of the pattern of discharge described as being from an intermediate type of receptor. Plots of the responses of receptors (Kappagoda, Linden & Snow, 1972) illustrate that a high discharge of these atrial receptors is not observed until a volume of 8–10 cm³ of saline has been injected into the balloon (see later that this is precisely the volume which has to be injected in order to obtain the reflex response of an increase in heart rate). That this order of distension was also within the physiological range was concluded from experiments in which infusions were given until the discharge from atrial receptors was the same as that obtained by the distension of the balloon with 8–10 cm³ of saline. It was found that a pressure of 0.8–1.2 kPa in the right atrium was required to obtain a distension equal to that caused by the distension of the balloon. Such pressures are well within the physiological range of right atrial pressures. Again, the pulsatile discharge observed during distension of the balloon has a typical pattern and thus, although the balloon stimulus may be 'unphysiological', the input to the medulla could be regarded as well within the physiological range of number and pattern.

It may be concluded from these experiments that stimulation of receptors by small balloons in the pulmonary vein–atrial junctions and by a larger balloon in the superior vena caval–atrial junction results in perfectly reasonable numbers and patterns of trains of impulses to the second order neurons in the medulla.

Reflex response: left atrial receptors

Ledsome & Linden (1964*b*) distended the small balloons in the pulmonary vein–atrial junctions and observed an increase in heart rate in every experiment; in 78 distensions in 24 dogs there was an average increase in heart rate of 24 beats/min (range 2–89). One of the best examples from this investigation is shown in Fig. 8.3; the increases in heart rate were calculated by subtracting the mean of the heart rate recorded in the control periods before and 3 min after distension from the heart rate during distension, giving an increase in this experiment of 77 beats/min. No change in mean arterial blood pressure was observed and the change in pulse pressure, pulmonary artery pressure and the small falls in left atrial pressure and right atrial pressure are assumed to result from the increase in heart rate caused by the stimulation of left atrial receptors. In experiments since then, and following the description of atrial receptors in the appendages (Floyd, Linden & Saunders, 1972), another small balloon has been placed in the left

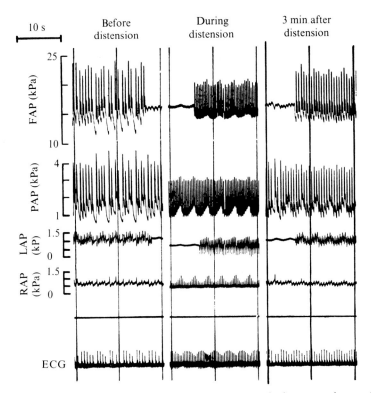

Fig. 8.3. Effects of stimulation of left atrial receptors on the heart rate in one dog, illustrating, not the best, but one of the better responses. Records from above downwards: femoral arterial pressure (FAP); pulmonary arterial pressure (PAP); left atrial pressure (LAP); right atrial pressure (RAP); datum line; ECG. The three panels show records before, during, and at 3 min after distension of the balloons. The change in heart rate from 80–170 to 105 beats/min gave an increase of 77 beats/min by averaging the heart rates before and after the distension and subtracting the result from the heart rate during distension (Ledsome & Linden, 1964*a,b*).

atrial appendage in addition to one in the upper pulmonary vein and one in the middle pulmonary vein; distension of these balloons again caused an increase in heart rate (Kappagoda, Linden & Mary, 1975) and an example from this work is shown in Fig. 8.4. Here it can be seen that the heart rate before distension of the balloons was 94 beats/min, during distension was 162 beats/min and 3 min after distension was 116 beats/min, giving a mean increase in heart rate of 57 beats/min. Ten distensions of three balloons in five dogs resulted in a mean increase in heart rate of 35 beats/min in these

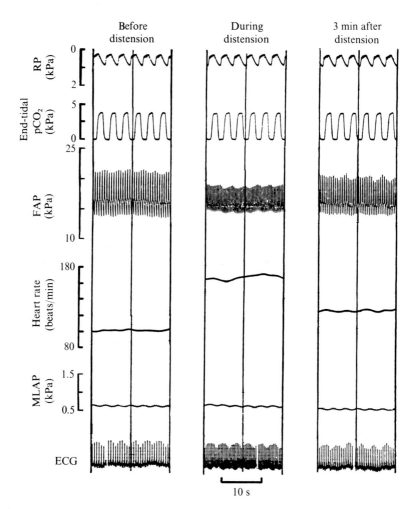

Fig. 8.4. Effect of stretching the left upper and middle pulmonary vein–atrial junctions and the left atrial appendage (three balloons). From above downwards, tracheal pressure (RP), end-tidal pCO_2, femoral arterial pressure (FAP), heart rate, mean left atrial pressure (MLAP) and the ECG. Records obtained immediately before distension of the balloons (heart rate, 94 beats/min), during distension (heart rate, 162 beats/min) and 3 min after distension (heart rate, 116 beats/min). Mean increase in heart rate, 57 beats/min (Kappagoda, Linden & Mary, 1975).

animals with a range of initial heart rates from 48 beats/min to 133 beats/min.

If all the investigations in Leeds of which reports have been published are considered, together with two reports from Vancouver (Albrook, Bennion & Ledsome, 1972; Burkhart & Ledsome, 1974), then over 700 distensions of the balloons in the vein–atrial junctions of over 200 dogs have always resulted in an increase in heart rate and never a bradycardia even though the initial control heart rates have ranged from 39 to 217 beats/min.

Afferent limb of reflex. Sectioning or cooling the vagi in the chest and neck has always reduced or abolished the response. By sectioning or cooling the vagal nerves at various levels within the chest and neck, so as to block the nerve impulses from the atrial receptors, it was shown (Ledsome & Linden, 1964*b*) that the response to distension of pulmonary vein–atrial junctions was first reduced and then abolished. Sectioning the vagal nerves at the level of the right lung root and at the level of the upper border of the aorta on the left eliminated afferent nerves emanating from atrial receptors and abolished the responses. The fact that the control heart rate was unchanged and the carotid sinus reflexes were intact allowed the conclusion that the efferent vagal nerves had not been sectioned. In 16 dogs, during four separate investigations, 25 distensions of the left atrial balloons following sectioning or cooling of both vagi in the chest or neck resulted in a response of only 0.8 beats/min (average of the means of the four investigations; range 0–2); the control response to distension before the sectioning or cooling of the vagi was 32 beats/min (average; range 25–36).

It may be concluded that the afferent limb of this reflex is in the vagi.

Efferent limb of reflex. Because the response of an increase in heart rate took 30 s to 5 min to decay following the removal of the distension of the balloons it was considered that the efferent limb of the reflex could be in the sympathetic nerves to the heart. By sectioning the right and left ansae subclavae (which contain all the sympathetic efferent nerves to the heart) and the use of sympathetic blocking drugs (propranolol, bretylium tosylate) the efferent limb of the reflex was shown to be solely in the efferent sympathetic nerves to the heart. During five different investigations there were 20 distensions of the balloons in 18 dogs following section of both ansae subclavae; the control increase in heart rate was 25 beats/min (average of means; range 16–31) and following section of the ansae subclavae the response was reduced to 4.5 beats/min (average of means;

range 0–16). In 41 dogs 55 distensions of the balloon in ten separate investigations resulted in a control increase in heart rate of 23 beats/min (average of means; range 11–60) which, after adrenoreceptor blocking drugs (propranolol or atenolol) had been administered to the animal, was reduced to 3.8 beats/min (average of means; range 0–9.8). An explanation of this residual response lies in the fact that these drugs are competitive antagonists and the release of noradrenaline in sufficient concentration, even in the presence of blocking drugs, still has an effect. The administration of bretylium tosylate (10 mg·kg^{-1}) abolishes all remaining increases in heart rate; bretylium tosylate blocks the action of all efferent sympathetic nerves without affecting central connections of cardiac reflexes or efferent vagal nerves (Ledsome & Linden, 1964a). In 54 dogs 100 distensions of the balloons in eight separate investigations caused an increase in heart rate of 21 beats/min (average of the means; range 11–35) in the control period before the administration of bretylium tosylate, and 0.6 beats/min (average of means; range 0–2) after bretylium tosylate. In spite of the claims to the contrary there is little evidence to support the assertion (Burkhart & Ledsome, 1974; Ledsome, Chapter 20 of this volume) that there is a vagal component to this reflex; this controversy has been discussed elsewhere (Kappagoda, Linden, Scott & Snow, 1975; Linden, 1975a, 1976).

Because the efferent limb of this reflex is in the sympathetic nerves it was also expected that there would be a reflex effect on the heart muscle as well as on the heart rate. From investigations involving stimulation of sympathetic nerves to the heart (e.g. Linden, 1975b) it is known that in addition to the increase in heart rate there is a concomitant positive inotropic response of the ventricles. Reflex responses involving efferent sympathetic nerves to the heart, for example a fall in pressure in the carotid sinus, can cause not only an increase in heart rate, but also a positive inotropic response (e.g. Sarnoff *et al.*, 1960; Hainsworth & Karim, 1973). Therefore it was expected that this reflex response to stimulation of left atrial receptors, which could be completely abolished by sympathetic blockade or section of the ansae subclavae, would involve a positive inotropic response. In the event, there was no positive inotropic response, there was only an increase in heart rate. The changes in the force of contraction of the muscle were measured by recording the maximum increase in the rate of change of pressure in the left ventricle (dP/dt$_{max}$) which had been shown to be the best index to use when observing inotropic responses in this dog preparation (Furnival, Linden & Snow, 1970). The mean change in dP/dt_{max} as a result of stimulating the atrial receptors in 37 experiments in 12 dogs was 25.7 kPa/s (mean; range 4.8–110.4); the increase in heart rate in this series of experiments was 21 beats/min (mean; range 6–90). Notice that even though there was an increase in heart rate in this investigation of as much as 90 beats/min in response to stimulation of left atrial receptors, there was increase in dP/dt_{max} of only 36.8 kPa/s, even less than would be obtained by simple pacing, presumably explained by concomitant increase in baroreceptor stimulation with a consequent withdrawal of activity in sympathetic nerves. It was confidently concluded that there was no concomitant inotropic effect when atrial receptors were stimulated (Furnival, Linden & Snow, 1971).

Peripheral resistance and respiration. In two other series of experiments (Ledsome & Hainsworth, 1970; Carswell, Hainsworth & Ledsome, 1970) it was shown that stimulation of left atrial receptors did not affect either respiration or peripheral resistance. The finding of no effect on peripheral resistance was supported by the work of Karim, Kidd, Malpus & Penna (1972) who found that on stimulating the left atrial receptors there was an increased discharge in nerves in the right ansa subclavia (presumably to the sino-atrial node), a decrease in discharge in the nerves to the kidney, but no change in any other sympathetic nerves. In spite of these very definite findings when discrete portions of the pulmonary vein–atrial junctions are stimulated by inflation of small balloons, Edis, Donald & Shepherd (1970), using larger balloons, found that the distension of the balloons at the pulmonary vein–atrial junctions caused not only increases and decreases in heart rate but a sustained reflex systemic hypotension (a criticism of their conclusion has been given previously by Linden (1975*a*, 1976)). Recently, Mason & Ledsome (1974) have observed that inflation of small balloons in the pulmonary vein–atrial junctions caused an increase in blood flow to the kidney with a fall in resistance in the kidney but no change in resistance in an isolated perfused limb. It is difficult to avoid the conclusion that the atrial receptors when stimulated have a response in the nerves to the heart and to the kidney but do not effect peripheral resistance.

Effect of acidaemia. It is important to realise that acidaemia depresses the reflex response from atrial receptors. In five dogs the preparation was set up complete with atrial balloons following which controlled responses of an increase in heart rate were elicited. The dogs were then made acidaemic by either ventilation with carbon dioxide or the infusion of hydrochloric acid, and the distensions were repeated. Finally, the blood chemistry was brought back to normal and the distensions were repeated. It was observed that increasing the acidaemia depressed the magnitude of the responses obtained from the stimulation of left atrial receptors (Harry, Kappagoda & Linden, 1971).

Other reviews (Paintal, 1972, 1973; Coleridge & Coleridge, 1972; Shepherd, 1973; Pelletier & Shepherd, 1973) have commented on the relationship between the atrial receptors and heart rate. Some of the reviewers accept that stimulation of the atrial receptors results in a biphasic response, i.e. an increase in heart rate when the initial heart rate is low and a decrease in heart rate when the initial heart rate is fast, and again that the efferent limb of the reflex is not solely in the sympathetic nerves but partially in the vagal nerves. Comments disagreeing with these other opinions have already been published (Kappagoda, Linden, Scott & Snow, 1975; Linden, 1975*a*, 1976).

However, to answer a main criticism that the response is somehow related to less arterial baroreceptor activity following a transient fall in blood pressure (Paintal, 1973) it is necessary briefly to make two points: *(a)* the transient response of a fall in blood pressure (commented on by Ledsome & Hainsworth, 1970; Linden, 1975*a*) is observed in less than 30% of distensions, and *(b)* when there is no transient response the steady-state increase in heart rate observed during stimulation of the atrial receptors is attained in less than 1 min (from 5–30 s). An example of the latter response is given in Fig. 8.5; in the experiment to obtain this record the balloons were deliberately distended for only 1 min so that the whole sequence could be shown in one figure (the heart rate responses usually reported are counted after 2–3 min of distension so that at least 1 min of steady state is obtained).

Distension of a pouch of left atrium. In anaesthetised dogs a pouch of the left atrium was constructed by placing a specially made clamp across the left atrium. By sliding the posterior limb of the clamp in front of the oesophagus but behind the left atrium and placing the anterior limb of the clamp lateral to the appendage but medial to the pulmonary vein–atrial junctions and thrusting the whole clamp towards the pulmonary artery, it was possible to construct a pouch containing only the left pulmonary vein–atrial junctions; the pulmonary veins were tied off as far into the left lung as possible (Ledsome & Linden, 1967). The pouch was then distended by means of pulsatile changes in pressure. It was shown that the same reflex response of an increase in heart rate could be elicited. High pressures were

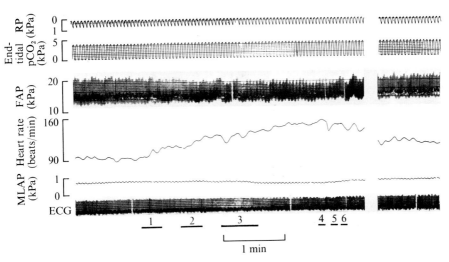

Fig. 8.5. Continuous record during distension of small balloons at (1) upper pulmonary vein–atrial junction; (2) middle junction, each with 1 cm³ saline, and (3) left atrial appendage with 2 cm³; (4, 5 and 6) volumes of saline removed from balloons. Right-hand panel recorded after an interval of 3 min. RP = tracheal pressure; FAP = femoral arterial pressure; MLAP = mean left atrial pressure. Heart rate during first control, 96 beats/min; during distension, 158 beats/min; and 3 min after distension removed, 118 beats/min. In order to make this published record relatively larger the distensions were terminated after 1 min in this experiment (Linden, 1975a).

required in the pouch but it was later demonstrated (Kidd *et al.*, 1966), by recording action potentials in slips of vagal nerves and by invoking the Law of La Place, that the tensions in the wall of the pouch and therefore the stretch of receptors would be much less even though the pressure in the lumen was higher, since the pouch had a radius of curvature about four times smaller than that of the whole atrium. This investigation served to provide more circumstantial evidence that the atrial receptors were involved in this reflex response. Again no bradycardia was observed during this investigation. This reflex elicited from distension of the pouch also had its afferent pathway in the vagal nerves and its efferent pathway only in the sympathetic nerves, identical findings to those when the balloons were used in the atrial venous junctions and in the appendage.

Reflex response: right atrial receptors
Reference above to stimulation of atrial receptors includes a description of

the techniques used in stimulating right atrial receptors; also the overall results quoted above include those resulting from the stimulation of right atrial receptors. However, because of the previous comment that there exists little evidence for the so-called Bainbridge reflex it is necessary to emphasise the evidence elicited solely from distension of the superior vena caval–right atrial junction. As pointed out above, on evidence of the response of action potentials in myelinated fibres in the vagus the stimulus caused by distension of the terminal balloon of the complex catheter with about 10 cm³ of warm saline resulted in reasonable trains of impulses in the vagus within the physiological range of pattern, frequency and number. Thus the technique is not 'unphysiological' in that trains of impulses within the normal range enter the medulla.

Distension of the terminal balloon to stimulate right atrial receptors always resulted in an increase in the heart rate (Kappagoda, Linden & Snow, 1972). In 65 distensions in 16 dogs there was an increase in heart rate of 18 beats/min (mean; range 5–73); these increases in heart rate were observed to occur from an initial control heart rate of 133 beats/min (mean; range 60–202). There was no relationship between the small changes in arterial pressure and superior vena caval pressure and the changes in heart rate. By cooling each vagus in turn the afferent pathway of the reflex was again shown to be in the vagi. Sectioning both vagi simultaneously was not considered to be a valid test of the afferent pathway such as the one postulated here because a totally different animal base line, which would not be comparable with the state in the previous control periods, is observed after bilateral vagal section. Cooling the nerve to 6 °C (thermode temperature 5 °C) blocks conduction in myelinated nerve fibres of the size attached to atrial receptors, and reduces this response of an increase in heart rate; the response of an increase in heart rate was less when a vagal nerve was cooled than the response either before cooling or after rewarming the nerve. From these results it was possible to conclude that the afferent pathway of this reflex is at least partly in the vagi.

The efferent pathway of this reflex was also shown to be in the sympathetic nerves. Crushing both ansae subclavae or blocking the action of sympathetic nerves with drugs completely abolished the response of an increase in heart rate. No bradycardia was observed after sympathetic blockade; it may be significant that in the three dogs in which the right ansa alone was sectioned the response was completely abolished, though more experiments would obviously have to be completed before any definite conclusions could be made.

There was no evidence to conflict with the hypothesis that the response to stimulation of right atrial receptors was the same as stimulation of left atrial receptors and that the afferent and efferent pathways were in the vagal and sympathetic nerves, respectively.

Summary. Stimulation of right and left atrial receptors results in a reflex increase in heart rate, the afferent nerves of this reflex being in the vagi and the efferent limb solely in the sympathetic nerves, changes in which are observed in the cardiac efferent sympathetic nerves; there is no concomitant positive inotropic effect (atrial receptors also affect sympathetic nerves to the kidney and urine flow, see Kappagoda, Chapter 9 of this volume).

Detailed examination of afferent limb of reflex

Graded responses by recruiting receptors. We had, however, always been aware that the weak link in our argument to substantiate this reflex was the degree of certainty with which we could claim that the atrial receptors were involved. A point of criticism was that there was no correlation between the magnitude of the stretch and the increase in heart rate and it was suggested that this reflex mechanism was just a 'trigger' response such that once the threshold to the receptors has been reached the response mounts so rapidly that in effect this reflex could not contribute very much to the physiological control of the circulation. Because of the peculiar geometry of the vein–atrial junctions it has not been possible to obtain a graded increase in the stimulus by varying the volume of warm saline injected into each balloon and explanations have been offered for this failure (Ledsome & Linden, 1964*b*; Linden, 1975*a*).

However, it was possible to answer this criticism another way. We found that there were about ten atrial receptors per atrial appendage in the dog (Floyd *et al.*, 1972) and that distension of the atrial appendage caused the same reflex increase in heart rate (Kappagoda, Linden & Saunders, 1972). It was therefore possible to stimulate successively different areas containing receptors and therefore successively greater numbers of receptors by distensions of balloons, first, at one atrio-venous junction, secondly, at two atrio-venous junctions, and thirdly, at two vein–atrial junctions together with the distension of the left atrial appendage. The pulmonary vein–atrial junctions were each distended with 1 cm^3 of warm saline and the atrial appendage with 2 cm^3. An example of the results of one experiment is shown in Fig. 8.6. In six dogs 24 distensions of a single pulmonary

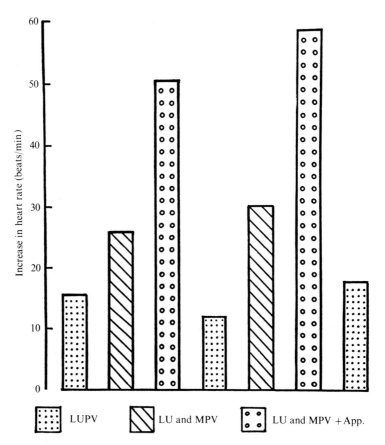

Fig. 8.6. Gradation of the responses to distension of the balloons in one dog: LUPV, distension of the balloon in left upper pulmonary vein alone; LU and MPV, distension of the two balloons in left upper and middle pulmonary veins, respectively; LU and MPV + App., distension of the three balloons in the left upper and middle pulmonary veins and in the left atrial appendage, respectively. The responses are expressed as increases in heart rate in beats/min (Kappagoda, Linden & Mary, 1975).

vein–atrial junction resulted in an increase in heart rate of 10.8 beats/min (mean; range 6–18); 12 distensions of the two pulmonary vein–atrial junctions resulted in an increase in heart rate of 22.2 beats/min (mean; range 12–37); and by distending the appendage in addition (i.e. 10 distensions of three balloons simultaneously) an increase of 35.2 beats/min (mean; range 22–57) was observed. Single balloon distensions were obtained both before and after the multiple balloon distensions. The differences between groups are statistically significant. Thus, when the stimulus to the atrial receptors is graded by successively recruiting groups of atrial receptors the response of an increase in heart rate is also graded (Kappagoda, Linden & Mary, 1975). The area stimulated by these balloons is small compared with the total area in which the atrial receptors are to be found, e.g. the right atrial receptors, atrial receptors on the right side of the left atrium and an unknown number on the left side of the left atrium would not be stimulated in this series of experiments.

However, whatever conclusions one can draw from this investigation there is no doubt at all that we could recruit receptors responsible for the reflex increase in heart rate and no doubt at all we could recruit them from areas known to contain atrial receptors. But again, we still had the conclusion that the most likely receptors to be involved were the receptors attached to the myelinated fibres because we had never properly eliminated receptors attached to C fibres from the argument.

Graded response by blocking conduction in nerves in vagi. It has been reported during several investigations (e.g. Sleight & Widdicombe, 1965; Coleridge *et al.*, 1973; Thorén, 1976; Thames, this volume) that there are also receptors in the atrium which discharge into C fibres (conduction velocity less than 2.5 m/s) in the vagal nerves. Thorén (1976) has shown an increase in C-fibre discharge of up to 20 imp/s when pressures were raised in the atria up to 2.7–4.0 kPa. Thus it was probable that the distension of the balloons was also exciting these receptors and possible that the C fibres could be involved in the heart rate–reflex response. Therefore in three series of investigations, the responses of the increase in action potentials in myelinated fibres, in C fibres and the increase in heart rate during stimulation of the atrial receptors were examined whilst progressively cooling the vagi in the neck (Kappagoda *et al.*, 1977). First, various cooling thermodes were examined and finally one constructed 2 cm long, which when cooled, resulted in the temperature in the centre of the nerve (mid-thermode) being at the same temperature as the surface of the thermode (mid-thermode). Because cooling was slow this relationship remained the same during the cooling. At least 1 min was allowed at each temperature before records were taken.

From previous experiments we knew that the atrial receptors discharged during the control period at a mean frequency of 30–40/s and that during distension of the small balloon in the pulmonary vein–atrial junctions with 1 cm^3 of warm saline this discharge was increased to a frequency of

80–120/s. We also knew that these fibres had a conduction velocity in our experiments of the order of 10–35 m/s. A diagram published by Paintal (1972) shows that the maximum frequency of impulses which a myelinated nerve fibre can conduct depends on the temperature and the conduction velocity, i.e. in any one fibre the effect of cooling is frequency dependent. Part of Paintal's figure is shown re-drawn in Fig. 8.7; the results of cooling two myelinated fibres, one with conduction velocity of 35 m/s and another

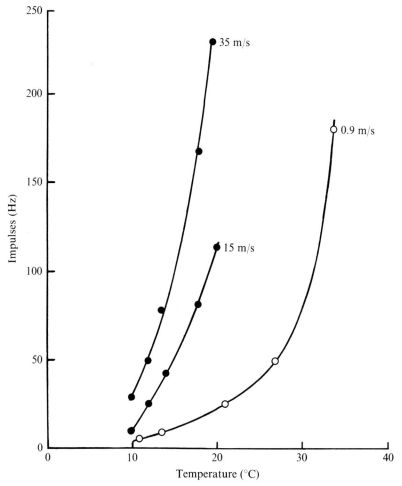

Fig. 8.7. Maximum frequency of impulses conducted by myelinated fibres with conduction velocities of 35 m/s and 15 m/s (from Paintal, 1972); and of a C fibre of conduction velocity 0.9 m/s (from Franz & Iggo, 1968). Ordinate: impulses in nerve fibre, Hz; abscissa: temperature of nerve on cooling thermode.

with 15 m/s are shown. As the nerve is cooled the maximum frequency at which impulses can pass through the cold block is indicated by each curve. Note the events which occur at cooling temperatures of 16 and 12 °C. At 16 °C it can be seen that a frequency of 120 imp/s would just get through the cold block in a fibre conducting at 35 m/s and at 12 °C a frequency of 30 imp/s would just not pass the block in a fibre conducting at 15 m/s. Because the range of activity from 120 imp/s to 30 imp/s covers about the range of activity in nerve fibres from atrial receptors, which we had previously encountered, the cooling temperatures of 16 and 12 °C were important reference points in our subsequent investigations into the effects of cooling the vagal nerves on the response to stimulation of atrial receptors.

We therefore first tested the hypothesis that the increases in numbers of action potentials brought about by distending the balloons at the pulmonary vein–atrial junctions would be gradually blocked as we cooled the vagi; the results of the examination of 11 units in nine dogs are shown in Fig. 8.8. It can be seen that, in general, as the vagus nerve was cooled from 18 °C to 10 °C the response to distension of the balloons decreased with each 2 °C fall in temperature. Examining the two reference points it can be seen that at 16 °C about 80% of the response of the increase in action potentials was still present and at 12 °C only about 20% of the response was still present. It could be concluded from this evidence that the fibres which we were examining were attached to atrial receptors and therefore we were stimulating atrial receptors.

In the next series of experiments we examined the effect of cooling (in 2 °C steps) the vagus nerves on the heart rate response to distension of the balloons; the results from 105 distensions in eight dogs are shown in Fig. 8.9. There is a general similarity in the effect of cooling the vagi on the heart rate response to that of cooling the vagi on the increase in action potentials in myelinated fibres discharging as a result of stimulating atrial receptors; two continuous lines drawn on Fig. 8.9 enclose all the responses of an increase in heart rate on cooling the vagi and the same lines on Fig. 8.8 enclose all but three responses of increases in action potentials on cooling. In particular it may be seen that at the reference points the same proportional effects were observed at 16 °C, about 80–85% of the response still remained, and at 12 °C only about 20% of the response remained. Thus, it may be concluded that the effects of cooling the vagal nerves in a stepwise fashion are the same on both the response of the increase in action potentials to distension of the balloons and the response of an increase in heart rate to distension of the balloons. It may therefore be concluded

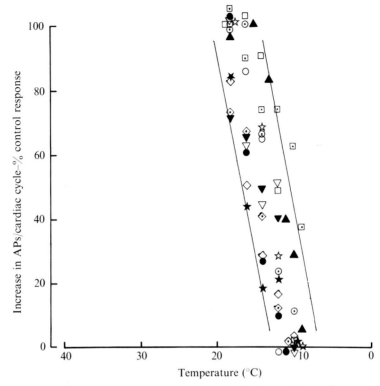

Fig. 8.8. Response of the increase in action potentials (APs) in myelinated vagal fibres resulting from distension of balloons in pulmonary vein–atrial junctions to cooling of the vagal nerve in 2 °C steps. Each point represents the response of the increase in action potentials per cardiac cycle, at each temperature of the vagus, expressed as a percentage of the response with the vagus at 36 °C. Each symbol represents the response in one nerve fibre; in all, 11 fibres with conduction velocities ranging from 9 to 37.5 m/s were examined in nine dogs. The responses at 36 °C were all rated 100% response and not shown in the diagram. The two continuous lines, which are the same as in Fig. 8.9, enclose all but 3 of the 59 responses to cooling (Kappagoda, Linden & Sivananthan, unpublished).

again, now with a little more conviction, that the atrial receptors are the most likely receptors to be involved in the reflex.

However, this does not completely eliminate the C fibres as being partly responsible. We had noticed in passing that when we had a C fibre on the electrodes and we distended the balloons then there was an increase in discharge to a frequency of about 20 imp/s which suggested that with this sort of discharge in slowly conducting fibres (less than 2 m/s) the block

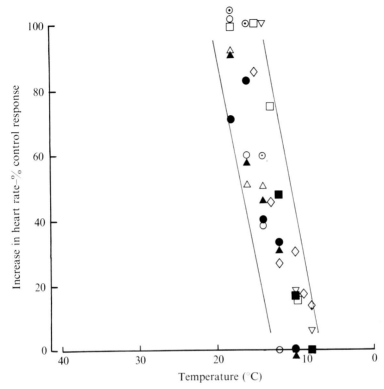

Fig. 8.9. Response of the increase in heart rate resulting from distension of balloons in the pulmonary vein–atrial junctions, to cooling of the vagi in 2 °C steps. Each point represents the response of the increase in heart rate at each temperature of the vagi expressed as a percentage of the control responses obtained with the vagi at 36 °C; the control responses were obtained before and after the cooled response. Each symbol represents the results from one dog. The responses at 36 °C are not shown in the diagram. The continuous lines enclose all the responses to cooling (cf. Figs 8.8 and 8.10) (Kappagoda, Linden & Sivananthan, unpublished).

should be complete at a fairly high temperature, around 20 °C. Unfortunately there is no investigation comparable with that of Paintal (1972) in which the effect of cooling has been systematically examined on a C fibre population to record the maximum frequency which will pass the cold block at each temperature in C fibres. However, we found in the literature (Franz & Iggo, 1968) a report of one C fibre, the impulse traffic in which was examined while cooling; this response has been plotted alongside those of Paintal in Fig. 8.7. The C fibre had a conduction velocity of 0.9 m/s, and it can be seen from Fig. 8.7 that a frequency of about 20 imp/s would just not get through the cold block at 19 °C – well above the blocking temperature of our reflex response.

The next investigation, therefore, consisted of obtaining C fibres which increased their discharge during distension of the small balloons in the vein–atrial junctions and systematically noting the temperature of the cold block which prevented this increase in discharge passing the block. Once a C fibre was found the control responses to distension were obtained at 36 °C, then the nerve was cooled to 18 °C. If the increase in discharge to distension of the balloon was the same as the control the nerve was cooled in 2 °C steps and the response to distension obtained at each step; if the increase in discharge was blocked then the nerve was returned to 30 °C and cooled in 2 °C steps as above. At the time of the symposium only three fibres had been examined, all discharging at more than 20 imp/s during distension of the balloon and in each case the response of an increased discharge to the distension was abolished above a temperature of 18 °C during the distension; such an effect would be predicted from Fig. 8.7. Since then we have investigated nine other fibres and all results from the examination of the 12 C fibres in 22 dogs are shown in Fig. 8.10; it seems that there is little similarity between the response in C fibres to progressive cooling and that of the heart rate response. It can be seen now that the two continuous lines drawn on Fig. 8.10, which are the same lines as drawn in Figs 8.8 and 8.9, enclosed hardly any of the responses of C fibres to cooling. There was a large range of response in receptors attached to C fibres in that the discharge during the control period was 3.12 imp/s (mean; range 0–11.8) and this discharge rose to 13.9 imp/s (mean; range 1.3–33.6) during distension of the balloons with the vagi warm. However, it does seem that there are possibly two groups of receptors within this large group, four with relatively high discharge rates of 7.9 imp/s (mean; range 0.1–11.2), rising during distension to 28.5 imp/s (mean; range 23.1–33.6) and eight with the usually reported discharge rates of 0.7 imp/s (mean; range 0–1.8), rising during distension to 6.7 imp/s (mean; range 1.3–12.5). Thus, when considering our two reference points, at both 16 °C and 12 °C the response of the high frequency group is fully blocked while less than 5% of response of the low frequency group is blocked. Thus, there is no relationship between the response in C fibres and the response of the increase in heart rate by cooling the vagi. When considering these responses of C fibres to distension of the balloons during progressive cooling of the vagi it is important to remember that to block the response of an *increase* in discharge it is not necessary wholly to block the trains of impulses in C

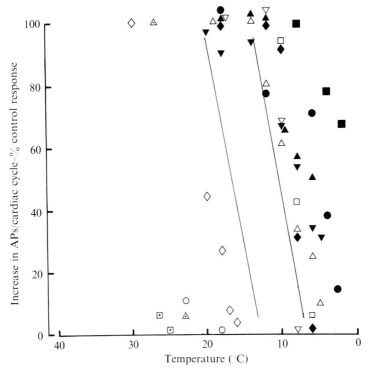

Fig. 8.10. Response of the increase in action potentials (APs) in vagal C fibres resulting from distension of balloons in the pulmonary vein–atrial junctions, to cooling of the vagal nerve in 2 °C steps (see text for description of technique). Each point represents the response of the increase in action potentials per cardiac cycle, at each temperature of the vagus, expressed as a percentage of the response with the vagus at 36 °C. Each symbol represents the responses in one nerve fibre; in all, 12 fibres were obtained in 22 dogs. The responses at 36 °C were all rated 100% response and not shown in the diagram. The continuous lines are the same as those in Figs 8.8 and 8.9; note that nearly all the responses to cooling lie outside these lines (Kappagoda, Linden & Sivananthan, unpublished).

fibres, i.e. it is not necessary to cool to 2 °C. Thus impulses with a frequency of the order of those observed during the control period will still be occurring when impulses at the higher frequency caused by distension of the balloons (the *increase* in discharge) have been blocked. It is the effect of distending balloons which is being examined. It is probable that the rates of discharge in the four fibres, which are higher than those usually observed in C fibres, are a result of the artificial presence of the balloons causing local abnormal increases in tension – and even greater increases in tension during distension of the balloons, which would account for the high discharge rates.

However, it *is* the effect of the balloons in position during the control period and during distension which is being examined when the reflex response of action potentials and/or increases in heart rate are being investigated, therefore it must be concluded that it is highly unlikely that the C fibres are concerned in the reflex response of an increase in heart to stimulation of receptors in the atrium.

In *conclusion* it seems that the atrial receptors which have been stimulated by balloons and which are causing the reflex increase in heart rate are the Paintal-type atrial receptors which are attached to myelinated fibres of a conduction velocity of the order of 10–35 m/s. It seems that the afferent nerves are solely in the vagi and the efferent nerves solely in the sympathetic nerves to the sino-atrial node and there is no positive inotropic response.

References

Ahmed, G. & Nicoll, P.A. (1963). Chronotropic response to intravenous infusion in the anaesthetized dog. *Am. J. Physiol.*, **204**, 423–6.

Albrook, S.M., Bennion, G.R. & Ledsome, J.R. (1972). The effects of decerebration on the reflex response to pulmonary vein distension. *J. Physiol.*, **226**, 793–803.

Aviado, D.M., Jr & Schmidt, C.F. (1955). Reflexes from stretch receptors in blood vessels, heart and lungs. *Physiol. Rev.*, **35**, 247–300.

Aviado, D.M., Jr & Schmidt, C.F. (1959). Cardiovascular and respiratory reflexes from the left side of the heart. *Am. J. Physiol.*, **196**, 726–30.

Avidao, D.M., Jr, Li, T.H., Kalow, W., Schmidt, C.F., Turnbull, G.L., Peskin, G.W., Hess, M.E. & Weiss, A.J. (1951). Respiratory and circulatory reflexes from the perfused heart and pulmonary circulation of the dog. *Am. J. Physiol.*, **165**, 261–77.

Bainbridge, F.A. (1915). The influence of venous filling upon the rate of the heart. *J. Physiol.*, **50**, 65–84.

Burkhart, S.M. & Ledsome, J.R. (1974). The response to distension of the pulmonary vein–left atrial junctions in dogs with spinal section. *J. Physiol.*, **237**, 685–700.

Carswell, F., Hainsworth, R. & Ledsome, J.R. (1970). The effects of distension of the pulmonary vein–atrial junctions upon peripheral vascular resistance. *J. Physiol.*, **207**, 1–14.

Coleridge, H. & Coleridge, J. (1972). Cardiovascular receptors. In *Modern trends in physiology*, ed. C. B. B. Downman, 1st edn, pp. 245–67. Butterworths: London.

Coleridge, H.M., Coleridge, J.C.G., Dangel, A., Kidd, C., Luck, J.C. & Sleight, P. (1973). Impulses in slowly conducting vagal fibres from afferent endings in the veins, atria and arteries of dogs and cats. *Circulation Res.*, **33**, 87–97.

Coleridge, H.M., Coleridge, J.C.G. & Kidd, C. (1964). Cardiac receptors in the dog, with particular reference to two types of afferent endings in the ventricular wall. *J. Physiol.*, **174**, 323–39.

Coleridge, J.C.G., Hemingway, A., Holmes, R.L. & Linden, R.J. (1957). The location of atrial receptors in the dog. A physiological and histological study. *J. Physiol.*, **136**, 174–97.

Coleridge, J.C.G. & Linden, R.J. (1955). The effect of intravenous infusions upon the heart rate of the anaesthetized dog. *J. Physiol.*, **128**, 310–19.

Daley, I. de B., Ludany, G., Todd, A. & Verney, E.B. (1937). Sensory receptors in the pulmonary vascular bed. *Q. Jl exp. Physiol.*, **27**, 123–46.

Donald, D.E. & Shepherd, J.T. (1963). Changes in heart rate on intravenous infusion in dogs with chronic cardiac denervation. *Proc. Soc. exp. Biol. Med.*, **113**, 315–17.

Edis, A.J., Donald, D.E. & Shepherd, J.T. (1970). Cardiovascular reflexes from stretch of pulmonary vein–atrial junctions in the dog. *Circulation Res.*, **27**, 1091–1100.

Floyd, K., Linden, R.J. & Saunders, D.A. (1972). Presumed receptors in the left atrial appendage of the dog. *J. Physiol.*, **227**, 27–8P.

Franz, D.N. & Iggo, A. (1968). Conduction failure in myelinated and non-myelinated axons at low temperature. *J. Physiol.*, **199**, 319–45.

Furnival, C.M., Linden, R.J. & Snow, H.M. (1970). Inotropic changes in the left ventricle: the effect of changes in heart rate, aortic pressure and end-diastolic pressure. *J. Physiol.*, **211**, 359–87.

Furnival, C.M., Linden, R.J. & Snow, H.M. (1971). Reflex effects on the heart of stimulating left atrial receptors. *J. Physiol.*, **218**, 447–63.

Goetz, K.L. (1965). Effect of increased pressure within a right heart cul-de-sac on heart rate in dogs. *Am. J. Physiol.*, **209**, 507–12.

Hainsworth, R. & Karim, F. (1973). Left ventricular inotropic and peripheral vasomotor responses from independent changes in pressure in the carotid sinuses and cerebral arteries in anaesthetized dogs. *J. Physiol.*, **228**, 139–55.

Harry, J.D., Kappagoda, C.T. & Linden, R.J. (1971). Depression of the reflex tachycardia from the left atrial receptors by acidaemia. *J. Physiol.*, **218**, 465–75.

Heymans, C. & Neil, E. (1958). *Reflexogenic areas of the cardiovascular system*. J. & S. Churchill: London.

Hirsch, L.J., Boyd, E. & Katz, L.N. (1964). Effect of intravenous volume infusion on heart rate in unanaesthetized dogs. *Am. J. Physiol.*, **206**, 992–6.

Kappagoda, C.T., Linden, R.J. & Mary, D.A.S.G. (1975). Gradation of the reflex response from atrial receptors. *J. Physiol.*, **251**, 561–7.

Kappagoda, C.T., Linden, R.J. & Saunders, D.A. (1972). The effect on heart rate of distending the atrial appendages in the dog. *J. Physiol.*, **225**, 705–9.

Kappagoda, C.T., Linden, R.J., Scott, E.M. & Snow, H.M. (1975). Atrial receptors and heart rate: the efferent pathway. *J. Physiol.*, **249**, 581–90.

Kappagoda, C.T., Linden, R.J. & Sivananthan, N. (1977). The receptors which mediate a reflex increase in heart rate. *J. Physiol.*, **266**, 89–90P.

Kappagoda, C.T., Linden, R.J. & Snow, H.M. (1972). The effect of stretching the superior vena caval–right atrial junction on right atrial receptors in the dog. *J. Physiol.*, **227**, 875–87.

Karim, F., Kidd, C., Malpus, C.M. & Penna, P.E. (1972). The effects of stimulation of the left atrial receptors on sympathetic efferent nerve fibres. *J. Physiol.*, **227**, 243–60.

Kidd, C., Ledsome, J.R. & Linden, R.J. (1966). Left atrial receptors and the heart rate. *J. Physiol.*, **185**, 78P.

Ledsome, J.R. & Hainsworth, R. (1970). The effects upon respiration of distension of the pulmonary vein–atrial junctions. *Resp. Physiol.*, **9**, 86–94.

Ledsome, J.R. & Linden, R.J. (1964a). The effect of bretylium tosylate on some cardiovascular reflexes. *J. Physiol.,* **170,** 442–55.

Ledsome, J.R. & Linden, R.J. (1964b). A reflex increase in heart rate from distension of the
pulmonary vein–atrial junctions. *J. Physiol.,* **170,** 456–73.

Ledsome, J.R. & Linden J.R. (1967). The effect of distending a pouch of the left atrium on the
heart rate. *J. Physiol.,* **193,** 121–9.

Linden, R.J. (1973). Function of cardiac receptors. *Circulation,* **48,** 463–80.

Linden, R.J. (1975a). Reflexes from the heart. In *Progress in cardiovascular diseases,* ed. E. H.
Sonnenblide, pp. 201–21. Stratton: New York.

Linden, R.J. (1975b). Sympathetic nerves to the heart. In *The contraction and relaxation of the
myocardium,* ed. W. G. Nayler, pp. 191–266. Academic Press: London.

Linden, R.J. (1976). Reflexes from receptors in the heart. *Cardiology,* **61,** (suppl. 1), 7–30.

Mason, J.M. & Ledsome, J.R. (1974). Effects of obstruction of the mitral orifice or distension
of the pulmonary vein–atrial junctions on renal and hind-limb vascular resistance in the
dog. *Circulation Res.,* **35,** 24–32.

Nonidez, J.F. (1937). Identification of the receptor areas in the venae cavae and pulmonary
veins which initiate reflex cardiac acceleration (Bainbridge's reflex). *Am. J. Anat.,* **61,**
203–31.

Paintal, A.S. (1972). Cardiovascular receptors. In *Handbook of sensory physiology,* vol. III/I,
Enteroreceptors, pp. 1–45. Springer-Verlag: Berlin.

Paintal, A.S. (1973). Vagal sensory receptors and their reflex effects. *Physiol. Rev.,* **53,**
159–227.

Pathak, C.L. (1959). Alternative mechanism of cardiac acceleration in Bainbridge's infusion
experiments. *Am. J. Physiol.,* **197,** 441–4.

Pelletier, C.L. & Shepherd, J.T. (1973). Circulatory reflexes from mechanoreceptors in the
cardio-aortic area. *Circulation Res.,* **33,** 131–8.

Sarnoff, S.J., Gilmore, J.P., Brockman, S.K., Mitchell, J.H. & Linden, R.J. (1960).
Regulation of ventricular contraction by the carotid sinus. Its effect on atrial and
ventricular dynamics. *Circulation Res.,* **8,** 1123–36.

Sassa, K. & Miyazaki, H. (1920). The influence of venous pressure upon the heart-rate. *J.
Physiol.,* **54,** 203–12.

Sleight, P. & Widdicombe, J.G. (1965). Action potentials in fibres from receptors in the
epicardium and myocardium of the dog's left ventricle. *J. Physiol.,* **181,** 235–58.

Shepherd, J.T. (1973). Intrathoracic baroreflexes. In *Mayo Clinic proceedings,* **48,** 426–37.

Thorén, P. (1976). Atrial receptors with non-medullated vagal afferents in the cat – discharge
frequency and pattern in relation to atrial pressure. *Circulation Res.,* **38,** 357–62.

Discussion

(a) Some discussion ranged round the temperatures at which conduction in the nerves was blocked, it being pointed out that myelinated fibres
are completely blocked at 6 °C and C fibres at 2 °C. Linden agreed but
pointed out that in the 'balloon' experiments the nerve fibres were *not*
completely blocked and emphasised that: (1) it was important to remember

that it was a response to distension of balloons in the atrio-venous junctions that was examined; (2) the resting discharge of the receptors while the balloon was not distended was such that there were trains of impulses in the myelinated fibres and in the C fibres which passed the cold block; (3) it was the difference between the resting discharge and the discharge during distension, i.e. the increase in distention (i.e. the response) which was being blocked. Thus, attention must only be paid to the relationship of the *increase* in heart rate and the *increase* in numbers of impulses to the various blocking temperatures. Linden agreed with a member of the audience that this was an unusual way of using the cold block techniques.

(b) There was a question of whether type A or type B receptors were being stimulated by distension of the balloons. Linden replied that in a separate series of experiments they had shown that the receptors stimulated by distension of the balloons were found to be type A and type B, and they were found in the same proportion as found in a random investigation of receptors in the atrium.

9. Atrial receptors and urine flow

C. T. KAPPAGODA

In 1924, Bazett, Thurlow, Crowell & Stewart observed that immersion of human subjects in water resulted in an increase in the flow of urine. This observation, which has been confirmed by several groups of investigators (e.g. Graveline & Jackson, 1962; Kaiser, Eckert, Gauer & Linkenbach, 1969), formed the basis of the speculation regarding the role of receptors in the low pressure areas of the circulation on the regulation of extracellular fluid volume. This speculation was formally expressed in a review published by Gauer & Henry in 1956 in which they suggested that these nerve endings could sense the 'fullness of the circulation' and thereby regulate the extracellular fluid volume. They considered the atrial receptors discharging into myelinated nerves in the vagi as the ones most likely to be involved in this process. Thus these receptors became identified in the contemporary physiological literature as the *volume receptors*. In this review an attempt will be made to assess some of the evidence in support of this proposition. For the purpose of this analysis, the term atrial receptor will be taken to refer solely to the complex unencapsulated nerve endings in the atrial endocardium which discharge into the myelinated nerves in the vagi. When other receptors are referred to, they will be defined separately.

Almost all the investigations performed in this field have been undertaken on the basis that any mechanism which regulates extracellular fluid volume will exert an influence on the rate of flow of urine. Hence, many investigators have sought to establish a link between the activity of these atrial receptors and urine flow. The techniques which have been used to study the effect of stimulating the atrial receptors on urine flow can be classified as follows.

(a) Negative pressure breathing
(b) Centrifugation

(c) Immersion

(d) Alteration of vascular and extravascular fluid volumes by dialysis

(e) Distension of the atria

(f) Stretching localised areas of the atrium.

While all these procedures result in changes in urine flow it is often not possible to define either the nature of the stimulus to the circulation or the receptors involved in any reflex because of the protean nature of the effects of many of these techniques on the cardiovascular system. Against this background, the experiments of Gauer, Henry and their colleagues, who attempted to stimulate the atria directly, stand out as one of the major advances in this field.

Stimulation of the left atrium

The basic experimental technique developed by them is illustrated in Fig. 9.1*a*. In anaesthetised dogs a balloon was inserted through the atrial appendage into the lumen of the left atrium. Distension of the balloon resulted in partial obstruction of the mitral valve with a concurrent rise in pressure in the left atrium. Henry, Gauer & Reeves (1956) observed that this procedure resulted in an increase in left atrial pressure and a small fall in mean arterial pressure. These changes were accompanied by an increase in urine flow. Snaring the pulmonary veins to increase the pressure in the pulmonary vasculature alone resulted in no significant changes in urine

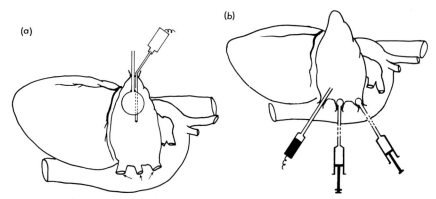

Fig. 9.1. Methods of stimulating atrial receptors. *(a)* Schematic diagram of preparation for producing partial obstruction of the mitral orifice. The balloon is inserted through the atrial appendage. *(b)* Schematic diagram of preparation for stretching the pulmonary vein–atrial junctions. Small balloons are positioned at the entrances of the left pulmonary veins into the left atrium.

flow. This response was also obtained in a second investigation reported by Henry & Pearce (1956) who confirmed the salient features of the response. Over the next 20 years experiments essentially similar to those described above have been performed in several laboratories and the nature of the response has been established beyond doubt (e.g. Ledsome, Linden & O'Connor, 1961; Lydtin & Hamilton, 1964; Ledsome & Linden, 1968; Kinney & DiScala, 1972; Lawrence, Ledsome & Mason, 1973; Kappagoda, Linden, Snow & Whitaker, 1974).

In the majority of the investigations listed above the protocol adopted was such that the urine was collected and measured at 10-min intervals. After an initial control period of 30–40 min the balloon was distended for 30 min following which the distension was released. The urine was collected for a further 40 min. The initial and final 30-min periods formed the two control periods. The test period was taken to be the final 20 min of the distension and the first 10 min after the release of the distension. An example of one such distension is illustrated in Fig. 9.2. In this experiment, distension of a balloon in the left atrium resulted in an increase in mean left atrial pressure of 1.3 kPa and no significant change in mean arterial pressure. Distension of the balloon was followed 5–10 min later by an increase in urine flow. Thereafter the urine volume increased over the next 20 min. After the release of the distension the urine flow began to diminish after 5–10 min. It should be noted that the urine flow during the 10 min immediately following the release may on occasion be equal to or even greater than that immediately before the release. Over the next 30–40 min the urine flow gradually returns to control values. Thus, there is clearly a lag in the onset and decay of the response of an increase in urine flow when compared with the distension of the balloon. At a very superficial level this time course can be taken to suggest the involvement of a humoral agent in this response.

Many of the investigations which have used this particular method of stimulating the left atrium yielded a consistent pattern of changes in the urine. The urine produced during distension of the balloon in the left atrium is, in the main, dilute when compared with that produced during the control periods. This dilution is indicated by an increase in free water clearance (Fig. 9.3). In addition there is also a small increase in the excretion of solutes. Ledsome *et al.* (1961) found an increase of approximately 20% in the solutes excreted – which presumably consists mainly of urea and sodium and potassium salts. Considering the excretion of sodium alone, there is a considerable measure of agreement between the results

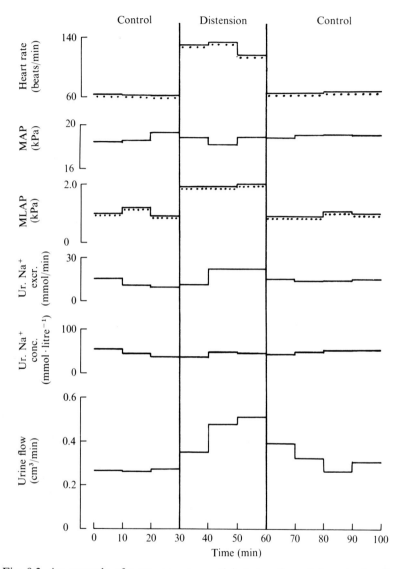

Fig. 9.2. An example of a response to partial obstruction of the mitral orifice. MAP=mean arterial pressure; MLAP=mean left atrial pressure; Ur.Na$^+$ excr. = urinary sodium excretion; Ur.Na$^+$ conc. = urinary sodium concentration. Distension of the balloon resulted in an increase in LAP of 1.3 kPa which was accompanied by an increase in urine flow.

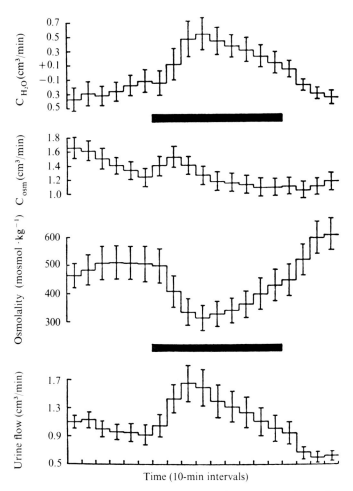

Fig. 9.3. Effect of partial obstruction of the mitral orifice. C_{H_2O} = free water clearance; C_{osm} = osmolar clearance. Distension of the balloon was maintained for 90 min and was accompanied by an increase in free water clearance (Lawrence, Ledsome & Mason, 1973). Bars = s.e.m.

from laboratories which have adopted comparable experimental protocols. For instance, Arndt, Reineck & Gauer (1963) obtained very small and inconsistent changes in sodium excretion while Ledsome & Linden (1968) commented that sodium excretion was relatively constant during balloon distension. In 11 experiments Ledsome & Mason (1972) obtained a mean increase of 8%, while Kappagoda *et al.* (1974) obtained a mean increase of 8% in four experiments.

The only experiments in which the excretion of sodium was widely different was the series reported by Lydtin & Hamilton (1964) who observed a mean increase of 217%. There are, however, considerable differences in the manner in which these responses were calculated which could account for this result.

Distension of a balloon in the lumen of the left atrium clearly results in major changes in the cardiovascular system. The procedure is invariably accompanied by an increase in heart rate. It is also usual to observe a reduction in mean arterial pressure. Henry *et al.* (1956) found that the cardiac output was also reduced. On the basis of these findings it must be concluded that partial obstruction of the mitral orifice probably results in the activation of a multiplicity of cardiovascular reflexes. Thus, in order to conclude that the changes observed in urine flow were caused by activation of structures in the left atrium it was necessary to devise an alternative method of providing a discrete stimulus to the atrium. Since it was already established that the majority of the atrial receptors was located in the endocardium of the pulmonary vein–left atrial junctions (Nonidez, 1937) Ledsome & Linden (1968) attempted to stimulate this region by distending balloons located at the entrances of the pulmonary veins into the left atrium (Fig. 9.1*b*). Unlike the technique described previously, this procedure did not obstruct the flow of blood through the left atrium.

Ledsome & Linden (1968) observed that distension of balloons positioned at the vein–atrial junctions resulted in an increase in urine flow in anaesthetised dogs. In addition it was found that when mitral obstruction and distension of the pulmonary vein–atrial junctions were applied alternately in the same animal both procedures resulted in an increase in urine flow (Fig. 9.4). The response to distension of the pulmonary vein–atrial junctions alone was somewhat smaller than that evoked by the larger balloon. Recently, Kappagoda *et al.* (1974) have confirmed these observations. The time course of the response and the concurrent changes in sodium excretion were similar to that observed in the earlier studies with the bigger stimulus. Another important feature of the study reported by Ledsome & Linden (1968) was that the application of local anaesthetic to the pericardial sac (a procedure likely to block transmission in nerves originating from the heart) abolished the response to mitral obstruction. From these studies it was concluded that both mitral obstruction and stretching of the pulmonary vein–atrial junctions resulted in an increase in urine flow which was qualitatively the same and that activation of structures within the left atrium was responsible for this response.

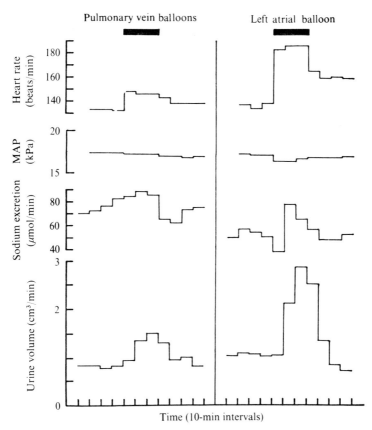

Fig. 9.4. Effect of stretching the pulmonary vein–atrial junctions *(a)* and partial obstruction of the mitral orifice *(b)* on urine flow. MAP = mean arterial pressure. Both procedures result in increases in urine flow which have the same time course.

Stimulation of the right atrium

Since the endocardium of the right atrium contains receptors which are morphologically similar to those found in the left atrium, Kappagoda, Linden & Snow (1973) attempted to extend the techniques described above to the right atrium. It was found that distension of a balloon in the lumen of the right atrium so as to cause partial obstruction of the tricuspid valve also resulted in an increase in urine flow. This response of an increase in urine flow had the same characteristics as that evoked by obstruction of the mitral orifice. Distension of a balloon positioned at the superior vena caval–right atrial junction resulted in a much smaller response than that produced by tricuspid obstruction – probably because of the very small number of receptors in this position (Kappagoda *et al.*, 1973).

Stimulation of atrial appendages

The evidence discussed so far indicates that stimulation of structures in the atrial wall results in an increase in urine flow. This assertion has been further supported by experiments in which the atrial appendages alone were stretched. Kappagoda, Linden & Snow (1972) found that distension of small balloons located in the atrial appendages also resulted in an increase in urine flow and that this response was abolished by crushing the bases of both atrial appendages.

Thus, in summary, it is concluded that there are structures within the walls of the atria which when stimulated result in an increase in the rate of urine flow. The evidence which suggests that this response of an increase in urine flow is a reflex mediated by the stimulation of atrial receptors is presented below.

Evidence for the reflex nature of the response

The reflex nature of any response affecting the cardiovascular system can only be established by demonstrating at least some of the components of the reflex arc mediating the response. In the context of the response of an increase in urine flow, the components of the arc which have been investigated are *(a)* the afferent pathway, *(b)* the receptors and *(c)* the efferent pathway.

Afferent pathway

One of the earliest attempts at demonstrating that this response was reflex in nature was the experiments described by Henry & Pearce (1956). They found that cooling the cervical vagi abolished the response produced by partial obstruction of the mitral orifice. Ledsome *et al.* (1961) confirmed this effect (Fig. 9.5).

Due to the diffused nature of the stimulus provided by obstruction of the mitral valve Ledsome & Linden (1968) re-examined the problem by distending balloons positioned at the vein–atrial junctions. The resulting increase in urine flow was abolished by cutting the vagi in the chest. Cooling one cervical vagus also diminished the responses produced by stimulating the atrial appendages (Kappagoda *et al.*, 1972) and by partial obstruction of the tricuspid orifice (Kappagoda *et al.*, 1973).

Sectioning the ansae subclaviae did not abolish the response to either mitral obstruction or distension of the pulmonary vein–atrial junctions (Ledsome & Linden, 1968).

These findings, taken collectively, provided strong evidence in support of the claim that the response of an increase in urine flow which followed stimulation of the atria was reflex in nature. They also indicated that the afferent pathway of this reflex was at least partly in the vagi.

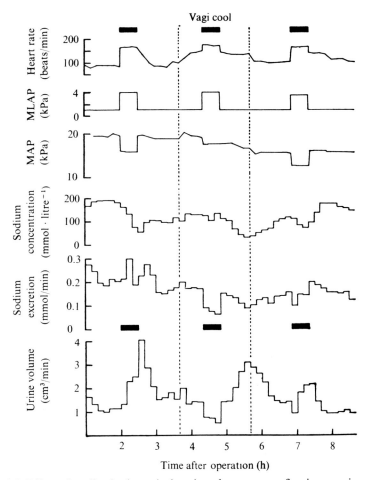

Fig. 9.5. Effect of cooling both cervical vagi on the response of an increase in urine flow which results from partial mitral obstruction. MLAP = mean left atrial pressure; MAP = mean arterial pressure. Cooling both vagi abolished the response which was restored after rewarming the nerves (Ledsome, Linden & O'Connor, 1961).

Receptors

The atrial receptors (as previously defined) have long been implicated as the receptor mechanism of the reflex arc mediating this response of an increase in urine flow. This proposition has been questioned in recent years because of the demonstration of more than one type of afferent nerve ending in the atrium. There is electrophysiological evidence of three such types of nerve endings: *(a)* the atrial receptors which discharge into the myelinated nerves of the vagi; *(b)* receptor endings which discharge into non-myelinated nerves in the vagi; and *(c)* receptor endings which discharge into the sympathetic nerves. All three types of nerve endings respond to moderate increases (1.0–2.0 kPa) in atrial pressure (e.g. Paintal, 1973; Kostreva *et al.,* 1975; Thorén, 1976). Thus it is likely that the techniques which are used to stimulate the atria so as to elicit changes in urine flow would also stimulate all three types of nerve endings.

In order to resolve the controversy regarding the receptors involved in this response, Kappagoda, Linden & Sivananthan (1977) studied the effect of graded cold blockade of the cervical vagi on the response of an increase in urine flow and on the discharge from receptors in the atrium. They found that distension of a balloon to occlude the mitral orifice activated both types of receptors which discharge into vagal fibres. Cooling the vagi resulted in attenuation of the responses to balloon distension in both types of fibres (Fig. 9.6). Most of the responses in myelinated nerve fibres were abolished over a range of 8–12 °C. The responses in non-myelinated fibres were abolished over a wide range of temperatures. When these effects were compared with the effect of cold blockade of the cervical vagi on the response of an increase in urine flow it was found that the latter was reduced in the same manner as the response in the myelinated fibres (Fig. 9.6).

Graded cold blockade of the cervical vagi also yielded another item of evidence. It was found that cooling both vagi to approximately 10 °C resulted in abolition of the response of an increase in urine flow to partial obstruction of the mitral orifice. Whilst this finding confirmed the earlier observations of Henry & Pearce (1956) it also rendered it unlikely that the receptors discharging into the sympathetic nerves were involved in this response.

Thus it is concluded that the atrial receptors discharging into the myelinated fibres in the vagi are the ones most likely to be involved in the reflex increase in urine flow.

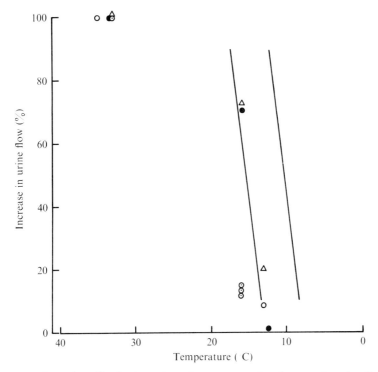

Fig. 9.6. Effect of cooling both vagi on the response of an increase in urine flow. Ordinate: the response at the various temperatures was expressed as a percentage of the response at 34–36 °C. Abscissa: temperature of the vagi. The parallel lines represent the effect of cooling on the responses to balloon distension in myelinated fibres.

Efferent pathway

The efferent pathway of this reflex remains a source of considerable controversy. The first attempt at resolving it was that of Ledsome *et al.* (1961) who studied the effect of denervating one kidney. Atrial stimulation by partial obstruction of the mitral valve resulted in an increase in urine flow in both kidneys (Fig. 9.7). This observation was corroborated by the investigations of Carswell, Hainsworth & Ledsome (1968) who observed that the response was also not abolished by the administration of the ganglion-blocking drug hexamethonium bromide. On the basis of these findings it is unlikely that the renal nerves were primarily involved in the response.

Investigations into the efferent mechanism of this response were then directed towards the involvement of humoral agents. Carswell, Hainsworth & Ledsome (1970) perfused an isolated kidney with blood from a dog undergoing atrial stimulation by partial obstruction of the mitral orifice. They observed that an increase in urine flow occurred in the isolated kidney. These results led to the inevitable conclusion that a humoral agent

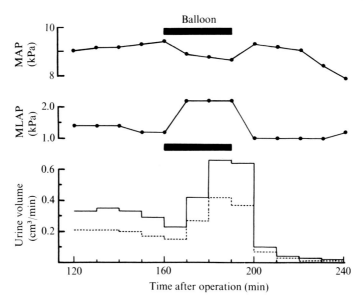

Fig. 9.7. Effect of denervating one kidney on the response to mitral obstruction. MAP = mean arterial pressure; MLAP = mean left atrial pressure. Interrupted line represents the intact kidney and the solid line is the denervated kidney. Balloon distension results in an increase in urine flow in both kidneys (Ledsome, Linden & O'Connor, 1961).

was involved in the response. It also seemed possible, though unlikely, that the humoral agent was released by a neurally mediated mechanism from the kidneys of the dog undergoing atrial stimulation.

In order to explore this possibility a series of experiments in which the right kidney alone was denervated was performed (Kappagoda, Linden, Snow & Whitaker, unpublished observations). After the response to balloon distension of an increase in urine flow was demonstrated the intact kidney was removed. Thus only a denervated kidney was linked to the circulation of the dog. Distension of a balloon to cause partial obstruction of the mitral valve resulted in an increase in urine flow in the denervated kidney (Fig. 9.8). The innervated kidney was clearly not the source of the humoral agent.

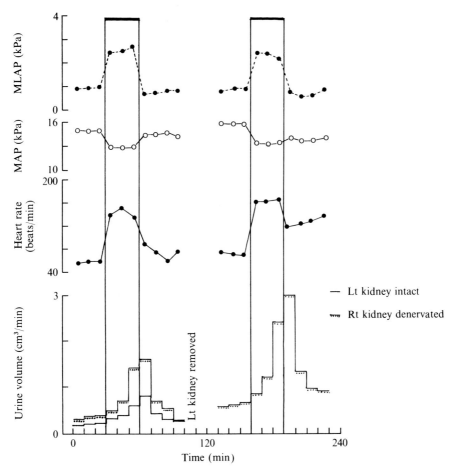

Fig. 9.8. Effect of denervation of one kidney. MLAP = mean left atrial pressure; MAP = mean arterial pressure. Distensions of the balloon are indicated by the horizontal bars. Removal of the intact kidney does not abolish the response in the denervated kidney.

However, the role of the renal nerves cannot be dismissed readily because of the observations of Karim, Kidd, Malpus & Penna (1972). They observed that stretching the pulmonary vein–atrial junctions in the manner of Ledsome & Linden (1968) diminished the discharge in the sympathetic nerves to the kidney (Fig. 9.9). More recently, Mason & Ledsome (1974) have shown that both forms of atrial stimulation resulted in an increase in renal blood flow and a decrease in renal vascular resistance. It is therefore possible that the renal nerves may yet be shown to play a role in the increase in urine flow which follows atrial stimulation.

The identity of the humoral agent responsible for the increase in urine flow remains in dispute. Much of the controversy centres around the role of the antidiuretic hormone (ADH) which in the popular view is believed to mediate this response (Gauer & Henry, 1976). There are several objections

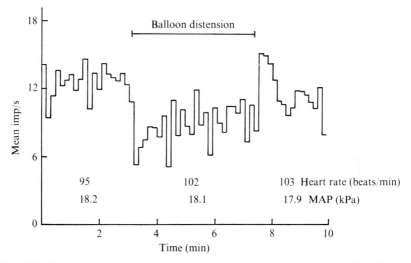

Fig. 9.9. Continuous record of the mean frequency of a renal sympathetic efferent strand. Balloon distension resulted in a significant decrease in the mean frequency without significant changes in mean arterial blood pressure (MAP) and heart rate (Karim, Kidd, Malpus & Penna, 1972).

to this point of view. The main one is concerned with the sensitivity of anaesthetised dogs to ADH.

Using the suppression of a water diuresis as the index it is possible to compare the sensitivity to ADH of anaesthetised dogs (Perlmutt, 1961) and conscious dogs (Orloff, Wagner & Davidson, 1957). From this comparison it can be concluded that the sensitivity of anaesthetised dogs is less than that of conscious ones. This conclusion is borne out by the finding that the concentration of ADH in the plasma of anaesthetised dogs is relatively high – approximately 10 $\mu IU \cdot cm^{-3}$ (e.g. De Torrente, Robertson, McDonald & Schrier, 1975).

Using the data from studies on the clearance of vasopressin from plasma (Lauson & Bocanegra, 1961) it could be predicted that, in conscious animals, the suppression of a water diuresis would be maximal when the concentration of ADH in plasma is of the order of 2 μIU·cm^{-3}. The comparable value for anaesthetised dogs is 10 μIU·cm^{-3}. It must be emphasised that these sensitivities have been demonstrated only in the presence of an unequivocal water diuresis – a condition which is wholly irrelevant to the situation prevailing in the experiments in which the atrial receptors have been stimulated.

One investigation which does bear some relevance to the latter experiments is the study by Mason & Ledsome (1971). In dogs anaesthetised with chloralose they found that altering the rate of infusion of vasopressin from 0.4 mIU·kg^{-3}/min to 0.04 mIU·kg^{-1}/min resulted in an increase in urine flow. After 30 min at the lower rate of infusion, the urine continued to show a negative free water clearance. After a comparable period of atrial stimulation, the urine tends to show a positive free water clearance (see Fig. 9.3). Even if this discrepancy is ignored, one cannot ignore the probable concentrations of ADH in the plasma during these infusions. Using the clearance data of Lauson & Bocanegra (1961), the concentration of ADH in the plasma during the higher rate of infusion can can be predicted to be approximately 60 μIU·cm^{-3} and the lower rate of infusion to be 6.0 μIU·cm^{-3}. These changes in concentration are likely to involve the pressor effects of vasopressin. In fact, changes in arterial pressure were observed during the infusions (Mason & Ledsome, 1971).

One could, however, extrapolate from this study to the work of De Torrente *et al.* (1975). They observed that distension of a balloon in the left atrium resulted in an increase in urine flow which was accompanied by a change in the concentration of ADH in plasma from 11 μIU·cm^{-3} to 5 μIU·cm^{-3}. The claim was made that the latter was responsible for the increase in urine flow. Closer observation of the data reveals that a positive free water clearance was not achieved during balloon distension even after 1 hour. If the physiological functions of ADH (in terms of its antidiuretic activity) are operational in anaesthetised dogs, this finding must indicate that the animals were 'at the top of the dose–responsive curve' for ADH. If this is the case much greater changes in the concentration of ADH could reasonably be expected so as to account for the increase in urine flow (cf. Mason & Ledsome, 1971). On the basis of this analysis, the data presented by De Torrente *et al.* (1975) does not permit the conclusion that the changes in the concentration of ADH were responsible for the increase in urine flow. Indeed, this possibility would seem somewhat remote.

In an effort to re-examine the role of ADH in this response, Kappagoda, Linden, Snow & Whitaker (1975) studied the effect of partial obstruction of the mitral orifice in anaesthetised dogs after destruction of the pituitary gland by electrocoagulation (Fig. 9.10). In spite of the destruction of the

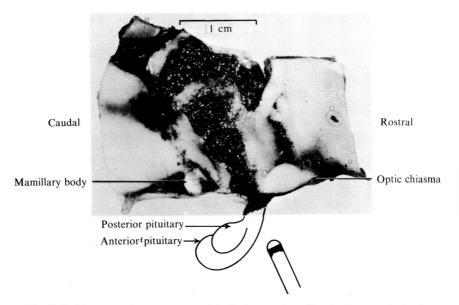

Fig. 9.10. Macroscopic appearance of the lesion produced by electrocoagulation in one dog. The position of the pituitary gland and the path of the probe are also illustrated. The probe severed the pituitary stalk and entered the adjacent areas of the hypothalamus to a depth of 1.5 cm producing a cylindrical lesion approximately 1.5 cm in diameter (Kappagoda, Linden, Snow & Whitaker, 1975).

pituitary gland, the response of an increase in urine flow was observed (Fig. 9.11). Fig. 9.11 shows a summary of these responses along with some of the responses previously reported in intact dogs by Ledsome & Linden (1968).

For the reasons outlined above it would appear unlikely that the humoral agent responsible for the increase in urine flow which follows atrial stimulation is ADH. It has been suggested that it could be an agent which is primarily diuretic in nature (Kappagoda *et al.,* 1975). On this basis, Kappagoda, Knapp, Linden & Whitaker (1976) have attempted to identify alternative humoral agents by examining the plasma from dogs on a variety of bioassay preparations. One such preparation was the

Malpighian tubule of *Rhodnius prolixus*. These Malphigian tubules were suspended *in vitro* in drops of 'test' and 'control' dog plasma in the manner of Madrell (1969). It was found that the tubules suspended in 'test' plasma secreted at a slower rate than those suspended in 'control' plasma. These differences were preserved when the plasma was extracted with butanol and reapplied to the tubules. Cutting the cervical vagi abolished the difference in both plasma and the extracts. These observations indicate that this preparation may serve as a useful technique for establishing the identity of the humoral agent responsible for the diuresis. It is of interest to note that ADH does not influence the rate of secretion of these tubules.

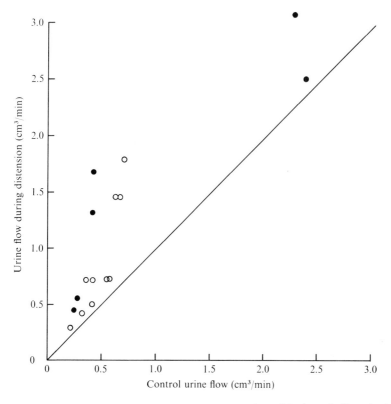

Fig. 9.11. Plot of increases in urine flow during distension of the large balloon in the left atrium. Ordinate: urine flow during period of distension; abscissa: average urine flow during control periods before and after period of distension. Responses were obtained in dogs with posterior pituitary gland intact (●) and with the posterior pituitary completely destroyed (o).

Summary

In this brief review evidence has been discussed to show that stimulation of the atria of anaesthetised dogs results in an increase in urine flow which is reflex in nature. The receptors most likely to mediate this response are the complex unencapsulated nerve endings located in the atrial endocardium which discharge into the myelinated fibres in the vagi. The efferent path of this reflex is indubitably humoral in nature and its identity remains controversial.

References

Arndt, J.O., Reineck, H. & Gauer, O.H. (1963). Ausscheidungsfunktion und Hamodynamik der Nieren bei Dehnung des linken Vorhofes am narkotisierten Hund. *Pflügers. Arch. ges. Physiol.*, **227**, 1–15.

Bazett, H.C., Thurlow, S., Crowell, C. & Stewart, W. (1924). Studies on the effects of baths in man. II. The diuresis caused by warm baths together with some observations on urinary tides. *Am. J. Physiol.*, **70**, 430–52.

Carswell, F., Hainsworth, R. & Ledsome, J.R. (1968). Elimination of agents possibly responsible for the diuretic response to left atrial distension. *J. Physiol.*, **198**, 23–4P.

Carswell, F., Hainsworth, R. & Ledsome, J.R. (1970). The effects of left atrial distension upon urine flow from the isolated perfused kidney. *Q. Jl exp. Physiol.*, **55**, 173–82.

De Torrente, A., Robertson, G.L., McDonald, K.M. & Schrier, R.W. (1975). Mechanism of diuretic response to increased left atrial pressure in the anaesthetized dog. *Kidney Int.*, **8**, 355–61.

Gauer, O.H. & Henry, J.P. (1956). Circulatory basis of fluid volume control. *Physiol. Rev.*, **43**, 423–81.

Gauer, O.H. & Henry, J.P. (1976). Neurohormonal control of plasma volume. *International review of physiology II*, **9**, 145–90.

Graveline, D.E. & Jackson, M.M. (1962). Diuresis associated with prolonged water immersion. *J. appl. Physiol.*, **17**, 519–24.

Henry, J.P., Gauer, O.H. & Reeves, J.L. (1956). Evidence of the left atrial location of receptors influencing urine flow. *Circulation Res.*, **4**, 85–90.

Henry, J.P. & Pearce, J.W. (1956). The possible role of cardiac atrial stretch receptors in the induction of changes in urine flow. *J. Physiol.*, **131**, 572–85.

Kaiser, D., Eckert, P., Gauer, O.H. & Linkenbach, H.J. (1969). Die diurese bei immersion in ein thermoindefferentes Vollbad. *Pflügers Arch. ges. Physiol.*, **306**, 247–61.

Kappagoda, C.T., Linden, R.J. & Sivananthan, N. (1977). The receptors mediating reflex increase in urine flow. *J. Physiol.* (In Press.)

Kappagoda, C.T., Linden, R.J. & Snow, H.M. (1972). The effect of distending atrial appendage in the dog. *J. Physiol.*, **227**, 233–44.

Kappagoda, C.T., Linden, R.J. & Snow, H.M. (1973). Effect of stimulating right atrial receptors on the urine flow in the dog. *J. Physiol.*, **235**, 493–502.

Kappagoda, C.T., Linden, R.J., Snow, H.M. & Whitaker, E.M. (1974). Left atrial receptors and the antidiuretic hormone. *J. Physiol.*, **237**, 663–83.

Kappagoda, C.T., Linden, R.J., Snow, H.M. & Whitaker, E.M. (1975) Effect of destruction of the posterior pituitary on the diuresis from left atrial receptors. *J. Physiol.*, **244**, 757–70.

Kappagoda, C.T., Knapp, M.F., Linden, R.J. & Whitaker, E.M. (1976). A possible bioassay for the humoral agent responsible for the diuresis from atrial receptors. *J. Physiol.*, **254**, 59P.

Karim, F., Kidd, C., Malpus, C.M. & Penna, P.E. (1972). The effect of stimulation of left atrial receptors on sympathetic efferent nerve activity. *J. Physiol.*, **27**, 243–60.

Kinney, M.J. & DiScala, V.A. (1972). Renal clearance studies of effect of left atrial distension in the dog. *Am. J. Physiol.*, **222**, 1000–3.

Kostreva, D.R., Zuperku, E.J., Purtock, R.V., Coon, R.L. & Kampine, J.P. (1975). Sympathetic afferent activity of right atrial origin. *Am. J. Physiol.*, **229**, 911–15.

Lauson, H.D. & Bocanegra, M. (1961). Clearance of exogenous vasopressin from plasma of dogs. *Am. J. Physiol.*, **200**, 493–7.

Lawrence, M.J.R., Ledsome, J.R. & Mason, J.M. (1973). The time course of the diuretic response to left atrial distension. *Q. Jl exp. Physiol.*, **58**, 219–27.

Ledsome, J.R. & Linden, R.J. (1968). The role of the left atrial receptors in the diuretic response to left atrial distension. *J. Physiol.*, **198**, 487–503.

Ledsome, J.R., Linden, R.J. & O'Connor, W.J. (1961). The mechanisms by which distension of the left atrium produces diuresis in anaesthetized dogs. *J. Physiol.*, **159**, 87–100.

Ledsome, J.R. & Mason, J.M. (1972). The effects of vasopressin on the diuretic response to left atrial distension. *J. Physiol.*, **221**, 427–40.

Lydtin, H. & Hamilton, W.F. (1964). Effect of acute changes in left atrial pressure on urine flow in anaesthetized dogs. *Am. J. Physiol.*, **207**, 530–6.

Maddrell, S.H.P. (1969). Secretion by the Malpighian tubules of *Rhodnius*. The movements of ions and water. *J. exp. Biol.*, **51**, 71–97.

Mason, J.M. & Ledsome, J.R. (1971). The effects of changes in the rate of infusion of vasopressin in anaesthetized dogs. *Can. J. of Physiol. Pharmac.*, **49**, 933–40.

Mason, J.M. & Ledsome, J.R. (1974). Effects of obstruction of the mitral orifice or distension of the pulmonary vein–atrial junctions on renal and hind limb vascular resistance in the dog. *Circulation Res.*, **35**, 24–32.

Nonidez, J. (1937). Identification of the receptor areas in the vena cavae and pulmonary veins which initiate reflex cardiac acceleration (Bainbridge's reflex). *Am. J. Anat.*, **61**, 203–31.

Orloff, J., Wagner, H.N. & Davidson, D.G. (1957). The effect of variations in solute secretion and vasopressin dosage on the excretion of water in the dog. *J. clin. Invest*, **37**, 458–64.

Paintal, A.S. (1973). Vagal sensory receptors and their reflex effects. *Physiol. Rev.*, **53**, 159–227.

Perlmutt, J.H. (1961). Renal activity of vasopressin in anaesthetized dogs. *Am. J. Physiol.*, **200**, 400–4.

Thorén, P. (1976). Atrial receptors with nonmedullated vagal afferents in the cat – discharge frequency and pattern in relation to atrial pressure. *Circulation Res.*, **38**, 62.

Discussion

(a) Answering a question about the time of onset of responses in relation to the distension of the balloon at the mitral orifice, Kappagoda pointed

out that 10–20 cm³ of saline could not be injected into the balloon immediately – the injections were over 1–2 min. At the end of the injection the heart rate had already increased and the blood pressure had fallen 1–3 kPa. Using the small balloons there was no fall in blood pressure; the heart rate, being a result of increased sympathetic activity, increased about 4–5 s after the injection of 1 cm³. The urine flow started to increase, with both balloon systems, about 5 min after distension.

(b) In answer to a question suggesting that mammalian tissue would have been better than the Malpighian tubule of *Rhodnius prolixus,* Kappagoda pointed out that this tubule was used as a test to differentiate plasma obtained during a period of stimulation of atrial receptors from plasma obtained during the control period. Extracts were made of each plasma and the differences between the extracts were then followed by using the tubule – any piece of tissue would be adequate for this purpose as long as only minute volumes of test substance could be used.

(c) Kappagoda pointed out that there was no way of knowing how much of an increase in blood volume there would have to be to increase the pressure in the atria so as to produce responses similar to those that he described.

There were suggestions from the floor that the response of an increase in urine flow could result from the haemodynamic changes caused by distension of the balloons, e.g. there were changes in heart rate. Kappagoda replied that the haemodynamic changes were not causative because blocking the effect of efferent sympathetic nerves and parasympathetic nerves by sectioning the nerves or blocking with pharmacological agents did not alter the response of an increase in urine flow.

10. A search for evidence linking atrial receptors to renal regulatory mechanisms

K. L. GOETZ

Our attempts to elicit reflex effects from atrial receptors began with experiments designed to alter atrial transmural pressure in the conscious dog. In those experiments we employed a technique that enabled us to produce a selective increase in pressure outside of the atrial walls. This was accomplished by surgically modifying the pericardium; the pericardium lying over the ventricles was trimmed away and the remaining edge was sewn to the atrio-ventricular groove to produce a pericardial pouch that enclosed the atria but not the ventricles (Fig. 10.1). Vascular catheters were implanted during the operation to facilitate subsequent haemodynamic monitoring.

Our initial experiments were performed on the fully conscious animal after it had recovered from the operation. Cardiovascular variables were measured continuously, and urine was collected at 10-min intervals. After a constant urine flow had been established, saline was infused into the pericardial pouch to increase pressure on the outside of the atrial walls and thereby reduce atrial transmural pressure. Because infusion of the saline produces a mild, controlled tamponade of the atria, we refer to this procedure as 'atrial tamponade'. The tamponade has little effect on haemodynamics as shown in Fig. 10.2. Most of our original experiments included the measurement of left atrial pressure but not central venous pressure, and left atrial pressure does not increase during tamponade. Consequently, we believed that a tracing such as the one depicted in Fig. 10.2 demonstrated that atrial tamponade had no effect on systemic haemodynamics (Goetz, Hermreck, Slick & Starke, 1970); we later learned that this was not true. When we began measuring central venous pressure routinely during tamponade experiments we learned that increasing intrapericardial pressure by 0.8–1.1 kPa causes about a 0.3–0.4 kPa

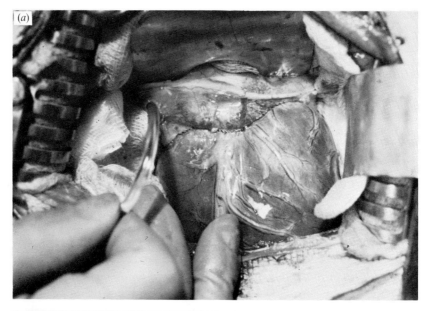

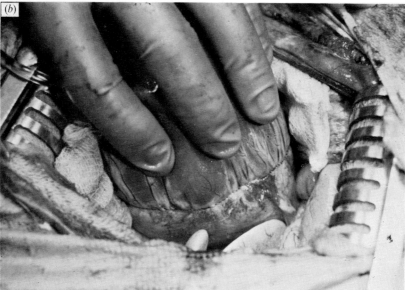

Fig. 10.1. An atrial pericardial pouch viewed through a left lateral thoracotomy incision. *(a)* 'Front' view with anterior descending artery visible in centre of ventricles. Suture line marks lower limits of atrial pouch. A large plastic catheter enters the pouch. *(b)* 'Back' view with the apex of the heart lifted up and retracted away from the field of view.

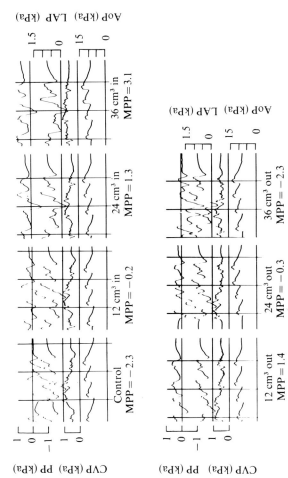

Fig. 10.2. Effect of atrial tamponade on systemic haemodynamics. Pressure tracings from the pericardial pouch (PP), left atrium (LAP), right atrium (CVP), and aorta (AoP) are shown. The addition of successive 12 cm³ increments of saline to the pouch produced the expected increase in mean pouch pressure (MPP), but the rest of the variables were quite stable. Removal of the saline reduced pouch pressure to control levels. (From Goetz, Hermreck, Slick & Starke, 1970.)

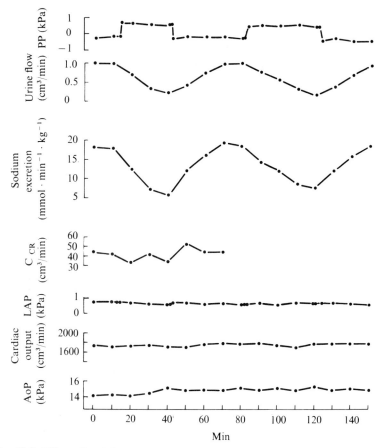

Fig. 10.3. Effect of atrial tamponade on renal function. PP = pericardial pressure; C_{CR} = creatinine clearance; LAP = left atrial pressure; AoP = aortic blood pressure. (From Goetz, Hermreck, Slick & Starke, 1970.)

increase in central venous pressure (Goetz & Bond, 1971). Actually a slight change in central venous pressure is discernible in this experiment (Fig. 10.2), especially if one compares the central venous pressure during peak tamponade with the recovery period. Measuring the areas under the two tracings indicates that the central venous pressure during peak tamponade is 0.2 kPa higher than during the recovery period.

Atrial tamponade produces obvious changes in renal function. One of our earliest experiments is summarised in Fig. 10.3. Intrapericardial pressure is shown at the top of the graph. Atrial tamponade was produced by infusing enough saline into the pouch to increase intrapericardial pressure by about 0.8 kPa. The resultant decreases in urine flow and sodium excretion are readily apparent. After 30 min the saline was withdrawn from the pouch and both urine flow and sodium excretion returned to control levels. A second period of atrial tamponade produced comparable results. Measured haemodynamic variables were virtually constant throughout the experiment as indicated by levels of left atrial pressure, cardiac output and aortic blood pressure. Because atrial tamponade produces a decrease in atrial transmural pressure and because we detected no change in systemic haemodynamics in these early experiments, we suggested that the decreases in urine flow and sodium excretion that occur during atrial tamponade are produced by a reflex initiated from atrial

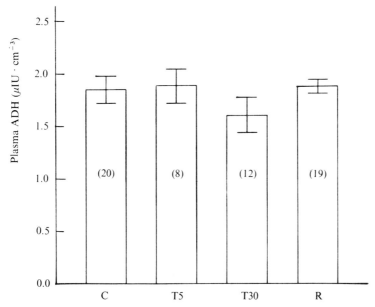

Fig. 10.4. Effect of atrial tamponade on the concentration of antidiuretic hormone (ADH) in plasma. C is the control concentration of ADH; T5 is the concentration after 5 min of tamponade; T30 is the concentration after 30 min of tamponade; R is the recovery concentration. Bars are ±s.e.m. (From Goetz, Bond, Hermreck & Trank, 1970.)

receptors (Goetz, Hermreck, Slick & Starke, 1970). In other words, these results appeared to be compatible with the atrial volume receptor hypothesis as formulated by Gauer & Henry (1963).

   In other experiments, we measured plasma antidiuretic hormone (ADH) levels with a bioassay in the hydrated rat (Goetz, Bond, Hermreck &

Trank, 1970). Results are shown in Fig. 10.4. The left bar denotes the control concentration of ADH; the next two, the values after 5 and 30 min of tamponade, and the right bar represents the recovery value. We were unable to detect any effect of atrial tamponade on plasma ADH levels in these experiments.

We had performed these early experiments on the assumption that the decrease in atrial transmural pressure would decrease the discharge activity of atrial receptors with a type B firing pattern even though we recognised that receptors located along the extreme limits of the pouch where the pericardium reflects back on itself conceivably might be little affected or even stretched by the tamponade (Goetz, Bond, Hermreck & Trank, 1970). We later obtained direct evidence regarding this point (Fig. 10.5). In each panel, from above downward, are shown the electrocardiogram, central venous pressure, aortic pressure and the electroneurogram. A typical B-type firing pattern from an atrial receptor appears in the control panel. Addition of saline in 10 cm³ increments to the pericardial pouch caused progressive decreases in the amount of discharge per cardiac cycle. When a total of 30 cm³ had been added, the impulses disappeared completely. Removal of the saline in 10 cm³ steps restored the discharge pattern to normal. There is little evidence for any haemodynamic response to this

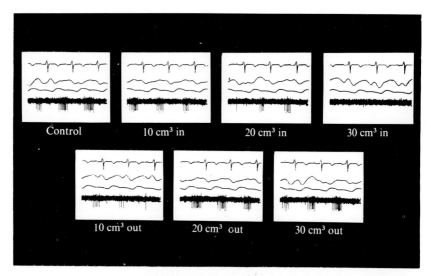

Fig. 10.5. Changes in atrial type B receptor discharge induced by graded atrial tamponade. In each panel, from above downwards, are shown the ECG, central venous pressure, aortic pressure and the electroneurogram.

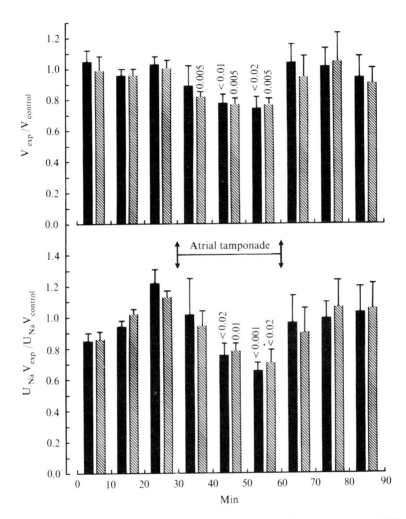

Fig. 10.6. Effect of atrial tamponade in dogs with one kidney denervated. The period of atrial tamponade is marked by the arrows. The data from the intact and the denervated kidneys were 'normalised' by the following procedure. In each experiment, the values obtained during the three 10-min control periods were averaged to give the mean control value. Each individual value was then expressed as the ratio of the experimental value over the mean control value. Urine flow is shown in the top panel and sodium excretion in the lower panel. Solid histograms, intact kidney; cross-hatched histograms, denervated kidney.

mild atrial tamponade although careful measurement indicates that there was a slight decrease in heart rate when 30 cm³ of saline were in the pouch. This was not a consistent change; in other experiments there was either no change or a slight increase in heart rate. Sixteen fibres were studied in this manner. Eleven decreased their rate of discharge during atrial tamponade although most not as markedly as the example in Fig. 10.5, three increased their rate of discharge and two were unchanged. Thus, although there are exceptions, atrial tamponade generally decreases the activity of atrial type B discharge in accordance with our original hypothesis.

Because of the prominent renal response elicited by atrial tamponade, we attempted to identify the efferent pathway of this presumed reflex. In one series of experiments the left kidney was denervated by carefully stripping the nerves and adventitia from the renal artery, renal vein, ureter and capsule. As an additional precaution, the renal artery was transected and re-anastomosed to ensure that all sympathetic fibres in the arterial wall were disrupted. The maximum time interval between the denervation procedure and the subsequent experiment was 15 d. On the day of the experiment the dog was anaesthetised with pentobarbital and the ureters were exposed retroperitoneally through bilateral flank incisions.

The effect of atrial tamponade on urine flow and sodium excretion in seven unilateral denervation experiments is shown in Fig. 10.6. The period of atrial tamponade is indicated by the arrows. Urine flow, shown in the top panel, decreased significantly in both the intact kidney and in the contralateral denervated kidney during atrial tamponade. Sodium excretion, shown in the bottom panel, also decreased significantly in both the innervated and denervated kidneys during tamponade. The response of the denervated kidneys did not differ significantly from that of the innervated kidneys.

Haemodynamic data from these experiments are summarised in Fig. 10.7. The period of atrial tamponade is indicated by the period of increased intrapericardial pressure, shown on top. Pericardial pressure was increased by about 0.8 kPa for 30 min and then was returned to control levels. Mean central venous pressure increased approximately 0.3 kPa during atrial tamponade. Mean atrial pressure, pulse pressure (S–D), and heart rate showed no significant alterations throughout these experiments (Goetz & Bond, 1971).

These experiments made it evident that the renal nerves are not required to produce the renal response to atrial tamponade, so we performed additional studies which I will summarise briefly. Both kidneys were

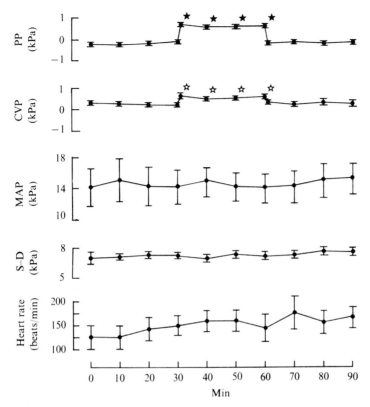

Fig. 10.7. Effect of atrial tamponade on haemodynamics in unilateral denervation experiments. PP = pericardial pressure; CVP = central venous pressure; MAP = mean arterial pressure; S-D = pulse pressure.

denervated surgically during a preliminary operation. After the operation, all animals were given 15 mg of deoxycorticosterone acetate (DOCA), intramuscularly, daily for at least 8 d to allow them to escape from the effects of salt-retaining steroids. On the morning of the experiment five units of vasopressin in oil were given intramuscularly as was one half of the daily dose of DOCA. In addition, pharmacological blockade of the sympathetic nervous system was produced with the intravenous adminis-tration of propranolol (1 mg·kg^{-1}) and phenoxybenzamine (5 mg·kg^{-1}) to eliminate potential effects on the denervated kidneys from changes in circulating catecholamines. Experiments in the conscious dog after these manoeuvres, however, continued to cause comparable decreases in urine flow and sodium excretion. We therefore concluded that an unidentified blood-borne substance was responsible for the antidiuresis and anti-natriuresis elicited by atrial tamponade (Goetz & Bond, 1971). This conclusion is comparable to that reached by the Leeds group when they attempted to identify the efferent pathway of a renal regulating reflex thought to arise from atrial receptors (Carswell, Hainsworth & Ledsome, 1970), and also somewhat reminiscent of the work of Keeler (1974, 1975), who demonstrated that neither renal denervation nor vasopressin infusion prevents the diuretic response to unilateral stimulation of the carotid sinus in rats.

The work described so far had not established that the renal response elicited by atrial tamponade actually was attributable to a reflex arising from atrial receptors although we assumed that it was. An alternate possibility was that the subtle changes in systemic haemodynamics were responsible for part or all of the renal response. In an attempt to differentiate between these two possibilities we produced tamponade of the atria after the known afferent pathways from the heart had been interrupted. Dogs for these experiments were prepared somewhat differently. The chest was opened through a mid-line sternotomy and a pericardial pouch was prepared. In addition, the sympathetic trunk from the stellate ganglion to T5 was removed bilaterally. The animals were allowed to recover from the operation for at least 1 week. In one series of experiments the animals were anaesthetised with pentobarbital and the vagi were divided in the mid-cervical region. Each animal was prehydrated with saline and a stable urine flow was established. In spite of the sympathectomy and vagotomy, atrial tamponade continued to produce the typical antidiuresis and antinatriuresis. With this degree of autonomic denervation, however, there was an additional change in haemodynamics noted; aortic blood pressure decreased during tamponade (Goetz, Bloxham, Bond & Sharma, 1976). Although the mean maximum decrease was only 1.1 kPa, this finding provided additional evidence that atrial tamponade does influence systemic haemodynamics. Apparently the carotid sinus baroreceptors are unable to maintain a stable blood pressure during tamponade in an animal after cervical vagotomy and a partial thoracic sympathectomy.

The decrease in arterial blood pressure complicated the interpretation of these experiments because it was conceivable that a simple fall in perfusion pressure of the kidneys was responsible for the observed response. In order to eliminate this possibility, we conducted an additional series of experiments in which renal perfusion pressure was controlled. This was

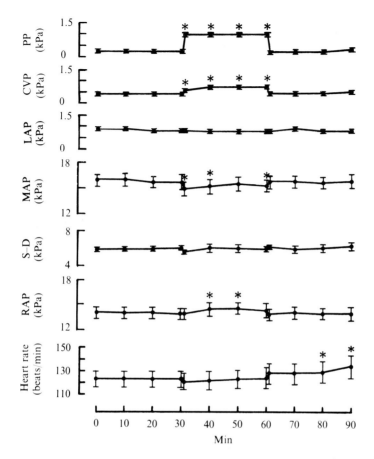

Fig. 10.8. Effect of atrial tamponade on haemodynamics in a group of sympathec-tomised, vagotomised dogs. Renal artery pressure (RAP) was controlled during the tamponade period. PP=pericardial pressure; CVP=central venous pressure; LAP=left atrial pressure; MAP=mean arterial pressure; S-D=pulse pressure. (From Goetz, Bloxham, Bond & Sharma, 1976.)

accomplished by placing an inflatable occluder around the descending aorta during the preliminary thoracotomy. At the beginning of the experiment, the aorta was partially constricted with the occluder to produce a pressure gradient of about 2.0–2.7 kPa between the aortic arch and renal arteries. As aortic arch pressure decreased slightly during atrial tamponade, the aortic constriction was reduced in order to prevent renal artery pressure from falling.

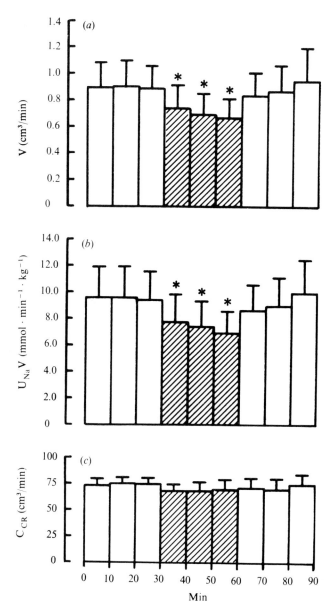

Fig. 10.9. Effect of atrial tamponade on renal function in a group of sympathec-tomised, vagotomised dogs. *(a)* Urine flow; *(b)* sodium excretion; *(c)* creatinine clearance (C_{CR}). (From Goetz, Bloxham, Bond & Sharma, 1976.)

We studied eight sympathectomised, vagotomised dogs in this manner. To eliminate another potential afferent pathway from the heart the recurrent laryngeal nerves had been sectioned above the origin of the cardiac branches in three of these dogs. The haemodynamics recorded from all eight dogs are summarised in Fig. 10.8. Central venous pressure increased slightly during tamponade and mean aortic arch pressure decreased. Renal artery pressure was not allowed to fall, however; in fact there was a slight rise during tamponade. Left atrial pressure, pulse pressure and heart rate were not changed significantly during these experiments.

The renal variables recorded in these experiments are summarised in Fig. 10.9. The three dogs with divided recurrent laryngeal nerves responded in a manner indistinguishable from that of the other five dogs; the results of all eight, therefore, were combined for analysis. Urine flow and sodium excretion decreased during atrial tamponade, and the magnitude of the decrease did not differ significantly from our earlier results in intact, unanaesthetised dogs. The clearance of creatinine was not changed significantly during tamponade. It is possible that some afferent neural pathways from the heart remained intact in these dogs but the vagotomy and thoracic sympathectomy doubtless removed the great bulk of afferent innervation. Moreover, we did demonstrate that the efferent pathways to the heart were effectively eliminated because occlusion of the common carotid arteries bilaterally for 15 s caused no change in heart rate in any of these animals (Goetz *et al.*, 1976).

The presence of an unabated renal response to atrial tamponade after cardiac denervation may be interpreted in various ways. One interpretation would be that atrial receptors did not contribute significantly to the renal response elicited by atrial tamponade in our earlier experiments on the intact, conscious dog. In any event, the results of the tamponade experiments in general and these experiments in particular imply that the renal excretion of salt and water is quite sensitive to subtle changes in systemic haemodynamics. Moreover, this sensitivity was not noticeably altered following rather extensive denervation of the heart in these experiments.

I would like to describe another method that we have used to stimulate selected groups of atrial receptors. We are able to distend selected portions of either atrium of the conscious dog by implanting a small cup such as those depicted in Fig. 10.10. The cup, which is about 2.5 cm in diameter, is made of fine stainless steel mesh that is fashioned into a dome, fitted with a

Fig. 10.10. Photograph of two atrial cups.

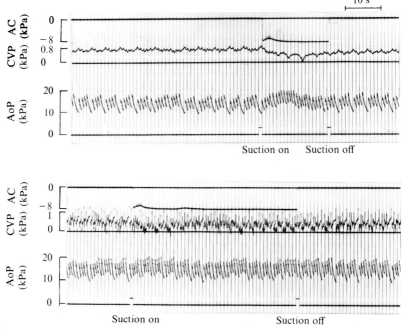

Fig. 10.11. Effect of localised atrial distension on cardiovascular variables.
AC = atrial cup; CVP = central venous pressure; AoP = aortic blood pressure.

dacron edge and sealed with liquid silastic. A thick-walled silastic tube is fitted to the cup for application of negative pressure that tends to pull the underlying atrial tissue into the cup. When small portions of atria were stretched in the conscious dog, we at times observed haemodynamic effects such as those illustrated in Fig. 10.11. In this particular dog the cup was located on the inferior vena cava and adjacent right atrium. A reduction of pressure inside the atrial cup by suction (top panel) elicited an increase in heart rate and blood pressure, and a decrease in central venous pressure. Shortly afterward, a longer period of distension (bottom panel) produced a lesser response. The heart rate and blood pressure changes appeared to adapt quite quickly, and we were unable to produce long-term changes in these variables. In some of these animals we measured urine flow and sodium excretion. No detectable change in renal function was noted during 30 min of continuous distention of either the right or left atrium in these experiments.

There are, of course, limitations to these studies. First, excessive stretch may have been applied to the atrial segment, and the distended receptors could have adapted quite rapidly. Secondly, and perhaps more importantly, the application of a localised, artificial stimulus to the cardiovascular system does not ensure that this procedure will necessarily elucidate the normal function of these receptors. As Korner (1971) has pointed out, the autonomic responses elicited in response to changes in the cardiovascular system are the result of integration of numerous individual receptor inputs and central nervous system interactions. A single input does not invariably produce the same response.

For example, increased pressure within the carotid sinus usually is considered to elicit, among other things, a reflex bradycardia, yet heart rate usually increases after the intravenous infusion of fluid even though there is a concomitant increase in carotid sinus pressure. Experiments of this type led to the concept of the Bainbridge reflex (Bainbridge, 1915). One might argue, therefore, that a reflex tachycardia elicited by atrial receptors overpowers any reflex cardiac slowing from carotid sinus baroreceptors in this situation. Distension of atrial receptors, however, does not invariably cause a tachycardia. Evidence for this statement was obtained from our experiments (Goetz, Saba & Geer, 1974) on dogs in which the sino-aortic regions were denervated by the technique of Edis & Shepherd (1971). These dogs, when conscious, usually show large changes in blood pressure and heart rate as shown in Fig. 10.12. This recording was made from a conscious dog resting on a padded table. The top two panels are continuous

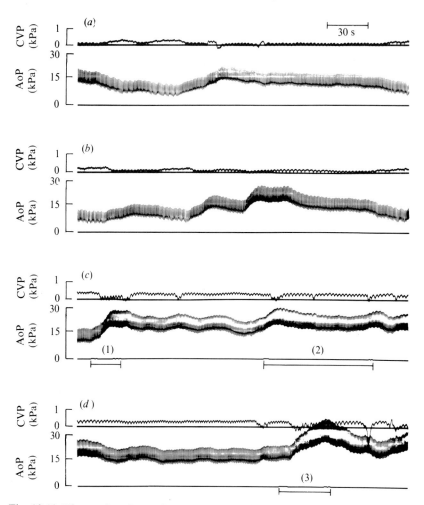

Fig. 10.12. Fluctuations in aortic and central venous pressures and heart rate in a conscious dog 24 d after its carotid and aortic baroreceptors were denervated. CVP = central venous pressure; AoP = aortic blood pressure.

and demonstrate spontaneous cardiovascular changes that are character-istic of many of these animals. Note that there tends to be a direct relationship between blood pressure and heart rate. Changes in these variables also can be elicited by certain stimuli as shown in the bottom two panels. While this tracing was being recorded the experimenter tapped on the table that the dog rested on (1) and patted the dog's head (2). Also shown is the response elicited when another person knocked loudly on the

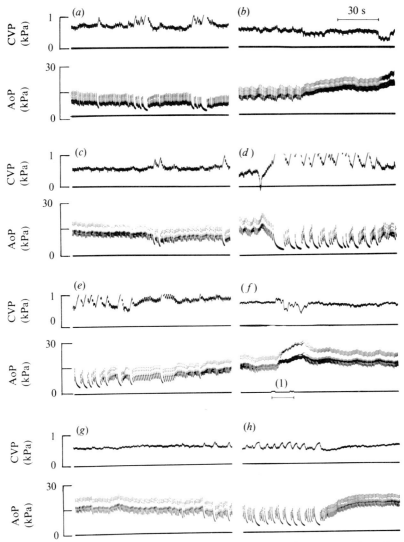

Fig. 10.13. Selected sections of a continuous 2-h recording from a conscious dog with chronic sino-aortic denervation (32 d post-operative). CVP = central venous pressure; AoP = aortic blood pressure.

outside of the door of the experimental room (3). The dog raised its head in response to the knocking and also obviously raised its blood pressure. Peak systolic blood pressure reached 41.6 kPa during this time.

Now to come to the point regarding atrial receptors and reflex cardiac acceleration. Note that increases in central venous pressure tend to correlate with decreases in heart rate. This is better demonstrated in Fig. 10.13 which depicts a record taken from the same animal at a later date. Panels *(a)–(h)* depict representative samples from a continuous 2-h recording in the awake, resting animal. All changes were spontaneous except for the pressor response in *(f)* which was induced by tapping on the dog's table (1). The periods of highest central venous pressure are associated with slow heart rates. Therefore a Bainbridge-type reflex was not apparent in these conscious dogs after sino-aortic denervation even though the vagal pathways were intact. These illustrations may have deviated a bit from my main theme, but they serve to illustrate that peripheral receptors do not invariably produce a given reflex response.

I would like to close on a somewhat philosophical note. In two early reviews on the importance of atrial volume receptors in fluid volume control Gauer & Henry (1963) and Gauer, Henry & Sieker (1961) discussed mechanisms that might be capable of sensing the fullness of the bloodstream. They noted that at one extreme there could be a generalised distribution of sensing elements throughout the circulation, with an integrating computer in the central nervous system. At the other extreme the integration could be performed mechanically in the cardiovascular system itself if distention of one compartment, amply supplied with volume-sensing elements, were representative of the total vascular bed. The authors concluded that both modes of control were used, but their reviews were concerned largely with the second alternative and the possibility that atrial receptors were the volume-sensing elements.

In the work I have described here, as well as in other experiments (Goetz & Bond, 1972, 1973; Goetz, Bond & Smith, 1974), we searched rather unsuccessfully for evidence linking atrial receptors to renal regulatory mechanisms. We have been impressed, however, that subtle changes in the cardiovascular system are followed by changes in urine flow and sodium excretion, even after afferent pathways from the heart have been interrupted. Consequently, the first of the two extremes mentioned previously is to be favoured, namely that input from peripheral receptors located throughout the cardiovascular system is integrated within the central nervous system. Closely interrelated with the process of neural integration is the humoral network that responds to changes in salt and water balance through a variety of sensing systems. There is growing evidence that integration of body function depends upon intimate neuro-humoral interactions. Cardiovascular and renal homeostatic adjustments, therefore, are derived from a multitude of factors, many of them poorly understood or perhaps not even suspected. Keeping this complexity in mind as I evaluate our own work on atrial receptors (Goetz, Bond & Bloxham, 1975) as well as that of others (Gauer, Henry & Behn, 1970) I remain to be convinced that atrial receptors should be singled out for special attention as volume receptors that influence renal function.

It is a pleasure to acknowledge the contributions of Dr G. C. Bond and Mrs Pamela Geer to unpublished work discussed in this paper.

References

Bainbridge, F.A. (1915). The influence of venous filling upon the rate of the heart. *J. Physiol.*, **50**, 65–84.

Carswell, F., Hainsworth, R. & Ledsome, J.R. (1970). The effects of left atrial distension upon urine flow from the isolated perfused kidney. *Q. Jl exp. Physiol.*, **55**, 173–82.

Edis, A.J. & Shepherd, J.T. (1971). Selective denervation of aortic arch baroreceptors and chemoreceptors in dogs. *J. appl. Physiol.*, **30**, 294–6.

Gauer, O.H. & Henry, J.P. (1963). Circulatory basis of the fluid volume control. *Physiol. Rev.*, **43**, 423–81.

Gauer, O.H., Henry, J.P. & Behn, C. (1970). The regulation of extracellular fluid volume. *A. Rev. Physiol.*, **32**, 547–95.

Gauer, O.H., Henry, J.P. & Sieker, H.O. (1961). Cardiac receptors and fluid volume control. *Prog. cardiovasc. Dis.*, **4**, 1–25.

Goetz, K.L., Bloxham, D.D., Bond, G.C. & Sharma, J.N. (1976). Persistence of the renal response to atrial tamponade after cardiac denervation. *Proc. Soc. exp. Biol. Med.*, **152**, 423–7.

Goetz, K.L. & Bond, G.C. (1971). Evidence for a humoral link between atrial receptors and renal function. *Physiologist, Wash.*, **14**, 152.

Goetz, K.L. & Bond, G.C. (1972). Effects of intravenous veratridine on plasma antidiuretic hormone concentration and renal function in dogs. *Proc. Soc. exp. Biol. Med.*, **140**, 604–8.

Goetz, K.L. & Bond, G.C. (1973). Reflex diuresis during tachycardia in the dog: evaluation of the role of atrial and sinoaortic receptors. *Circulation Res.*, **32**, 434–41.

Goetz, K.L., Bond, G.C. & Bloxham, D.D. (1975). Atrial receptors and renal function. *Physiol. Rev.*, **55**, 157–205.

Goetz, K.L., Bond, G.C., Hermreck, A.S. & Trank, J.W. (1970). Plasma ADH levels following a decrease in mean atrial transmural pressure. *Am. J. Physiol.*, **219**, 1424–8.

Goetz, K.L., Bond, G.C. & Smith, W.E. (1974). Effect of moderate hemorrhage in humans on plasma ADH and renin. *Proc. Soc. exp. Biol. Med.*, **145**, 277–80.

Goetz, K.L., Hermreck, A.S., Slick, G.L. & Starke, H.S. (1970). Atrial receptors and renal function in conscious dogs. *Am. J. Physiol.*, **219**, 1417–23.

Goetz, K.L., Saba, G.S. & Geer, P.A. (1974). Blood pressure variations in conscious dogs with sinoaortic denervation. *Fed. Proc. Fedn Am. Socs exp. Biol.*, **33**, 359.

Keeler, R. (1974). Natriuresis after unilateral stimulation of carotid receptors in unanesthetized rats. *Am. J. Physiol.*, **226**, 507–11.

Keeler, R. (1975). Failure of vasopressin to prevent the natriuretic and diuretic response to unilateral stimulation of carotid baroreceptors. *Can. J. Physiol. Pharmac.*, **53**, 1193–7.

Korner, P.I. (1971). Integrative neural cardiovascular control. *Physiol. Rev.*, **51**, 312–67.

Discussion

(a) Gilmore wondered whether atrial receptors were important in primates and, particularly, man. He described experiments designed to study volume control in the monkey, using negative pressure breathing, immersion and balloon inflation to distend the atria. He found that type B atrial receptors were insensitive in that they showed little change in discharge rate with changes in atrial pressure. Negative pressure breathing resulted in a water diuresis which could be inhibited with ADH. Immersion of the monkey in a chair in a tank, however, resulted in an increase in urine flow and sodium excretion despite the animal being given ADH and DOCA; this response was still present after vagotomy. In other experiments in the monkey in which atrial pressure was raised by balloon inflation no diuresis was obtained although the same experiment in the dog resulted in a diuresis. He suggested that the reflexes described by Linden in the dog may be of importance to the dog but did not occur in the primate. He concluded that atrial receptors and their responses had little or no relevance to man.

(b) It was pointed out that Goetz's experiments were designed to cause a reduction in atrial transmural pressure but that most other investigators had caused an increase in atrial transmural pressure. Goetz's experiments produced a retention of sodium and water and those of other investigators had shown a loss of sodium and water. Was it possible that there were different mechanisms for conservation and excretion? Goetz said that such a possibility had been postulated by other investigators and that there had been speculation that different thresholds might allow one or another of these responses to occur. He went on to say that he and his colleagues had designed their experiments to confirm the atrial receptor hypothesis and that they had been disappointed that their results were not confirmatory. He said that the decreases in urine flow and sodium exretion that had occurred in animals with denervated hearts could be taken as evidence that the antidiuretic and antinatriuretic mechanism elicited by atrial tamponade did not originate in the heart, although these experiments had not identified where the mechanism did originate. It was pointed out from the floor that just because it was difficult to demonstrate a reflex effect did not necessarily exclude the possibility that such a reflex existed and that everybody knew

that there were multiple mechanisms involved within the kidney and circulation for compensating for various situations. So while some of the evidence that Goetz had presented should be viewed as being against the idea of atrial receptors being involved in blood volume control, at the same time these experiments did not exclude the possibility. Goetz agreed completely with this point of view. He added that his experiments had been designed to uncover possible atrial receptor involvement with several different techniques, all of which were unsuccessful. These results and other considerations had led him to conclude that atrial receptors were not particularly important in the reflex regulation of ADH and blood volume. He said this conclusion was consistent with some common observations. For example, one might predict from the atrial receptor hypothesis that a diuresis would occur during sleep because the supine position increases the central blood volume and hence stretches atrial receptors; actually an antidiuresis occurs during sleep. Paradoxically, patients with heart failure characteristically have a diuresis at night, yet there is evidence from Gilmore's laboratory that atrial receptors are damaged in heart failure. Neither of these observations appears to fit well with the atrial receptor hypothesis of blood volume control. Gauer closed the discussion by commenting that 'The heaviest attack against our volume receptor story came from my good friend Joe Gilmore, but if I may make a remark – we have suffered many heavy attacks in the last 20 years and have survived them – and if there were time now, we could survive the attack of Joe Gilmore.'

Tachycardia caused by left atrial stretch in the conscious dog

V. S. BISHOP & D. F. PETERSON

from Department of Pharmacology, Health Science Center,
University of Texas, San Antonio, Texas 78284, USA

Heart rate and peripheral vascular responses were investigated in nine conscious dogs during volume infusion (V), 1-min left circumflex coronary occlusion (O) and isolated stretch of the left atrial appendage (S). Mongrel dogs underwent sterile thoracotomy in order to chronically implant recording instruments and catheters for measurement of left atrial pressure, cardiac output and aortic pressure. Additionally, a jugular catheter was inserted for volume infusion, a cuff occluder was placed around the left circumflex coronary artery and an inflatable balloon was placed in the left atrial appendage. Ten days or more after surgery experiments were performed while the dog rested quietly on a hammock, unsedated and unrestrained. The following changes caused by each of the interventions were observed:

	Left atrial pressure (kPa)	Heart rate (beats/min)	Arterial pressure (kPa)	Cardiac output (cm^3/min)	Peripheral resistance (%)
V	+1.40	+32	+2.05	+1144	−27
O	+0.61	+27	−1.64	−312	−1
S	+0.05	+20	+0.08	+217	−10

It is concluded that, in the conscious dog, all three stimuli stretched left atrial receptors which contributed to the observed tachycardia. Earlier studies with conscious dogs have identified the peripheral reflex pathways mediating tachycardia during both volume infusion and coronary occlusion (Peterson, Kaspar & Bishop, 1973; Bishop & Peterson, 1976), while

234

pathways for isolated stretch have been studied only on anaesthetised dogs (Linden, 1973). The afferent pathways in all these studies was found to lie within the vagi. The efferent pathway during volume infusion is primarily in the vagi, whereas the right cardiac sympathetic nerves were the dominant pathway during both coronary occlusion and isolated stretch of the left atrial appendage. When the present results are compared, it is apparent that both coronary occlusion and volume infusion are less specific in activation of cardiopulmonary receptors than is direct atrial stretch. The location and adequate stimulation of other receptors activated during these abnormal physiological states have yet to be identified.

References

Bishop, V.S. & Peterson, D.F. (1976). Pathways regulating cardiovascular changes during volume loading in awake dogs. *Am. J. Physiol.*, **231**, 854–9.

Linden, R.J. (1973). Function of cardiac receptors. *Circulation*, **48**, 463–80.

Peterson, D.F., Kaspar, R.L. & Bishop, V.S. (1973). Reflex tachycardia due to temporary coronary occlusion in the conscious dog. *Circulation Res.*, **32**, 652–9.

The influence of exogenous antidiuretic hormone on diuresis produced by left atrial distension. Experiments on conscious dogs on different levels of sodium intake

B. ARNOLD, F. EIGENHEER, M. GATZKA, G. KACZMARCZYK, U. KUHL, U. LINDEMANN & H. W. REINHARDT

from Arbeitsgruppe Experimentelle Anaesthesie, Freie Universität Berlin, Klinikum Charlottenburg, 1 Berlin 19, Spandauer Damm 130, Germany

Experimental increase of left atrial pressure (LAP) leads to an augmentation of urinary output ($\dot{V}$) (Henry, Gauer & Reeves, 1956). Whether the increase of $\dot{V}$ is caused by a decrease of antidiuretic hormone (ADH)-secretion alone is still controversial (Kappagoda, Linden, Snow & Whitaker, 1974; Lydtin & Hamilton, 1964).

In the present studies conscious dogs (female beagles, 8–15 kg of weight) were examined under chronically controlled conditions of salt and water intake (high sodium intake, HSI, 14.0 mmol·kg^{-1} or low sodium intake, LSI, 0.5 mmol·kg^{-1}; 100 cm^3·kg^{-1} water were added to the food). Six dogs had been thoracotomised 3 weeks before the studies were started. They were fitted with a nylon purse string around the left atrium and a teflon catheter in the left atrium.

(1) After a fasting period of about 20 h, LAP was elevated by about 1 kPa for 60 min by tightening the purse string. The elevation of LAP augmented $\dot{V}$ in HSI and LSI groups between 100% and 600% in comparison with controls. If vasopressin (0.05 mIU/min·kg^{-1}) was infused simultaneously the rise in $\dot{V}$ was abolished in both groups.

(2) After a HSI meal LAP increased 0.4–0.6 kPa correlated to an increase of $\dot{V}$ and sodium excretion.

(3) In another series, postprandial $\dot{V}$ and sodium excretion were measured before surgery in 17 dogs on 126 d on various levels of dietary sodium intake (0.5 (LSI), 3.5, 7.0, 10.0, 14.0 (HSI) mmol·kg^{-1}). $\dot{V}$ always increased postprandially. Vasopressin (i.v.) did not diminish the postprandial diuresis after HSI, in contrast it decreased the postprandial diuresis if the sodium content of the diet was 10.0 mmol·kg^{-1} and less. Postprandial diuresis was abolished by vasopressin in dogs on LSI. The various diminutions or the suppression of left atrial diuresis by vasopressin

were probably caused by the necessity of sodium excretion competing with the vasopressin-dependent renal resorption of water.

References

Henry, J.P., Gauer, O. & Reeves, J.L. (1956). Evidence of the atrial location of receptors influencing urine flow. *Circulation Res.,* **4,** 85–90.

Kappagoda, C.T., Linden, R.J., Snow, H.W. & Whitaker, E.M. (1974). Left atrial receptors and the antidiuretic hormone. *J. Physiol.,* **237,** 663–83.

Lydtin, H. & Hamilton, W.E. (1964). Effect of acute changes in left atrial pressure on urine flow in unanaesthetized dogs. *Am. J. Physiol.,* **207,** 530–6.

Reflex effects of ventricular receptors

11. Possible physiological stimuli for ventricular receptors and their significance in man

P. SLEIGHT

In this paper I will consider only the receptors with unmyelinated fibres which are located throughout the muscle of the left ventricle. Recordings have been made from these fibres in dogs, cats and rabbits. The receptors appear to be very numerous and to be largely confined to the left ventricle. Receptors with myelinated fibres have been described from the right and left ventricles and from the coronary vessels (see reviews by Coleridge, Coleridge & Kidd, 1964; Brown, 1965; Paintal, 1972) but they are very much more sparse and their reflex effects unknown.

The receptors with C fibre afferents have sometimes been erroneously referred to as epicardial receptors. In an earlier paper I used this term to describe receptors stimulated by nicotine injected over the surface of the left ventricle in the dog (Sleight, 1964a). In a later paper (Sleight & Widdicombe, 1965) we found that the receptors were distributed throughout the wall of the left ventricle. At the same time, Coleridge et al. (1964) also recorded from these receptors; they too found they were stimulated both by mechanical events and by irritant chemicals.

It is this peculiarity of stimulation both by chemical and various mechanical stimuli, particularly distension, which has led to the widespread belief that the receptors have no physiological function. It is suggested that they exist as a nociceptive or protective mechanism which stops the heart in moments of extreme stress. I believe this view is no longer tenable, but first I will review the historical background to this belief.

Historical background

Bezold & Hirt (1867) in a classic paper gave their name to the depressor reflex which resulted from the intravenous injection of the hellebore and veratrum alkaloids, isolated some years earlier by two French chemists Pelletier and Cavantou. Bezold and his student colleague showed that the fall in blood pressure resulted from both a vasodilatation in the mesenteric bed and also a drop in heart rate. The effect was reflex and blocked by vagotomy. From observation that veratrum resulted in a tetanic contraction of a frog nerve–muscle preparation following a single electric stimulus they deduced that the alkaloid caused repetitive nerve discharge. They believed that the receptors signalled intracardiac pressure.

Since then many others, notably Jarisch and his colleagues, have added to our knowledge of this powerful reflex (Jarisch & Henze, 1937; Jarisch & Richter, 1939*a, b, c*; Jarisch, 1941*a, b*; Jarisch & Zotterman, 1948). Their main contributions were to show that the reflex was blocked by deep anaesthesia or by procaine in the pericardial sac. They also localised the reflex to the ventricles rather than the atria. They too believed the reflex was a protective or 'shelter' reflex. Dawes (1947) cannulated the coronary arteries and showed that the response was elicited from the distribution of the left but not the right coronary in the dog.

Sleight (1964*a*) used nicotine to stimulate the receptors directly since it was known that both nicotine and veratridine would elicit the response in the dog. Using pledgets soaked in nicotine solution (100 μg·cm^{-3}) it was found that the response was not particularly associated with the main coronary vessels but could be elicited from muscle without large surface vessels. It was also found that injection of nicotine into the pericardial sac

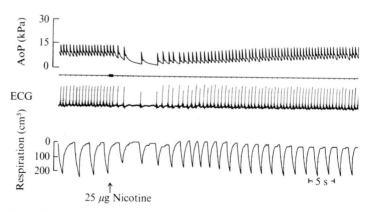

Fig. 11.1. Conscious dog: indwelling pericardial and internal mammary arterial catheters: respiration measured with a spirometer. Nicotine (25 μg) injected through the pericardial catheter. AoP is the aortic blood pressure. (From Sleight, 1964*a*.)

of conscious dogs (via previously implanted catheters) caused a more profound hypotension and bradycardia (see Fig. 11.1) with no apparent discomfort. It was therefore clear that the response was not the result of stimulation of pain fibres.

Electrophysiological studies

With the studies of Coleridge *et al.* (1964) and Sleight & Widdicombe (1965) there was then evidence that the receptors were mechanoreceptors with unmyelinated fibres with slow (*c.* 1 m/s) conduction velocity. Linden (1973) has stated with characteristic bluntness (but I believe with some inaccuracy) that the stimuli used to activate the receptors were 'assaults outside the physiological ranges' which provided no evidence yet of the natural physiological stimulus. I will take issue with this statement, particularly with regard to the effect of adrenaline on the receptor discharge. We found this to be the most powerful stimulus, one which produced pulse synchronous discharge in 28 out of the 33 fibres tested (Sleight & Widdicombe, 1965). The receptors have only a sporadic discharge under control conditions and this has prevented their acceptance as part of a normal cardiovascular control system. I believe that they could

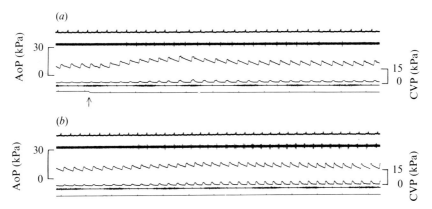

Fig. 11.2. Stimulation of a single-fibre preparation from a left ventricular epi-myocardial receptor by *(a)* aortic occlusion, *(b)* 20 μg adrenaline i.v. – injected 6 s beforehand. Records from above downwards: ECG, electroneurogram, aortic blood pressure (AoP), coronary venous pressure (CVP), diaphragmatic electroneurogram, marker. The adrenaline stimulated the fibre more vigorously at a lower mean arterial blood pressure. Both stimuli caused the fibre to discharge with a cardiac rhythm. (From Muers & Sleight, 1972*b*.)

signal intramyocardial tension and therefore monitor left ventricular contractility.

Effect of adrenaline

When we compared aortic occlusion, admittedly an unphysiological stimulus, with small doses of intravenous adrenaline which produced quite physiological changes in heart rate and aortic pressure we found that the cardiac synchronous discharge was much greater with adrenaline than with aortic occlusion, despite the fact that the latter produced very much larger changes in pressure. The receptors were excited during systole and we also found that small doses of adrenaline (2–3 μg) injected into a coronary artery produced vigorous receptor excitation but negligible change in aortic pressure or heart rate. These findings were also confirmed by Muers & Sleight (1972b) who found in addition (Muers, 1969) that the discharge occurred some 40 ms earlier during systole with adrenaline than during aortic occlusion.

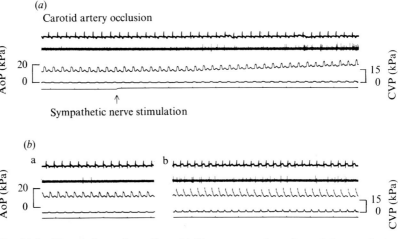

Fig. 11.3. Stimulation of small multi-fibre preparations from left ventricular epi-myocardial receptors by *(a)* carotid artery occlusion, *(b)* electrical stimulation of the left ventral ansa subclavia. *(a)* and *(b)* are separate preparations from the same experiment. Both runs were after bilateral vagotomy. Traces from above downwards: ECG, electroneurogram, aortic blood pressure (AoP) coronary venous pressure (CVP), marker. In *(a)* the arterial blood pressure rose slowly and as it did so fibres were recruited and others discharged faster. In *(b)* 'a' is the control discharge immediately before stimulation and 'b' the discharge immediately after stimulation. (From Muers & Sleight, 1972b.)

Left ventricular distension

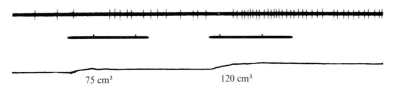

75 cm³   120 cm³

Fig. 11.4. Distension of the fibrillating left ventricle with a balloon. Figures below the marker (uncalibrated pressure) trace show the amount of saline injected. Top record = electroneurogram. Time in s, marker. Black bars = signal marker for beginning of successive balloon inflations. (From Sleight & Widdicombe, 1965.)

The ventricular receptors were also fired by a natural increase in sympathetic discharge caused by carotid artery occlusion. Of the ventricular receptors we studied, 85% were excited (again with a systolic discharge) by electrical stimulation of the ansa subclavia (10 imp/s for 5 s, Fig. 11.3). All these experiments resulted in an increase in cardiac contractility as measured by a Walton–Brodie strain gauge arc and were more effective in exciting the receptors than pressure increases or passive distension. Sleight & Widdicombe (1965) found that rather gross distension of the ventricle was needed to stimulate receptors in the fibrillating heart (Fig. 11.4).

The timing of stimulation within the cardiac cycle

I have emphasised that the ventricular receptors fire during systole when they are stimulated by mechanical events. This is of course a very important point when considering any possible physiological stimulus to the receptors.

There has in the past been some confusion on this point for, although Sleight & Widdicombe (1965) found that the receptors fired during systole in the dog, Öberg & Thorén (1972) found them to be excited during diastole in the cat. They further found a rather close relation between end-diastolic volume changes and discharge rate. The cat has a more rapid heart rate than the dog and this, together with the slow conduction from receptor to recording site, makes it very difficult to be sure of the exact time of excitation during the cardiac cycle. In later studies Thorén (1977) has now concluded that in the cat the fibres are in fact excited during systole. Öberg & Thorén (1972) also found that adrenaline was a powerful stimulus to the receptors. Their studies had been initiated in order to elucidate the cause of the sudden bradycardia and hypotension which occurs during acute haemorrhage in the cat. After the initial tachycardia and mobilisation of the sympathetic nervous system the extremely rapid circulatory collapse is very reminiscent of the fainting reaction in man.

Öberg & Thorén (1972) produced very convincing evidence to implicate the ventricular receptors in this response. It seems likely that the haemorrhage had activated the sympathetic system and that the effect of this on an empty heart was to stimulate the receptors strongly.

The relation of firing to end-diastolic pressure

Öberg & Thorén (1972) and Thorén (1977) found a variable but on the whole convincing increase in firing with increase in left ventricular end-diastolic pressure, despite the fact that the receptors fired in systole. I believe this could be explained by an increase in intramyocardial tension caused by the increase in force of contraction resulting from the Starling effect of increased end-diastolic volume.

The effect of myocardial congestion

Muers & Sleight (1972c) had set out to characterise the receptors responsible for the reflex bradycardia and hypotension which follows distension of the coronary sinus in the dog (Juhasz-Nagy & Szentivanyi, 1961). The latter authors claimed that this was responsible for the coronary chemoreflex in the dog. We were unable to find any reflex effects from the coronary sinus itself but did confirm that there was indeed a reflex caused by coronary venous occlusion. We located the receptors in the left ventricle and found they were the same receptors we have been discussing above (Muers & Sleight, 1972a, b). They were evidently not receptors in the walls of vessels since

(a) gross elevation of coronary venous pressure was needed to excite them (> 7 kPa, Fig. 11.5);

(b) there was a delay in the build up of discharge amounting to several beats after the elevation of pressure;

(c) there was a similar lag in cessation of discharge on releasing the sinus occlusion.

This suggested to us that the massive rise in coronary venous pressure had resulted in capillary transudation and thus an increase in the extracellular fluid pressure. This would be likely to sensitise the endings of the ventricular receptors.

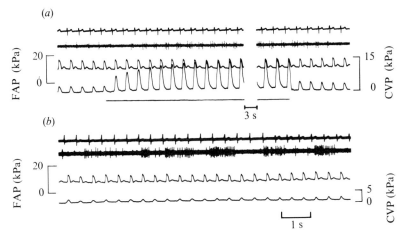

Fig. 11.5. Stimulation of a single-fibre afferent nerve preparation from the right recurrent cardiac nerve by *(a)* coronary sinus occlusion and *(b)* stroking of the left ventricular surface near the anterior descending coronary artery. Traces from above downwards: ECG, electroneurogram, femoral arterial blood pressure (FAP), coronary venous pressure (CVP), event marker. The upper record has been interrupted for 3 s during the occlusion. In *(a)* occluding the coronary sinus converted the slow irregular discharge of the fibre (0·6 imp/s) to a faster discharge (6.6 imp/s) with a regular cardiac rhythm. The regular discharge began when the peak coronary venous pressure had reached 9.9 kPa. At the end of the occlusion, the discharge returned to the prestimulation level within two to three heart beats. In *(b)* the high-frequency bursts of activity in the same fibre were produced by stroking the epicardium of the left ventricle near the anterior descending coronary artery. Stroking the ventricle elsewhere did not excite the fibre. (From Muers & Sleight, 1972*a, b.*)

The effect of asphyxia

Sleight & Widdicombe (1965) could find no good evidence that the receptors were primarily responsive to P_aO_2 or P_aCO_2 changes. This has been examined more carefully by Thorén (1972*a*). He found in cats that the receptors increased their discharge 40–70 s after coronary occlusion or asphyxia. The activation appeared to be caused by the mechanical bulging of localised dilatation rather than the anoxia. He went on to show that the bradycardia produced by severe hypoxia in the cat could be blocked by anaesthetising the ventricular receptors with local procaine, whilst the sinus node innervation by the vagus remained intact (Thorén, 1972*b*). Previously this bradycardia had been thought to be due solely to carotid chemoreceptor stimulation. Such a mechanism, from ventricular dilatation, could be responsible for the agonal bradycardia seen during severe asphyxia in man and during acute left ventricular dilatation occurring with severe cardiac insults such as myocardial infarction.

With regard to the exact mechanism of receptor excitation under these conditions an increase in ventricular size would lead to an increase in wall tension (by La Place's Law). A terminal sympathetic discharge and adrenaline release from the adrenal would augment this.

A physiological role for ventricular receptors

I have outlined the evidence which supports my belief that the receptors are primarily excited by increase in the force of contraction of the left ventricle or by intramyocardial tension. Before considering possible physiological roles for any reflex we should list the known reflex effects.

These are:

(a) Cardiac slowing.

(b) Hypotension which is partly caused by bradycardia and partly by peripheral vasodilatation. Bergel & Makin (1967) showed that in the dog this was caused by cholinergic vasodilatation in muscle. This is the same mechanism which is present in the fainting reaction in man (see below).

(c) Important effects upon renal blood flow and renin release. Öberg & Thorén (1973) showed that the ventricular receptors had a particularly strong action on lowering renal vascular resistance when compared with the action of the carotid baroreceptors. This suggests that the receptors could be important in blood volume control; Mancia, Romero & Shepherd (1975) have shown that they have a tonic effect on renin release.

(d) Reflex gastric dilatation and vomiting. Abrahamsson & Thorén (1973) showed that the receptors also caused reflex vomiting in cats (see below – *Myocardial infarction*).

Role in exercise

It seemed possible (Sleight, 1964*b*) that the reflex might have an important physiological role in exercise, by two mechanisms.

(a) In the anticipatory state before exercise, when sympathetic drive to the heart increases the force of contraction, the receptors would lead

to cholinergic muscle arteriolar dilatation and thus 'prime' the muscles with blood.

(b) During more severe muscular exercise high sympathetic drive to the myocardium might be accompanied by an inappropriately high and inefficient heart rate. A reflex vagal brake to the sinus node would enable maximal sympathetic inotropic effects without embarrassingly large chronotropic effects.

It is, however, difficult to find a role in this for reflex gastric dilatation, but it is probably the cause of the nausea, hypotension and collapse which affects atheletes immediately on *cessation* of extreme effort. At this time sympathetic drive is high but venous return diminishes abruptly on cessation of the muscle 'pump'. The situation is exactly akin to that seen during haemorrhage in the cat, when ventricular receptor stimulation is particularly strong. I understand that this type of post-exercise collapse is not seen when the athelete goes on jogging after the race so that the ventricle is not squeezing on an empty chamber (R. Bannister, personal communication).

It is possible also that the ventricular receptors together with the other low pressure receptors in the heart may 'tune' the gain of the carotid and aortic arterial baroreceptors and thus be involved in blood pressure control. There is some experimental evidence in favour of such a role.

Pathophysical implications in man

Myocardial infarction

Much of the unexplained circulatory disturbance seen during acute myocardial infarction could be attributed to stimulation of ventricular receptors. Webb, Adgey & Pantridge (1972) have reported that bradycardia is a prominent feature of the early stages of myocardial infarction. An illustration of this and its response to atropine is shown in Fig. 11.6. Under these conditions, with cardiac dilatation and high circulatory catecholamine stimulation, receptor excitation will be maximal.

Bennet *et al.* (1977) have recently demonstrated a significant increase in creatinine clearance and glomerular filtration in patients on the first day or two following myocardial infarction, particularly in those with evidence of raised end-diastolic pressure in the left ventricle. They speculated whether this could have resulted from stimulation of ventricular or atrial receptors (Kappogoda, Linden, Snow & Whitaker, 1975).

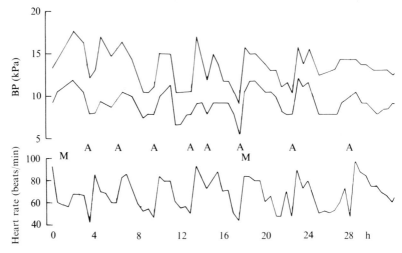

Fig. 11.6. Blood pressure (BP) and pulse of a 54-year-old man admitted with myocardial infarction. His shock-like state responded very well to atropine (A). M = morphine. (From Sleight, 1974.)

Stinson *et al.* (1976) have shown that coronary ligation in the dog gives rise to renal vasodilatation which is mediated via cholinergic nerves.

Coronary arteriography

Bradycardia and hypotension may occur following the injection of X-ray contrast media (but not saline) into the coronary arteries (Fig. 11.7). Zelis, Caudill, Baggette & Mason (1976) have shown in man that injection of the hypertonic contrast medium into the coronary arteries results in a reflex decrease in forearm vascular resistance of 32%. This effect was shown to be caused by cholinergic vasodilatation (cf. Bergel & Makin, 1967) since it was blocked by atropine injected into the brachial artery.

The fainting reaction

Ventricular receptors appear to be very likely candidates for the initiation of vasovagal syncope in man. The manifestations of this, bradycardia, hypotension and nausea very closely resemble those of ventricular receptor stimulation in animals. In the upright immobile position venous return falls off; this first triggers a reflex tachycardia from arterial baroreceptors. This inotropic drive to an empty heart is then a powerful stimulant to ventricular receptors which then initiate the faint. The parallel with the work on haemorrhage by Öberg and Thorén is very close.

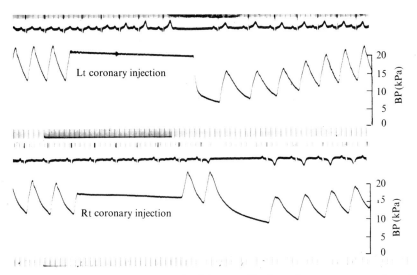

Fig. 11.7. The effect of injection of 76% urograffin into the coronary arteries of man. The coronary catheter pressure is temporarily interrupted during injection. BP = blood pressure. (From Sleight, 1974.)

Gastric syncope

In this condition vasovagal syncope is believed to follow gastric irritation or vomiting. In some instances I believe the true chain of events is different. I recently observed a patient who was reported to have this syndrome. She had experienced a number of attacks of sudden unexplained vomiting which rapidly led to syncope. I happened to be feeling her pulse when she began to have another attack. This time one could observe that the first event in the sequence was cardiac arrest; this was followed within 20 s by violent retching movements and simultaneous syncope. Without the knowledge of the cardiac standstill one would have been justified in calling this gastric syncope. It seems more probable that the gastric dilatation was secondary to the acute cardiac distension following standstill. I have since seen another similar patient. In both cases the vomiting attacks as well as the syncope were cured by a cardiac pacemaker.

Asphyxia bradycardia

Although the clinical response to hypoxaemia in man is usually a tachycardia this gives rise to a profound bradycardia terminally. In animals, Thorén (1972a, b, see above) has produced good evidence that ventricular dilatation is responsible for a large part of this. Frankel, Mathias & Spalding (1975) have published data from tetraplegic patients showing profound bradycardia and even cardiac arrest when the ventilator was disconnected. They speculated on different mechanisms for this but the time course of the response they saw suggests very strongly that ventricular receptors were activated secondarily to asphyxial cardiac dilatation. When oxygen was given before the respirator was disconnected the bradycardia and arrest was not seen; naturally the ventilator was not left off long enough for severe hypoxaemia to occur with this latter manoeuvre.

Digitalis bradycardia

There are many possible mechanisms for the sinus bradycardia which follows the administration of digitalis glycosides to animals or man. Amongst these are an enhancement of the effect of acetylcholine or vagal stimulation on the sinus node, a centrally mediated increase in vagal discharge and a stimulatory effect on the nodose ganglion. In addition, Sleight, Lall & Muers (1969) showed that in the dog a rapidly acting glycoside stimulated ventricular receptors and gave rise to bradycardia (Fig. 11.8). It is possible that part of the toxic nausea and vomiting caused by digitalis is mediated by way of stimulation of ventricular receptors.

Aortic stenosis syncope

Patients with severe aortic stenosis have a tendency to syncope on exercise. The reasons for this have been obscure and speculation has ranged from

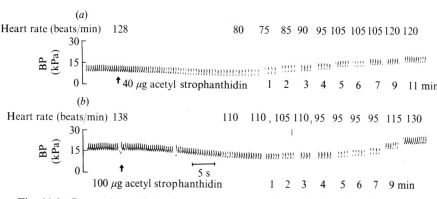

Fig. 11.8. Comparison of *(a)* intracoronary injection and *(b)* epicardial application of acetyl strophanthidin 30 min later; anaesthetised open chest dog. BP = blood pressure. (From Sleight, Lall & Muers, 1969.)

arrhythmias to an inadequate rise in cardiac output on effort. Mark *et al.* (1973) have produced very compelling evidence that the syncope is caused by the stimulation of ventricular receptors. They measured forearm blood flow and observed the change which occurred with exercise (Fig. 11.9). In normal subjects who performed leg exercise the forearm blood flow was reflexly diminished.

In patients with aortic stenosis syncope there was abnormal vasodilatation. In patients with mitral stenosis (where the rest of the cardiac pressures apart from left ventricular pressure would behave similarly) there was no vasodilatation. This appeared to localise the afferents to the ventricle. When the stenosed aortic valve was replaced, lowering left ventricular pressure, the response became normal too (Fig. 11.10).

A similar mechanism may be important in the syncope (and sometimes sudden death) which occurs in patients with subaortic stenosis caused by hypertrophic cardiomyopathy.

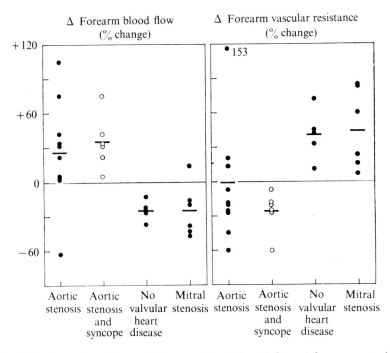

Fig. 11.9. Percentage change in forearm blood flow (left) and forearm vascular resistance (right) during the second minute of supine leg exercise. Dots represent responses in individual patients. The horizontal lines are the means of each group. (From Mark *et al.*, 1973.)

Before operation

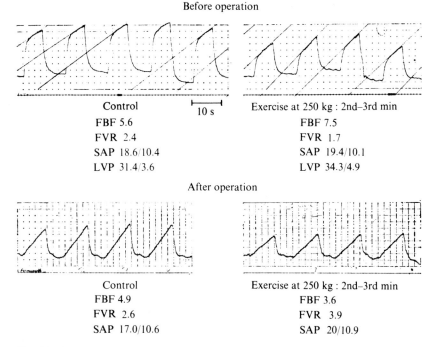

Control		Exercise at 250 kg : 2nd–3rd min
FBF 5.6	10 s	FBF 7.5
FVR 2.4		FVR 1.7
SAP 18.6/10.4		SAP 19.4/10.1
LVP 31.4/3.6		LVP 34.3/4.9

After operation

Control	Exercise at 250 kg : 2nd–3rd min
FBF 4.9	FBF 3.6
FVR 2.6	FVR 3.9
SAP 17.0/10.6	SAP 20/10.9

Fig. 11.10. Responses to supine leg exercise in a patient with aortic stenosis before and after aortic valve replacement. FBF = forearm blood flow; FVR = forearm vascular resistance; SAP = systemic arterial pressure; LVP = left ventricular pressure. (From Mark *et al.*, 1973.)

Prostaglandin syncope

Occasional collapse with bradycardia and hypotension has followed the use of prostaglandins for therapeutic abortion. It is therefore of interest that the Coleridges and their collaborators (see this volume) have shown that ventricular receptors can be excited by certain prostaglandins.

Hypotensive drugs: the veratrum alkaloids

It seems beyond question that the veratrum alkaloids used in the treatment of hypertension have their effect by stimulation of these receptors. Much effort has gone into attempts to find congeners with similar hypotensive effect but without the troublesome side effect of nausea and vomiting. The work of Abrahamsson and Thorén makes it equally certain that these effects stem from the same receptor and are therefore unlikely to be separable.

Summary

The hypothesis has been proposed that the major stimulus for ventricular receptor excitation is a rise in intramyocardial tension in the wall of the left ventricle. This may be brought about by several mechanisms of which catecholamine stimulation to an empty heart or an increase in ventricular distension are the most important.

The reflex may be physiologically important during and before exercise. It may be responsible for certain previously ill-understood syndromes in man. These range from some forms of cardiogenic shock in myocardial infarction, through aortic stenosis syncope, syncope during coronary angiography, digitalis bradycardia, prostaglandin syncope, to the simple faint or vasovagal attack.

References

Abrahamsson, H. & Thorén, P. (1973). Vomiting and reflex vagal relaxation of the stomach elicited from heart receptors in the cat. *Acta. physiol. scand.,* **88,** 433–40.

Bennett, E.D., Brooks, N.H., Keddie, J., Lis, Y. & Wilson, A. (1977). Increased renal function in patients with acute left ventricular failure and possible homeostatic mechanism. *Clin. Sci. mol. Med.,* **52,** 43–50.

Bergel, D.H. & Makin, G.S. (1967). Central and peripheral cardiovascular changes following chemical stimulation of the surface of the dog's heart. *Cardiovasc. Res.,* **1,** 80–90.

Bezold, A. von & Hirt, L. (1867). Über die physiologischen Wirkungen des essigsauren Veratrins. *Unters. Physiol. Lab. Wurzburg.,* **1,** 75–156.

Brown, A.M. (1965). Mechanoreceptors in or near the coronary arteries. *J. Physiol.,* **177,** 203–14.

Coleridge, H.M., Coleridge, J.C.G. & Kidd, C. (1964). Cardiac receptors in the dog with particular reference to two types of afferent ending in the ventricular wall. *J. Physiol.,* **174,** 323–39.

Dawes, G.S. (1947). Receptor areas in the coronary arteries and elsewhere as revealed by the use of veratridine. *J. Pharmac.,* **89,** 325–42.

Dawes, G.S. & Comroe, J.H., Jr (1954). Chemoreflexes from the heart and lungs. *Physiol. Rev.,* **34,** 167–201.

Frankel, H.L., Mathias, C.J. & Spalding, J.M.K. (1975). Mechanisms of reflex cardiac arrest in tetraplegic patients. *Lancet,* **2,** 1183–5.

Jarisch, A. (1941*a*). Der Einfluss der Vagusausschaltung auf den Blutdruck. *Arch. Kreislaufforsch,* **9,** 1–10.

Jarisch, A. (1941*b*). Die Bedeutung des Vagus für die Wirkung der Mistel und des Veratrins. *Arch. exp. Path. Pharmak.,* **197,** 266–70.

Jarisch, A. & Henze, C. (1937). Blutdrucksenkung durch chemische Erregung depressorischer Nerven. *Arch. exp. Path. Pharmak.,* **187,** 706–30.

Jarisch, A. & Richter, J. (1939*a*). Die Kreislaufwirkung des Veratrins. *Arch. exp. Path. Pharmak.,* **193,** 347–54.

Jarisch, A. & Richter, H. (1939*b*). Die Afferenten Bahnen des Veratrineffektes in den Hernznerven. *Arch. exp. Path. Pharmak.*, **193**, 335–71.

Jarisch, A. & Richter, H. (1939*c*). Der Bezold Effekt eine vergessene Kreislauf reaktion. *Klin. Wschr.*, **18**, 185–7.

Jarisch, A. & Zotterman, Y. (1948). Depressor reflexes from the heart. *Acta. physiol. scand.*, **16**, 31–51.

Juhasz-Nagy, A. & Szentivanyi, M. (1961). Localisation of the receptors of the coronary chemoreflex in the dog. *Archs int. Pharmacodyn. Thér.*, **131**, 39–53.

Kappagoda, C.T., Linden, R.J., Snow, H.M. & Whitaker, E.M. (1975). Effect of destruction of the posterior pituitary on the diuresis from left atrial receptors. *J. Physiol.*, **244**, 757–70.

Linden, R.J. (1973). Function of cardiac receptors. *Circulation*, **68**, 463–80.

Mancia, G., Romero, C.J. & Shepherd, J.T. (1975). Continuous inhibition of renin release in dogs by vagally innervated receptors in the cardiopulmonary region. *Circulation Res.*, **36**, 529–35.

Mark, A.L., Kioschos, M., Abboud, F.M., Heistad, D.D., Schmid, P.G. & Burr, J.W. (1973). Abnormal vascular responses to exercise in patients with aortic stenosis. *J. clin. Invest.*, **52**, 1138–46.

Muers, M.F. (1969). Ventricular mechanoreceptors and their reflex effects. D.Phil. thesis, Oxford.

Muers, M.F. & Sleight, P. (1972*a*). The reflex cardiovascular depression caused by occlusion of the coronary sinus in the dog. *J. Physiol.*, **221**, 259–82.

Muers, M.F. & Sleight, P. (1972*b*). Action potentials from ventricular mechanoreceptors stimulated by occlusion of the coronary sinus in the dog. *J. Physiol.*, **221**, 283–309.

Muers, M.F. & Sleight, P. (1972*c*). The cardiovascular effects of coronary sinus distension in the anaesthetised dog. *J. exp. Physiol.*, **57**, 359–70.

Öberg, B. & Thorén, P. (1972). Studies on left ventricular receptors, signalling in non-medullated vagal afferents. *Acta. physiol. scand.*, **85**, 145–63.

Öberg, B. & Thorén, P. (1973). Circulatory responses to stimulation of medullated and non-medullated afferents in the cardiac nerve in the cat. *Acta. physiol. scand.*, **87**, 121–32.

Paintal, A.S. (1972). Cardiovascular receptors in enteroceptors. In *Handbook of sensory physiology, vol. III/I, Enteroceptors*, ed. N. E. Berlin, pp. 1–45. Springer-Verlag: Basel.

Sleight, P. (1964*a*). A cardiovascular depressor reflex from the epicardium of the left ventricle in the dog. *J. Physiol.*, **173**, 321–43.

Sleight, P. (1964*b*). The coronary chemoreflex. M.D. thesis, Cambridge.

Sleight, P. (1974). Neural control of the cardiovascular system. In *Modern trends in cardiology*, vol. 3, ed. M. F. Oliver, pp. 1–43. Butterworths: London.

Sleight, P., Lall, A. & Muers, M.F. (1969). Reflex cardiovascular effects of epicardial stimulation by acetylstrophanthidin. *Circulation Res.*, **25**, 705–11.

Sleight, P. & Widdicombe, J.G. (1965). Action potentials in fibres from receptors in the epicardium and myocardium of the dog's left ventricle *J. Physiol.*, **181**, 235–58.

Stinson, J.M., Mootry, P.J., Jackson, C.G., Gates, H.O. & Scott, M.T. (1976). Renal vasodilation in response to coronary ligation in the dog. *Clin. exp. Pharmac. Physiol.*, **3**, 191–4.

Thorén, P. (1972*a*). Left ventricular receptors activated by severe asphyxia and by coronary artery occlusion. *Acta. physiol. scand.*, **85**, 453–63.

Thorén, P. (1972*b*). Reflex bradycardia elicited from left ventricular receptors during acute severe hypoxia in cats. *Acta physiol. scand.*, **87**, 103–13.

Thorén, P. (1977). Characteristics of left ventricular receptors with nonmedullated vagal afferents in the cat. *Circulation Res.*, **40**, 231–7.

Thorén, P., Donald, D.E. & Shepherd, J.T. (1976). Role of heart and lung receptors with non-medullated vagal afferents in circulatory control. *Circulation Res.*, **38** (supp. II), 2–9.

Webb, S.W., Adgey, A.A.J. & Pantridge, J.F. (1972). Autonomic disturbance at onset of acute myocardial infarction. *Br. med. J.*, **3**, 89.

Zelis, R., Caudill, C.C., Baggette, K. & Mason, D.T. (1976). Reflex vasodilation induced by coronary angiography in human subjects. *Circulation*, **53**, 490–3.

Discussion

(a) Sleight was questioned as to whether, when he said the endings were stimulated by increases in tension, he was including the epicardial receptors which responded exquisitely to touch, or only to the ones deeper in the myocardium. Sleight replied that they are all in the myocardium and the only difference between them is that those nearer to the surface can be stimulated easily. He said he had some evidence for this from cutting in from the endocardium outwards – it was possible to get a rough idea of where they were located and the ones that can be stimulated from the outside easily are just nearer the surface but they still would be signalling myocardial tension.

(b) It was suggested that the ones on the surface which are exquisitely sensitive to touch, e.g. blowing on them caused them to discharge, responded to drugs which do not affect those in the myocardium which discharge with a pulsatile rhythm; they seem to be quite a different category of ending. Sleight agreed that drugs act differently on different fibres but that, in his experience, 85% would discharge with a cardiac rhythm in response to small doses of adrenaline and that was not crazy behaviour but seemed to him like remarkably steadfast behaviour.

(c) In answer to a question about the relationship of the activity, Sleight said it correlated well with dP/dt and intramural tension. He referred to one of Thorén's slides where the impulses in one fibre had a linear relation to end-diastolic pressure. However, Sleight did not believe such a fibre was monitoring end-diastolic pressure. He believed that the explanation lay in the fact that as a result of the increase in fibre length and end-diastolic pressure there was a more forceful contraction and hence a greater stimulus to the receptors.

(d) In answer to another question Sleight said that during haemorrhage

the heart becomes empty and the receptors increased their discharge as a result of an increased force of contraction from increased catecholamines in the empty heart. This is probably the best stimulus for ventricular receptors.

(e) It was suggested that the altered chemical environment of these receptors could be responsible for the effects. Sleight thought the time course of the chemical changes was too long to allow this explanation.

12. Reflex effects of left ventricular mechanoreceptors with afferent fibres in the vagal nerves

P. THORÉN

The presence in the heart of receptors affecting the circulatory system was suggested more than 100 years ago by Bezold & Hirt (1867). They ascribed the hypotension and bradycardia produced by injections of veratrum alkaloids to stimulation of endings in the heart, although they never attempted to localise the receptors. That the veratrum-sensitive endings were, indeed, located in the heart and then mainly the left ventricle was clearly demonstrated by Jarisch & Richter (1939a, b) and by Dawes (1947). This reflex is now called the Bezold–Jarisch reflex.

Other early experiments suggested the presence of ventricular receptors. Thus Daly & Verney (1927) ascribed the reflex slowing of the heart following a pressure rise in the left ventricle to stimulation of left ventricular receptors.

The demonstration of receptors in the heart ventricle was greatly facilitated when adequate techniques for recordings of action potentials in afferent fibres from the heart became available. Thus Jarisch & Zotterman in 1948 described activity in slow-conducting vagal afferents emanating from receptors in the ventricles which responded to mechanical stimulation and to injection of veratrum alkaloids. The work by Paintal in 1955 where he recorded activity in ventricular receptors with medullated vagal afferents is also classical in the field.

Only recently has interest been focused on the characteristics (Coleridge, Coleridge & Kidd, 1964; Sleight & Widdicombe, 1965; Muers & Sleight, 1972b; Öberg & Thorén, 1972a; Thorén, 1976a, b) and the reflex effects (Thorén, Donald & Shepherd, 1976) of these atrial and ventricular receptors with non-medullated vagal afferents. The aim of this present paper is to describe the functional characteristics, the reflex effects and the possible physiological importance of the ventricular vagal C fibres.

Characteristics of ventricular C fibres

The existence of the left ventricular C fibres were first described in the dog by Coleridge *et al.* (1964). The ventricular C fibres normally had a low-frequency, irregular rhythm but could be markedly activated by stroking the epicardium or by injection of capsaicin and veratridine. The epi-myocardial type of left ventricular receptors in the dog have also been studied by Sleight & Widdicombe (1965) and Muers & Sleight (1972*b*). They examined the activity in the receptors during changes in ventricular pressure and contractility. A brief aortic occlusion stimulated 60% of the endings and the activity usually then displayed cardiac rhythmicity during systole. Adrenaline injection induces a more powerful activation of the receptors than aortic occlusion (Muers & Sleight, 1972*b*). The carotid occlusion reflex and electrical stimulation of the sympathetic nerves to the heart could also activate these receptors and several of these C fibre endings were markedly activated when the coronary venous pressure was increased during a partial occlusion of the coronary sinus.

The functional characteristics of the left ventricular epi-myocardial C fibres have also recently been studied in detail in the cat by Öberg & Thorén (1972*a*) and Thorén (1977). Normally, ventricular C fibres in the cat have a low-frequency activity (mean 1.4 imp/s), either irregular or with cardiac-modulated rhythm. The conduction velocity in the vagal nerve varies from 0.6 to 2.4 m/s and the conduction velocity within the heart is significantly lower than the conduction velocity in the vagal nerve (Thorén, 1977). The location of 26 receptors was established in the opened, non-beating heart by probing and the receptors were located throughout the entire left ventricle.

All the receptors were tested to occlusion of the ascending aorta. Fig. 12.1 shows an example of the response in one receptor during aortic occlusion. This receptor has a low tonic activity but becomes markedly activated with a maximal frequency of 12 Hz upon aortic occlusion. The maximal discharge during aortic occlusion ranged from 3.5 to 22 Hz in the 26 receptors. During a graded aortic occlusion the receptors do not respond to isolated changes in left ventricular systolic pressure. However, upon further occlusion and parallel with changes in left ventricular end-diastolic pressure (LVEDP) the receptor activity is markedly increased. This is illustrated in Fig. 12.2 where the curves relating the activity versus the systolic and end-diastolic pressure during a graded occlusion of ascending aorta is shown.

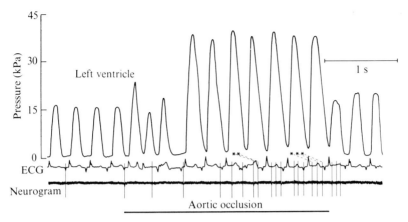

Fig. 12.1. Activity in a left ventricular C fibre during a short-lasting aortic occlusion. The receptor has a low sporadic spontaneous activity and is markedly activated during the aortic occlusion. Total conduction time was obtained by electrical stimulation over receptor area. The asterisks indicate corrected position of receptor activation in cardiac cycle after total conduction time is taken into account. During aortic occlusion, the receptor discharges throughout systole.

The receptors could also be moderately or markedly activated by a graded transfusion, and the receptor activity then increased in parallel with the change in LVEDP with a threshold of 0.3–1.6 kPa in LVEDP.

An interesting characteristic of these left ventricular C fibres is that the receptor response to aortic occlusion and transfusion is markedly influenced by changes in ventricular inotropism (Thorén, 1977). Fig. 12.2 shows the response in the cat of a left ventricular C fibre to graded aortic occlusion with normal inotropism, increased inotropism (isoprenaline infusion) and decreased inotropism (propranolol injection) of the ventricle. Notice that isoprenaline infusion increased the receptor response to aortic occlusion and that β-blockade markedly attenuates the response. The receptor response to transfusion is also markedly attenuated after injection of propranolol.

The fact that the receptors show a good correlation with changes in LVEDP both during aortic occlusion and transfusion might indicate that the receptors function as distension receptors. This is, however, not the case, because an unexpected finding is that the receptors are activated in systole during these circumstances (Muers & Sleight, 1972b; Thorén, 1977). This systolic activation of the receptor during aortic occlusion is shown in Fig. 12.1.

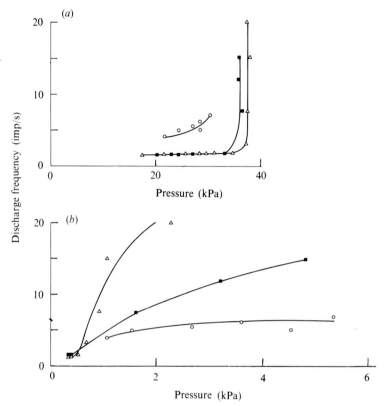

Fig. 12.2. Activity in one left ventricular C fibre plotted against *(a)* left ventricular systolic and *(b)* end-diastolic pressures during graded aortic occlusion before, during infusion of isoprenaline (1.25–2.5 g/min) and after administration of propranolol (0.2–0.3 mg·kg^{-1}) in the cat. Values for maximal rate of increase in left ventricular pressure (dP/dt$_{max}$) reflect changes in ventricular inotropism. ■, Control (dP/dt$_{max}$=613.2); ▵, isoprenaline (dP/dt$_{max}$=773.1), ○, propranolol (dP/dt$_{max}$=293.3).

The importance of ventricular systole in the activation of the receptors is also shown by the fact that during ventricular fibrillation the receptor discharge does not increase in spite of an increase in LVEDP but does so immediately after cardiac contraction has spontaneously become re-established. Also, depression of ventricular contractility impairs the receptor response to pressure and volume loading. In contrast, during graded aortic occlusion, receptor discharge at any given LVEDP was higher when inotropism was increased. Thus, we are left with the paradox that these left ventricular receptors are activated during systole but that the frequency of discharge appears related to diastolic events, and there is no good explanation for this finding.

There exists also a sparse population of receptors with C fibre afferents in the right ventricle in the dog (Muers & Sleight, 1972*b*) and the cat (Thorén, unpublished data). In addition, these receptors can respond to moderate pressure changes in the right ventricle (Thorén, unpublished data) but the receptor characteristics have not been studied in detail.

Selective electrical activation of cardiac vagal afferents

In order to analyse the cardiovascular responses to increased activity in medullated and non-medullated cardiac afferents the central end of the right cardiac nerve of the cat was stimulated electrically at different intensities and durations (Öberg & Thorén, 1973*a*). Stimulation of the low-threshold group of medullated fibres induced only weak excitatory responses. These increases were augmented if the buffering influences of the carotid baroreceptors were minimised by clamping both carotid arteries.

In contrast, high-intensity stimulation of both medullated and non-medullated fibres caused a pronounced bradycardia, a decrease in arterial blood pressure and a dilatation of the muscle and kidney vessels whether or not the arterial baroreceptors were functioning normally. These two types of cardiovascular responses during electrical activation of the cardiac vagal afferents are shown in Fig. 12.3.

The pattern of the reflex response induced by electrical stimulation of the afferent cardiac vagal C fibres have been studied in detail in the cat by Öberg & White (1970) and Little, Wennergren & Öberg (1975). The frequency–response characteristics of the reflex effects (Öberg & White, 1970; Little *et al.*, 1975) on aortic blood pressure, renal and muscle vascular bed and heart rate show that powerful reflex effects can be induced already at low-frequency stimulations (less than 2 imp/s). The kidney seems to be especially responsive to low-frequency stimulations of the right cardiac nerve. Thus only 1 imp/s induced about 30% of the total response in the kidney (Little *et al.*, 1975) and the maximal renal response is obtained at 10 imp/s. This powerful effect on the renal circulation at very low stimulation frequencies is of great interest because the spontaneous activity in the cardiac C fibres is normally low (see above).

If the *relation* between the renal vascular and muscle vascular responses

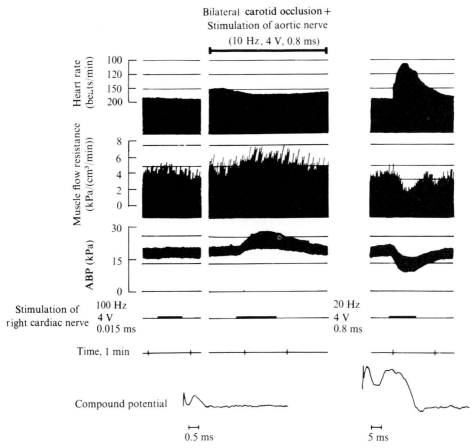

Fig. 12.3. Effects of stimulation of medullated (left and middle panel) and non-medullated fibres (right panel) in the main right cardiac nerve on blood pressure, heart rate and flow resistance in calf muscle. The evoked potentials recorded proximally from the same nerve are also shown. Note the augmented responses to stimulation of the medullated fibres when the buffering influence from arterial baroreceptors are largely removed (carotid occlusion) and a constant inhibitory restraint is placed on the vasomotor centre by stimulation of the aortic nerve (middle panel). ABP = arterial blood pressure. (From Öberg & Thorén, 1973*a*).

were compared over the whole range of stimulation frequencies, the cardiac receptors induced more powerful effects on the kidney than those elicited from the arterial baroreceptors. However, the *maximal* reflex response (Little *et al.*, 1975) of the renal vessels, obtained with right cardiac nerve stimulation, was equal to that caused by baroreceptor stimulation, but the maximal reflex response of the skeletal muscle vessels to cardiac nerve stimulation averaged only 66% of that resulting from baroreceptor stimulation. The cardiac C fibres also elicited a more pronounced reflex bradycardia, which was mainly caused by an increased vagal outflow to the heart.

Reflex effect elicited from ventricular C fibres during ventricular distension

Depressor reflexes elicited during left ventricular distension was first described by Daly & Verney in 1927, the later papers (Aviado & Schmidt, 1959; Doutheil & Kramer, 1959; Salisbury, Cross & Rieben, 1960; Ross, Frahm & Braunwald, 1961; Mark *et al.*, 1973*a*; Öberg & Thorén, 1973*b*; Chevalier *et al.*, 1974) have consistently confirmed this finding. The afferent pathway of the reflex in the vagal nerve and the dilatation is solely caused by inhibition of adrenergic vasoconstrictor activity. The bradycardia, particularly evident in the cat, is caused almost entirely by vagal activation (Öberg & Thorén, 1973*a*).

Distension of the right ventricle either has no effect on the cardiovascular system or the reflex responses are small and inconsistent.

The reflex effect of left ventricular distension on heart rate, kidney and hind-limb circulation was studied in a recent paper by Öberg & Thorén (1973*b*). The ventricular receptors were activated by aortic occlusion and nicotine injection into the pericardial sac. The contribution of left atrial receptors to the aortic occlusion response was not ruled out but data from others (Mark *et al.*, 1973*a*, *b*) indicate that the left ventricle is the main receptor station responsible for the depressor reflex seen during aortic occlusion. When the reflex effect from the left ventricular receptors was compared with reflex effects from carotid baroreceptors, the ventricular receptors have an especially marked effect on the vagal outflow to the heart and on the kidney circulation (as illustrated in Fig. 12.4). The vasodilatation in the hind limb was caused by a withdrawal of sympathetic tone and not by the activation of vasodilator pathways. Thus the pattern of reflex response observed during activation of the ventricular receptors was very similar to the pattern observed during electrical activation of the afferent vagal C fibres.

There are no data available to show whether the powerful reflex effects observed during left ventricular distension can be ascribed to ventricular C fibres or ventricular medullated fibres. However, indirect evidence indi-

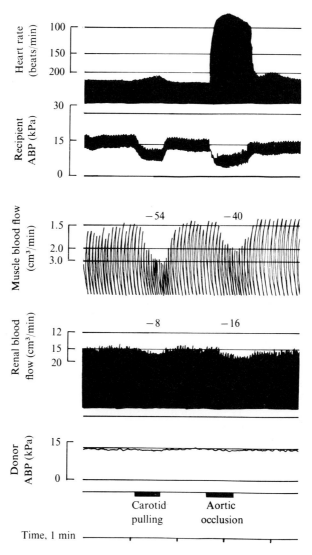

Fig. 12.4. Reflex effects of carotid pulling and aortic occlusion on heart rate, muscle and renal flow resistances. The skeletal muscle and kidney were cross-perfused from a donor at constant pressure. Note that the heart rate and renal vessel responses are more pronounced with aortic occlusion while the skeletal muscle response is more pronounced with carotid pulling. ABP = arterial blood pressure. (From Öberg & Thorén, 1973*b*.)

cates that the ventricular C fibres are the most important in this respect. First, ventricular distension induced a reflex response which is similar to the reflex response observed during electrical activation of the afferent C fibres from the heart (Öberg & Thorén, 1973*a*; Öberg & White, 1970; Little *et al.*, 1975).

Secondly, the ventricular C fibres by far outnumber the medullated ventricular receptors both in the cat and the dog. Only 20% of afferent vagal fibres in the heart are medullated fibres (Agostoni, Chinnock, Daly & Murray, 1957), and probably less than one-tenth of these medullated fibres are in the ventricles (Paintal, 1955; Coleridge *et al.*, 1964; Kolatat, Ascanio, Tallarida & Oppenheimer, 1967). In contrast, most of the afferent C fibres in the heart are in the left ventricle (Sleight & Widdicombe, 1965; Muers & Sleight, 1972*b*; Öberg & Thorén, 1972*a*; Thorén, 1977). This indicates that the medullated ventricular receptors are only a small fraction of the total afferent endings in the ventricle. This was also evident in the study by Thorén (1977) because only two filaments with ventricular medullated receptors were found in contrast with a large number of filaments with ventricular C fibres, despite the fact that recordings from medullated afferents have a more favourable signal to noise ratio.

Thirdly, right ventricular distension induced very weak or no reflex effects (Aviado *et al.*, 1951) in contrast with the powerful depressor reflex which can be elicited from the left ventricle. The vagal C fibres show a similar distribution and have their main receptor station in the left ventricle (Sleight & Widdicombe, 1965; Muers & Sleight, 1972*b*; Öberg & Thorén, 1972*a*; Thorén, 1977), with only a few endings in the right ventricle. In contrast, the medullated ventricular receptors are distributed equally in both ventricles (Coleridge *et al.*, 1964; Paintal, 1973*b*). Thus there is indirect evidence that left ventricular C fibres are the major receptor station responsible for the marked depressor reflex seen during left ventricular distension.

There are also other depressor reflexes originating in the left ventricle, such as the reflex elicited by local coronary sinus distension (Szentivanyi & Juhasz-Nagy, 1962), increased pressure in the coronary sinus (Szentivanyi & Juhasz-Nagy, 1962) and increased pressure in the left coronary artery in the cat (Brown, 1966). However, the response to local coronary sinus distension is probably of local origin and not a reflex (Muers & Sleight, 1972*c*), and the response to increased pressure in the coronary sinus is caused by the activation of ventricular C fibres throughout the ventricle (Muers & Sleight, 1972*a, b*) and not by receptors especially confined to the coronary venous system. The depressor reflex response to increased pressure in the coronary arteries (Brown, 1966) could very well be caused by activation of ventricular medullated fibres, as claimed by Brown (1965) but participation of ventricular C fibres in the response cannot be ruled out at present.

Bezold–Jarisch reflex

Extending the old observations of Bezold & Hirt (1867), Jarisch & Richter (1939*a*, *b*) showed that veratridine excites vagal receptors in the cardiopulmonary region and that this is accompanied by bradycardia and hypotension. Studies by Dawes (1947) indicate that the left ventricle is the main receptor station responsible for this reflex. The Bezold–Jarisch reflex or the coronary 'chemoreflex' can be elicited by a large number of drugs (Dawes & Comroe, 1954; Krayer, 1961; Zipf, 1966). Various receptors with medullated and non-medullated afferents in the lungs (Paintal, 1973*b*), ventricles (Paintal, 1955; Coleridge *et al.*, 1964; Sleight & Widdicombe, 1965; Paintal, 1973*b*; Öberg & Thorén, 1972*a*) and atria (Coleridge *et al.*, 1973; Paintal, 1973*b*) are activated by veratridine. Paintal claims that at least in the cat the ventricular medullated fibres are the receptors of main importance for the Bezold–Jarisch reflex (Paintal, 1955, 1973*a*). This statement has been questioned by other investigators who claim that the ventricular C fibres are mainly responsible for the Bezold–Jarisch reflex (Öberg & Thorén, 1972*a*, 1973*a*) for the following reasons. First, the circulatory adjustments in the Bezold–Jarisch reflex can be mimicked by afferent stimulation of the cardiac nerves only when such stimulation intensities are used which activate non-medullated afferents (Öberg & Thorén, 1973*a*). Secondly, the left ventricular C fibres by far outnumber the ventricular medullated endings (see above). Thirdly, the ventricular C fibre endings seem to be preferentially located in the left ventricle (Coleridge *et al.*, 1964; Sleight & Widdicombe, 1965; Muers & Sleight, 1972*b*; Öberg & Thorén, 1972*a*; Thorén, 1977), i.e. in that part of the heart from which the Bezold–Jarisch reflex is said to be elicited, at least in dogs (Dawes, 1947). Finally, the ventricular C fibres are activated by low doses of veratrum alkaloids both in dogs (Sleight & Widdicombe, 1965) and cats (Öberg & Thorén, 1972*a*). In contrast, the ventricular medullated fibres are activated by veratrum alkaloids in the cat (Paintal, 1955; Brown, 1965) but not in the dog (Coleridge *et al.*, 1964), and there is no reason to believe that the afferent mechanism for the Bezold–Jarisch reflex should be different in cats and dogs.

The epicardial 'chemoreflex' elicited by the injection of nicotine or other drugs into the pericardial sac (Sleight, 1964) can also be ascribed to activation of ventricular C fibres (Sleight & Widdicombe, 1965; Öberg & Thorén, 1972*a*) and is probably the same reflex mechanism as the Bezold–Jarisch reflex.

The word 'chemoreflex' in the terms 'coronary chemoreflex' and 'epicardial chemoreflex' is not adequate, because these reflexes are caused by chemical activation of *mechanoreceptors* and not by activation of 'chemoreceptors'. In fact, there is no evidence for *true* chemoreceptors in the myocardium (Mark *et al.*, 1974).

Reflex activation of the vomiting centre

Increased activity in cardiac C fibres in the cat can induce a marked relaxation of the stomach (Abrahamsson & Thorén, 1972) because of the activation of the vagal non-cholinergic relaxatory fibres to the stomach (Abrahamsson & Thorén, 1973). The relaxation can be elicited by electrical

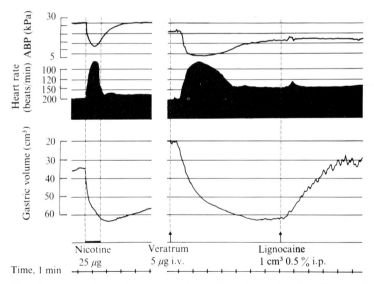

Fig. 12.5. Pronounced hypotension, bradycardia and gastric relaxation produced by local application of nicotine in the pericardium (left panel) and by intravenous injection of veratrum (right panel). The recovery of gastric volume after local administration of lignocaine in the pericardium suggests that the relaxation caused by veratrum is elicited from receptors localised in the heart. ABP = arterial blood pressure. (From Abrahamsson & Thorén, 1972.)

stimulation of the cardiac afferent C fibres, by aortic occlusion, by mechanical stroking of the epicardial surface, by coronary artery occlusion and by chemical activation (Fig. 12.5) of the cardiac C fibres with veratridin and nicotine (Abrahamsson & Thorén, 1972).

The reflex relaxation of the stomach forms part of a more complex activation of the vomiting centre (Abrahamsson & Thorén, 1973) because electrical stimulation of the vagal cardiac afferent fibres elicited a typical vomiting response in non-anaesthetised decerebrated cats. The reflex relaxation of the stomach observed in the anaesthetised animals is, therefore, probably a manifestation of a more complex vomiting response where the somatic component is suppressed by the anaesthesia of the animals.

This reflex relaxation of the stomach first occurs at considerably higher stimulation frequencies than the reflex cardiovascular events (Abrahamsson & Thorén, 1973). Even low stimulation frequencies (around 1 Hz) induce marked reductions of blood pressure, while the concomitant changes in gastric volume are slight or moderate. Thus, this reflex is probably not operating normally but may be of importance during pathophysiological and 'chemical' activation of the receptors. With stimulation frequencies above 5 Hz, pronounced changes in gastric volume occur, while a full vomiting response is elicited first with higher frequencies around 10 Hz. It therefore seems as if alternations in receptor activity in the lower frequency range mainly affects the cardiovascular system and that an all-out activation of the vomiting centre requires an intense afferent discharge of the receptors.

The functional role of left ventricular C fibres

Blood pressure control. Cardiac C fibres are probably not significantly involved in the tonic regulation of blood pressure because they cannot register small or moderate changes in arterial blood pressure (Muers & Sleight, 1972b; Öberg & Thorén, 1972a; Thorén, 1977). However, the receptors may be of importance for arterial pressure control in some situations of daily life, e.g. when blood pressure rises drastically to high values as a result of emotional influences, such as fear, anger, etc., and can then help to prevent the heart from acute overloading by inducing powerful depressor reflexes.

When the sinus and aortic nerves are cut in dogs the arterial pressure increases markedly but only for a short period of time. This return of the blood pressure towards normal might be caused by the tonic inhibition from cardiopulmonary receptors. This can also explain why these animals are so labile in their blood pressure because the cardiopulmonary C fibres are not able to sensor small or moderate changes in systolic blood pressure.

Exercise. The activity in left ventricular C fibres can be augmented when the contractility of the left ventricle is increased by infusion of isoprenalin (Muers & Sleight, 1972*b*; Thorén, 1977) or by stimulation of the sympathetic nerves to the heart (Muers & Sleight, 1972*b*). During exercise there is a marked augmentation of the sympathetic outflow to the heart, together with a rise in arterial blood pressure; and these factors together might activate these left ventricular C fibres. Accordingly, Sleight & Widdicombe (1965) have postulated that left ventricular C fibres play a role in the circulatory control during exercise and then act as a vagal brake on the heart.

Blood volume control. The role of cardiac receptors in blood volume control has been widely discussed for many years (Gauer, Henry & Behn, 1970; Goetz, Bond & Bloxham, 1975; Linden, 1975) and it is not the aim of the paper to discuss this in detail. However, some points are worth while making.

(1) Most of the vagal afferents from the heart are C fibres. The C fibre endings are located throughout the entire heart with the major receptor population in the left ventricle (Thorén, 1977) but also with a substantial population in the atria (Thorén, 1976*a*).

(2) Many of these C fibres can respond to changes in atrial pressure and LVEDP within the physiological pressure range and will therefore sense changes in cardiac filling within this normal range. Finally, these cardiac C fibres have a strong reflex inhibitory effect particularly on the sympathetic outflow to the kidney (Thorén *et al.*, 1976) which means that they probably will influence not only renal blood flow directly but also renin release (Zanchetti *et al.*, 1976) and sodium excretion. Electrical stimulation of cardiac vagal C fibres can also affect the antidiuretic hormone secretion (Harris & Spyer, 1974).

Thus, cardiac vagal C fibres constitute a receptor group which is likely to be involved in blood volume regulation. However, the definite role of these receptors in overall blood volume regulation is still not fully understood.

Pathophysiological aspects. These cardiac C fibres are probably also of importance during different pathophysiological conditions, such as myocardial infarction, severe asphyxia, vasovagal syncope and syncope in aortic stenosis.

During occlusion of a coronary artery the ventricular C fibres are markedly activated with a latency of about 20 s (Thorén, 1972, 1973*b*, 1976*a*). The stimulus for the receptor is the bulging of the ischaemic area and not the ischaemia *per se*. During a long-lasting occlusion of a coronary artery the receptor activity is maximal after about 60 s and then there is a progressive decline again in the receptor activity. This secondary decline in the activity is caused by an anoxic depression of the receptors. The pronounced vasodepressor reflex occurring early after occlusion of a coronary artery (Thorén, 1976*a*) is probably caused by this activation of the ventricular C fibres.

The ventricular C fibres are also markedly activated during an acute severe hypoxia or asphyxia (Thorén, 1972). The stimulus for the receptors is the ischaemic distension of the left ventricle. The marked bradycardia observed during these circumstances is at least in part triggered from these cardiac C fibres (Thorén, 1973*a*).

Acute severe haemorrhage can increase the activity in cardiac C fibres (Öberg & Thorén, 1972*b*). The stimulus for these mechanoreceptors in this situation seems to be the tearing and squeezing of the myocardium induced by the vigorous contraction of the left ventricle around the almost empty chamber.

It was recently shown by Mark *et al.* (1973*b*) that depressor reflexes from the left ventricle may be an important factor in the circulatory control in patients with severe aortic stenosis. These reflexes will then contribute to the syncope resulting from exertion in these patients. As discussed earlier the ventricular C fibres are then likely to be the major receptor station responsible.

References

Abrahamsson, H. & Thorén, P. (1972). Reflex relaxation of the stomach elicited from receptors located in the heart. An analysis of the receptors and afferents involved. *Acta physiol. scand.*, **84**, 197–207.

Abrahamsson, H. & Thorén, P. (1973). Vomiting and reflex vagal relaxation of the stomach elicited from heart receptors in the cat. *Acta physiol. scand.*, **88**, 433–9.

Agostoni, E., Chinnock, J.E., Daly, M. de B. & Murray, J.G. (1957). Functional and histological studies of the vagus nerve and its branches to the heart, lungs and abdominal viscera in the cat. *J. Physiol.*, **135**, 182–205.

Aviado, D.M., Jr, Li, T.H., Kalow, W., Schmidt, C.F., Turnbull, G.L., Peskin, G.W., Hess,

M.E. & Weiss, A.J. (1951). Respiratory and circulatory reflexes from the perfused heart and pulmonary circulation of the dog. *Am. J. Physiol.*, **165**, 261–77.

Aviado, D.M., Jr & Schmidt, C.F. (1959). Cardiovascular and respiratory reflexes from the left side of the heart. *Am. J. Physiol.*, **196**, 726–30.

Bezold, A. von & Hirt, L. (1867). Über die physiologischen Wirkungen des essigsauren Veratrins. *Unters. Physiol. Lab. Wurtzburg.*, **1**, 76–156.

Brown, A.M. (1965). Mechanoreceptors in or near the coronary arteries. *J. Physiol.*, **177**, 203–14.

Brown, A.M. (1966). The depressor reflex arising from the left coronary artery of the cat. *J. Physiol.*, **184**, 825–36.

Chevalier, P.A., Webber, K.C., Lyons, G.W., Nicoloff, D.M. & Fox, I.J. (1974). Hemodynamic changes from stimulation of left ventricular baroreceptors. *Am. J. Physiol.*, **227**, 719–28.

Coleridge, H.M., Coleridge, J.C.G., Dangel, A., Kidd, C., Luck, J.C. & Sleight, P. (1973). Impulses in slowly conducting vagal fibers from afferent endings in the veins, atria and arteries of dogs and cats. *Circulation Res.*, **33**, 87–97.

Coleridge, H.M., Coleridge J.C.G. & Kidd, C. (1964). Cardiac receptors in the dog, with particular reference to two types of afferent ending in the ventricular wall. *J. Physiol.*, **174**, 323–39.

Daly, E. de B. & Verney, E.B. (1927). The localization of receptors involved in the reflex regulation of the heart rate. *J. Physiol.*, **62**, 330–40.

Dawes, G.S. (1947). Studies on veratrum alkaloids. *J. Pharmac. exp. Ther.*, **89**, 325–42.

Dawes, G.S. & Comroe, J.H., Jr (1954). Chemoreflexes from the heart and lungs. *Physiol. Rev.*, **34**, 167.

Doutheil, U. & Kramer, K. (1959). Über die Differenzierung kreislaufregulierender Reflexe aus dem Linken herzen. *Pflügers Arch. ges. Physiol.*, **269**, 114–29.

Gauer, O.H., Henry, J.P. & Behn, C. (1970). The regulation of extracellular fluid volume. *A. Rev. Physiol.*, **32**, 547–95.

Goetz, K.L., Bond, G.C. & Bloxham, D.D. (1975). Atrial receptors and renal function. *Physiol. Rev.*, **55**, 157–205.

Harris, M.C. & Spyer, K.M. (1974). Inhibition of ADH release by stimulation of afferent cardiac branches of the right vagus in cats. *J. Physiol.*, **237**, 663–84.

Jarisch, A. & Richter, H. (1939a). Die Kreislaufwirkung des Veratrins. *Arch. exp. Path. Pharmak.*, **193**, 347–54.

Jarisch, A. & Richter, H. (1939b). Die afferenten Bahnen des Veratrinreflektes in den Herznerven. *Arch. exp. Path. Pharmak.*, **193**, 355–71.

Jarisch, A. & Zotterman Y. (1948). Depressor reflexes from the heart. *Acta physiol. scand.*, **16**, 31–51.

Kolatat, T., Ascanio, G., Tallarida, R.J. & Oppenheimer, M.J. (1967). Action potentials in the sensory vagus at the time of coronary infarction. *Am. J. Physiol.*, **213**, 71–8.

Krayer, O. (1961). The history of the Bezold–Jarisch effect. *Arch. exp. Path. Pharmak.*, **240**, 361–8.

Linden, R.J. (1975). Reflexes from the heart. *Prog. cardiovasc. Dis.*, **18**, 201–21.

Little, R., Wennergren, G. & Öberg, B. (1975). Aspects of the central integration of arterial baroreceptor and cardiac ventricular receptor reflexes in the cat. *Acta physiol. scand.*, **93**, 85–96.

Mark, A., Abboud, F.M., Heistad, D.D., Schmid, P.G. & Johannsen, U.J. (1974). Evidence against the presence of ventricular chemoreceptors activated by hypoxia and hypercapnia. *Am. J. Physiol.,* **227,** 178–82.

Mark, A.L., Abboud, F.M., Schmid, P.G., Heistad, D.D. & Johannsen, U.J. (1973*a*). Reflex vascular responses to left ventricular outflow obstruction and activation of ventricular baroreceptors in dogs. *J. clin. Invest.,* **52,** 1147–53.

Mark, A.L., Kioschos, J.M., Abboud, F.M., Heistad, D.D. & Schmid, P.G. (1973*b*). Abnormal vascular response to exercise in patients with aortic stenosis. *J. clin. Invest.,* **52,** 1138–53.

Muers, M.F. & Sleight, P. (1972*a*). The reflex cardiovascular depression caused by occlusion of the coronary sinus in the dog. *J. Physiol.,* **221,** 259–82.

Muers, M.F. & Sleight, P. (1972*b*). Action potentials from ventricular mechanoreceptors stimulated by occlusion of the coronary sinus in the dog. *J. Physiol.,* **221,** 283–309.

Muers, M.F. & Sleight, P. (1972*c*). The cardiovascular effects of coronary sinus distension in the anaesthetised dog. *Q. Jl exp. Physiol.,* **57,** 359–70.

Öberg, B. & Thorén, P. (1972*a*). Studies on left ventricular receptors signalling in nonmedullated vagal afferents. *Acta physiol. scand.,* **85,** 145–63.

Öberg, B. & Thorén, P. (1972*b*). Increased activity in left ventricular receptors during hemorrhage or occlusion of caval veins in the cat. A possible cause of the vasovagal reaction. *Acta physiol. scand.,* **85,** 164–73.

Öberg, B. & Thorén, P. (1973*a*). Circulatory responses to stimulation of medullated and nonmedullated afferents in the cardiac nerve in the cat. *Acta physiol. scand.,* **87,** 121–33.

Öberg, B. & Thorén, P. (1973*b*). Circulatory responses to stimulation of left ventricular receptors in the cat. *Acta physiol. scand.,* **88,** 8–22.

Öberg, B. & White, S. (1970). Circulatory effects of interruption and stimulation of cardiac vagal afferents. *Acta physiol. scand.,* **80,** 383–94.

Paintal, A.S. (1955). A study of ventricular pressure receptors and their role in the Bezold reflex. *Q. Jl exp. Physiol.,* **40,** 348–63.

Paintal, A.S. (1973*a*). Sensory mechanisms involved in the Bezold–Jarisch reflex. *Aust. J. exp. Biol. med.,* **51,** 3–15.

Paintal, A.S. (1973*b*). Vagal sensory receptors and their reflex effects. *Physiol. Rev.,* **53,** 159–227.

Ross, J. Jr, Frahm, C.J. & Braunwald, E. (1961). The influence of intracardiac baroreceptors on venous return, systemic vascular volume and peripheral resistance. *J. clin. Invest.,* **40,** 563–72.

Salisbury, P.F., Cross, C.E. & Reiben, P.A. (1960). Reflex effects of left ventricular distension. *Circulation Res.,* **8,** 530–4.

Sleight, P. (1964). A cardiovascular depressor reflex from the epicardium of the left ventricle in the dog. *J. Physiol.,* **173,** 321–43.

Sleight, P. & Widdicombe, J.G. (1965). Action potentials in fibres from receptors in the epicardium and myocardium of the dog's left ventricle. *J. Physiol.,* **181,** 235–58.

Szentivanyi, M. & Juhasz-Nagy, A. (1962). Two types of coronary vasomotor reflexes. *Q. Jl exp. Physiol.* **47,** 289–98.

Thorén, P. (1972). Left ventricular receptors activated by severe asphyxia and by coronary artery occlusion. *Acta physiol. scand.,* **85,** 455–63.

Thorén, P. (1973a). Reflex bradycardia elicited from left ventricular receptors during acute severe hypoxia in cats. *Acta physiol. scand.*, **87**, 103–12.

Thorén, P. (1973b). Reflex bradycardia elicited from left ventricular receptors during occlusion of one coronary artery in the cat. *Acta physiol. scand.*, **88**, 23–34.

Thorén, P. (1976a). Activation of left ventricular receptors with non-medullated vagal afferents during occlusion of a coronary artery in the cat. *Am. J. Cardiol.*, **37**, 1046–51.

Thorén, P. (1976b). Atrial receptors with nonmedullated vagal afferents in the cat – discharge frequency and pattern in relation to atrial pressure. *Circulation Res.*, **38**, 357–62.

Thorén, P. (1977). Characteristics of left ventricular receptors with nonmedullated vagal afferents in the cat. *Circulation Res.*, **40**, 231–7.

Thorén, P., Donald, D.E. & Shepherd, J.T. (1976). Role of heart and lung receptors with nonmedullated vagal afferents in circulatory control. *Circulation Res.*, **38** (Suppl. ii), 2–9.

Zanchetti, A., Stella, A., Leonetti, G., Morganti, A. & Teryoli, I. (1976). Control of renin release. A review of experimental evidence and clinical implications. *Am. J. Cardiol.*, **37**, 675–91.

Zipf, H.F. (1966). The pharmacology of visceroafferent receptors with special reference to endoanaesthesia. *Acta neuroveg.* **28**, 169–96.

Discussion

(a) In answer to a question from the floor Thorén replied that relaxation of the stomach was only observed if recording from the inserted balloons was isotonic and not isometric, probably because the gut normally works as a reservoir.

(b) Thorén was asked whether he thought the ventricular pressure receptors discharging into medullated fibres played any part in this response or if they were redundant to the control of the circulatory system. In reply Thorén said his argument that the responses following aortic occlusion were mainly caused by ventricular receptors discharging into C fibres was as follows: (1) There are considerably more ventricular C fibres than ventricular medullated fibres; he had recorded probably 150 C fibres altogether but had only found 2–3 medullated fibres from the ventricle. (2) He had shown that electrical stimulation of afferent nerves at an intensity which activates C fibres resulted in a pronounced bradycardia. (3) Ventricular pressure receptors with medullated afferents, as described by Paintal, were located mainly in the right ventricle and comparison of reflex responses evoked by occlusion of the pulmonary artery and aorta showed that the left ventricle is the receptive area largely responsible for the observed reflex response during bilateral ventricular distension. Thus, the receptors attached to C fibres were distributed in a way that would explain the reflex response to aortic occlusion or pulmonary artery occlusion. He

suggested that these facts indicated that left ventricular receptors associated with C fibres were the most important receptors involved.

Paintal commented that he thought the aim was to find out what each of these groups of receptors was doing and agreed that right ventricular fibres, whether 'C' or 'A', are not stimulated under these circumstances and do not produce such a cardiovascular reflex, as shown by the elegant experiments of Geoffrey Dawes (Dawes, 1947). However, what had to be found out was the proportions of the reflex effect produced by the medullated and non-medullated fibres.

From the floor it was pointed out that the philosophy which concluded that if one thing is more numerous than another it must be more important does not necessarily hold. Thorén was asked whether he was confident that the electrical stimuli activated only non-medullated ventricular fibres and if he was able to abolish all the effects by cutting the vagi. Thorén replied that when stimulating the cardiac vagal branch on the right side he probably also stimulated some fibres from the lung but the great majority of fibres originated from the heart. Secondly, although he stimulated C fibres he must also have stimulated medullated fibres because he was not using any technique for blocking fibres. However, if the nerves were stimulated with low-frequency bursts he felt sure that the reflex response was mainly caused by the C fibres, because at such low frequencies he did not expect the medullated fibres to have any effect. The effects of aortic occlusion and electrical activation were totally abolished by vagotomy.

(c) It was pointed out from the floor that there was considerable speculation about the physiological role of these receptors and it was difficult to understand how they could be involved in the responses to haemorrhage. In this situation, where the end-diastolic pressure is presumably falling, the systolic pressure falls, the output from these receptors falls and yet the receptors are said to be responsible for the bradycardia and hypotension. Thorén pointed out that during haemorrhage the heart beats more vigorously because of a large increase in sympathetic activity and then, when ventricular pressure reached a certain level, quite suddenly, very high ventricular pressures occurred, probably because the ventricle was virtually empty and vigorous contractions under these circumstances caused the development of very high pressures at a time when there is tremendous distortion of the ventricles. Thorén said that the fact that some receptors are activated by transfusion and, under these circumstances, show a good correlation with changes in end-diastolic pressure indicates that they might be responding to change in volume and

in stroke volume. The latter observations would fit with the isoprenaline data, where an increased stroke volume was observed after isoprenaline had activated the receptors. Some type of free nerve endings between the myocardial cells would form a basic structure which could be stimulated by distension.

Reference

Dawes, G.S. (1947). Studies on veratrium alkaloids; receptor areas in coronary arteries and elsewhere as revealed by use of veratridine. *J. Pharmac. exp. Ther.*, **89**, 325–42.

13. Ventricular receptors: some reflex effects and integrative mechanisms

B. ÖBERG

In the previous chapter by Thorén the reflex cardiovascular responses to ventricular receptor stimulation were outlined in general terms. It was pointed out that one characteristic feature in this response pattern is the relatively strong reflex inhibition of the sympathetic outflow to the kidneys, leading to a marked renal vasodilatation. Because of the great significance of this finding when discussing the possible role of the ventricular receptor reflex mechanism in body fluid volume regulation, a more detailed analysis of this particular reflex response seems of great interest. The first part of this presentation will deal with some aspects of this problem, and following this, some integrative aspects of the ventricular reflex mechanism will be discussed.

Analysis of the possible mechanisms behind the relatively strong renal vessel engagement in the ventricular receptor reflex

The idea of a particularly strong influence from the ventricular receptors on the renal circulation is based on observations from studies where the reflex renal and muscle vessel responses were compared when ventricular receptors on one hand and arterial baroreceptors on the other were excited. Such comparisons reveal that, for a given muscle vessel response, the reflex renal vessel effects were significantly more pronounced when the ventricular receptors were stimulated than those obtained when the arterial baroreceptors were excited (Öberg & Thorén, 1973).

In a series of experiments this non-uniform engagement of the two beds in the above-mentioned reflex mechanisms was analysed in more detail (Little, Wennergren & Öberg, 1975). The results of these studies show that even a supramaximal electrical stimulation of ventricular receptor afferents was not capable of inducing a complete inhibition of sympathetic constrictor fibre activity in the muscle, since a superimposed strong baroreceptor activation (high sinus pressure) induced a significant additional dilatation in the muscle (Fig. 13.1). On the other hand, when a cardiac afferent stimulation was added to an intense prevailing baroreceptor activity, no further muscle vasodilatation appeared. With regard to the renal vascular bed, it seemed as if a complete inhibition of sympathetic activity could be elicited from both the ventricular receptors and the baroreceptors since addition of one reflex to the other produced in neither case a further vasodilatation. It was further found that, if the reflex muscle and renal vessel responses (expressed as a percentage of the maximal reflex response) were compared when the ventricular receptor afferents were stimulated with varying frequencies (between 1 and 16 imp/s), the renal responses were significantly more pronounced than the muscle vessel responses in the lower frequency range (below 4 Hz). The normal, physiological variations of firing rate in the non-medullated afferents probably occur mainly within this narrow range. The finding of a particularly strong influence of the ventricular receptors on the renal vascular bed as compared with that on the muscle vessels therefore seems to be explained rather by the relatively modest reflex engagement of the muscle vascular bed.

The neuronal mechanisms responsible for the differentiated engagement of renal and muscle vasomotor neurons in response to an activation of the ventricular receptors are unknown. The reported findings suggest, however, that the ventricular receptor afferents are preferentially projected towards those central vasomotor neurons which control the sympathetic outflow to the kidneys, while the innervation of the muscle vasomotor neurons is more sparse and, with regard to a portion of these neurons, possibly entirely lacking. This preferential orientation of cardiac afferents to renal vasomotor neurons implies that the ventricular receptor reflex mechanism constitutes a sort of a cardiorenal reflex. This reflex arrangement, together with the observation of a good correlation between ventricular end-diastolic pressure and receptor activity (Thorén, 1974), strongly suggest a volume-regulating function of this control system.

Influence from the ventricular receptors on the renal urine excretion

If one assumes that the ventricular receptors are engaged in the control of body fluid volume, the question arises as to what extent the reflex vascular adjustments in the kidney affect the renal excretion of urine. This problem was studied in a series of experiments on anaesthetised cats in which urine formation and sodium excretion were followed before, during and after stimulation of the cardiac nerve (Wennergren, Henriksson, Weiss & Öberg, 1976). It was found that urine formation and sodium excretion increased significantly (on average, 24 and 38%, respectively) when the cardiac nerve was stimulated. Even modest circulatory adjustments in the kidney (a 10–15% increase in renal blood flow conductance) were often accompanied by drastic increases in urine flow and sodium excretion. The effects were similar in autoperfused kidney and kidneys perfused from a donor animal which indicates that an intra-renal mechanism was responsible for the observed effects. Furthermore, the response appeared and vanished rapidly suggesting that neurogenic rather than hormonal factors are involved, and then in all probability, the neurogenic adjustments on renal haemodynamics leading to an increased glomerular filtrate rate and/or a reduced tubular reabsorption in turn caused by either a redistribution of the renal blood flow or an increased hydrostatic pressure in peritubular capillaries. Besides these probably haemodynamically-induced changes in renal exretory function one may expect that the ventricular receptors, because of their strong influence on the sympathetic outflow to the kidneys, may also have a correspondingly marked effect on the renin secretion, with consequent effects on the renal handling of sodium (e.g. Mancia, Romero & Shepherd, 1975).

Vascular versus cardiac responses to ventricular receptor stimulation

When discussing the possible volume-regulating function of the ventricular receptor reflex, one naturally asks how the very marked bradycardia, which is a distinctive feature in the reflex response pattern, fits into the picture. This response does not seem very rational as far as a minute-to-minute volume control is concerned as it appears to constitute a positive feedback. It seems, however, as if in most studies dealing with the reflex effects of ventricular receptor stimulation and where marked bradycardia responses have been obtained, very powerful and non-discriminating stimuli have often been used. It is then a possibility that with more discrete and well-graded stimuli a sequential reflex activation of different cardiovascular target organs may take place, and that therefore the renal and cardiac responses are not necessarily always inseparable. One may, for instance, imagine that the ventricular receptors exhibit a spectrum of thresholds, with, e.g., a preferential orientation to the vagal motor nucleus from the high-threshold endings, so that the reflex bradycardia emerges in its most drastic form only in situations implying a very strong receptor activation. Such reflex arrangement fits with the old concept of a protective function of reflexes originating from cardiac receptors, a mechanism brought into play in the case of an imminent overloading of the heart.

Some still very preliminary experiments in our laboratories, designed to analyse the above-mentioned problem, seem to indicate that when there is a stepwise increase in the distention of the heart, and reasonably, of receptor activity, it is possible to dissociate the reflex vascular effects from the cardiac responses. In these experiments on chloralose-anaesthetised, thoracotomised cats a shunt was placed between the superior mesenteric artery and the left atrium. By opening the shunt in a stepwise fashion the

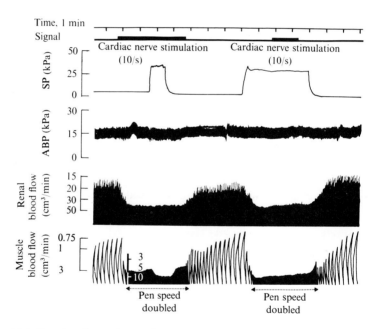

Fig. 13.1. Renal and muscle vessel responses to a strong baroreceptor stimulation superimposed on a 'maximal' cardiac nerve stimulation (left). A clear-cut muscle dilatation is induced when the sinus pressure is elevated while no additional dilator response is seen in the kidney. When the cardiac nerve stimulation is superimposed on a strong baroreceptor activation no additional vessel responses are seen (right). ABP = aterial blood pressure; SP = sinus pressure. (From Little, Wennergren & Öberg, 1975.)

inflow in the left side of the heart, and hence the distension of the left ventricle, could be varied. By means of ultra-sonic crystals sewn to the epicardial surface of the heart a rough estimate of cardiac dimensions was obtained. Heart rate, blood pressure, and muscle and renal blood flows were recorded. Fig. 13.2, which illustrates records from one experiment in this series, shows (left panel) that when the left ventricular volume is increased in a stepwise manner by opening the shunt gradually there is, with moderate cardiac distension, a dilatation in both vascular beds while the heart rate does not change noticeably. First, when the distension of the

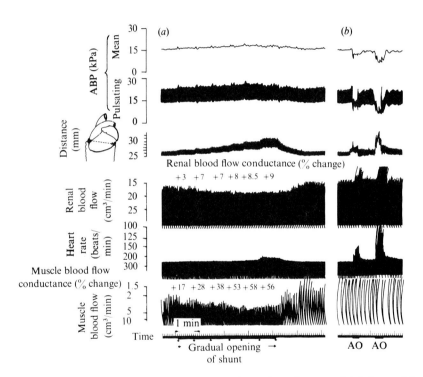

Fig. 13.2. *(a)* Left panel. Effects of a gradual opening of an arterial–left arterial shunt on blood pressure, dimension of the heart, heart rate and renal and muscle blood flow and flow conductances. Clear-cut heart rate responses are obtained first when the heart is distended enough to produce close to maximal (vascular) effects. *(b)* Right panel. Heart rate responses to partial occlusion of the ascending aorta (AO). Note the marked bradycardia obtained with moderate distensions of the heart in the first AO test. The aortic nerves were cut and both carotid arteries occluded throughout. ABP = aterial blood pressure.

heart is relatively pronounced and the vascular effects appear to have reached a maximum plateau a moderate reflex bradycardia appears. In contrast, when the ascending aorta is partially occluded (Fig. 13.2, right panel), there is a marked bradycardia indicating that the reflex pathways subserving the heart rate response were intact. It is interesting to note that the bradycardia following aortic occlusion emerges despite the fact that the distension of the heart is now less pronounced than during the arterial–left arterial shunting. Aortic occlusion is apparently more effective in producing a reflex bradycardia response than an increased diastolic filling.

The observations of a sequential activation of different viscero-motor neurons when the diastolic filling is increased may indicate that the non-medullated receptors do not in fact constitute a homogeneous group of endings, but that at least two groups exist, differing with regard to their natural stimulus and with regard to their central neurogenic connections. Such a possible functional differentation does not necessarily imply two morphologically distinct groups of endings; it may be simply a matter of differences with regard to their location and orientation in the ventricular wall. Endings in the innermost wall layers, i.e. in or close to the endocardium, which will experience a more pronounced distension when the diastolic filling is augmented than receptors located towards the epicardial surface may, for instance, have their afferents mainly oriented towards central *vasomotor* neurons, implying that *vascular* adjustments are preferentially induced when the diastolic filling is gradually increased. Receptors located towards the epicardial surface in the muscle layer and which, therefore, possibly become activated primarily by events occurring during the contraction process as, e.g., the tension development, may, on the other hand, have afferent projections also towards the *vagal* motor nucleus. This implies that a reflex bradycardia is most likely to occur in situations where the contraction process is changed, as when the afterload is increased, during positive inotropic influences, etc.

If there is a receptor arrangement along the above lines it means that at least two types of reflex mechanisms originate from ventricular non-medullated endings. First, one reflex elicited from 'endocardial' receptors, operating mainly via the *vasomotor* neurons and then preferentially engaging the renal vascular bed, which may be ascribed a volume-regulating function. Secondly, one reflex governed by 'myocardial' receptors, brought into play, e.g., when the tension development in the myocardium indicates that an overloading of the heart is imminent. By inducing a vagal brake on the pump, and a vasodilatation, and hence a blood pressure fall, this reflex will cause an effective unloading of the pump and may be assigned a protective function.

Aspects on the central integration of the ventricular reflex mechanism

Since the ventricular receptors are activated whenever the left ventricle becomes distended, following, e.g., an increased outflow resistance, augmented diastolic filling or a myocardial ischaemia, it is to be expected that these receptors contribute to the circulatory adjustments arising in different physiological and pathophysiological states. One problem is then, if, and to what extent the response patterns, induced reflexly or from higher centres in these situations, are modified by a simultaneous input from the ventricular receptors. One may expect that the arterial baroreceptors and ventricular receptors work in concert to restore the blood pressure in situations when the pressure is severely elevated, although the two reflex mechanisms, as earlier mentioned, seem to utilise somewhat different cardiovascular effector organs to produce a blood pressure fall. It may be more interesting in this context to analyse how different *excitatory* influences on the cardiovascular system are modified when there is a simultaneous *inhibitory* input from the ventricular receptors. In a series of experiments the interaction between the ventricular receptor reflex mechanism on one side and the excitatory effects of the arterial *chemoreceptors* stimulation and of stimulation of the *hypothalamic defence area* on the other was explored.

Interaction with the chemoreceptor reflex

As shown by Thorén (1972) the ventricular receptors become activated during severe asphyxia and anoxia and then contribute significantly to the reflex bradycardia observed in these situations (Thorén, 1973). The profound slowing of the heart, seen during submersion in diving species, also seems to be in part caused by a stimulation of ventricular receptors (Blix, Wennergren & Folkow, 1976). In these situations the ventricular receptors and the arterial chemoreceptors thus seem to work synergistically in producing a powerful reflex bradycardia.

With regard to the vascular beds the two reflex mechanisms evidently work in opposite directions. Since a reduced tissue blood flow and oxygen consumption seems essential in situations with limited oxygen supply the reflex vasoconstriction from chemoreceptors is of advantage, while the vasodilator influence from the ventricular receptors is detrimental. To analyse the net effect of these two oppositely operating mechanisms on the vascular bed and to explore the capacity of the chemoreceptor reflex to maintain a strong vasoconstriction in face of the inhibitory input from ventricular receptors a study was performed on cats (Wennergren, Little & Öberg, 1976). In one series of experiments the heart rate and muscle vessel responses to hypoxia (inspired gas containing 3–6% oxygen) were studied before and after interruption of the afferent impulse traffic from the cardiac area. It was found that while the bradycardia response to hypoxia was essentially abolished after cardiac deafferentation, the vasoconstrictor responses were not significantly influenced by this procedure. This finding suggests that although endings in the heart, and then probably the ventricular receptors, are activated during hypoxia, they are not capable of modifying the vascular effects.

To explore this problem further another series of experiments were done in which cardiac and vascular responses to afferent cardiac nerve stimulation were studied and compared under normal conditions and when there was a simultaneous strong chemoreceptor excitation present. During chemoreceptor activation the vasodilator responses to cardiac nerve stimulation were found to be markedly attenuated (Fig. 13.3). With a strong chemoreceptor activation the vasodilator response was even totally abolished. In contrast, the reflex vasodilator effects of baroreceptor stimulation were much less affected by a simultaneous chemoreceptor activity (Fig. 13.3). A strong baroreceptor stimulation was, in fact, found to be capable of breaking through even an intense chemoreceptor activation so that identical reflex vascular responses were obtained, irrespective of whether the chemoreceptors were activated or not. The small reflex vascular effects to stimulation of the cardiac nerve during chemoreceptor stimulation is therefore not explained by a mere competition of excitatory and inhibitory influences on the peripheral effector but seems to indicate a central blocking mechanism. This blockade of the ventricular receptor influence on the vascular bed by a simultaneous chemoreceptor activation was, however, not seen in cases where the primary chemoreceptor bradycardia response was absent. It therefore seems as if the primary bradycardia response and the blockade of the reflex vasodilatation of ventricular receptor origin are two chemoreceptor reflex effects which are in some way intimately connected with each other. The net result of this interference between the chemoreceptor and ventricular receptor mechanisms is that when they operate jointly there is a powerful

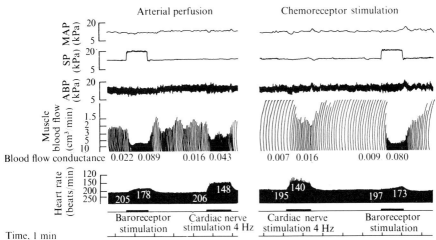

Fig. 13.3. Muscle vessel and heart rate responses to baroreceptor stimulation and cardiac nerve stimulation *without* (left panel) and *with* (right panel) a concomitant chemoreceptor stimulation. The blood pressure is kept essentially constant throughout. Note that the baroreceptor responses are basically similar in both situations, while the vessel response to cardiac nerve stimulation is markedly attenuated in the presence of a simultaneous chemoreceptor activity. MAP = mean aterial blood pressure; SP = sinus pressure; ABP = aterial blood pressure. (From Wennergren, Little & Öberg, 1976.)

bradycardia and a strong peripheral vasoconstriction, which is an advantageous circulatory adjustment during hypoxic states.

Interaction between the hypothalamic defence reaction and the ventricular receptor reflex mechanism

The complex adjustments of the cardiovascular system associated with the so-called *defence reaction* which can be elicited, e.g., by stimulation of the hypothalamic defence area, generally results in a prompt and marked increase of the arterial blood pressure. This pressure rise, in combination with the strong positive inotropic influence on the heart of a defence area stimulation, creates a situation where the ventricular receptors can be expected to be strongly excited. Recent preliminary experiments suggest that this is also the case (G. Wennergren, B. Lisander & P. Thorén, personal communication). Two essentially counteracting mechanisms are thus operating simultaneously, which raises the question of whether, and to what extent, the inhibitory ventricular receptor mechanism is capable of modifying the generally excitatory effects of a defence area stimulation.

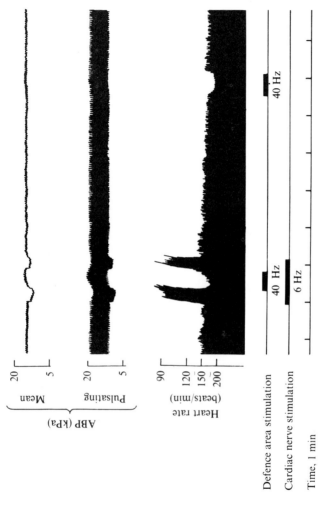

Fig. 13.4. Blood pressure and heart rate responses to a defence area stimulation, performed with and without a simultaneous afferent stimulation of the cardiac nerve. The cat is spinalised, leaving only the vagal control of heart rate intact. Note that the bradycardia response to cardiac nerve stimulation is reversed to a tachycardia by the defence area stimulation. ABP = arterial blood pressure. (From Wennergren, Lisander & Öberg, 1976.)

This problem was recently analysed in a series of experiments on anaesthetised cats (Wennergren, Lisander & Öberg, 1976). It was found that with regard to the heart rate responses a defence area stimulation, superimposed on a cardiac nerve stimulation, was capable of markedly attenuating or even eliminating the reflex bradycardia produced by cardiac ventricular receptor nerve stimulation so that the heart rate attained essentially the same level as when a defence area stimulation alone was present. This effect was most clearly seen in spinalised animals, where only the vagal control of heart rate remained (Fig. 13.4). This finding suggests that the vagal component of the reflex bradycardia elicited from cardiac receptors is effectively blocked by a concomitant defence area stimulation and that this blockade occurs at the central level and is not a mere competition between vagal and sympathetic influences on the peripheral effector, i.e. the sinus node.

The reflex inhibition of the sympathetic adrenergic outflow to the heart and blood vessels elicited from ventricular receptors was, on the other hand, still effectively operating during a defence area stimulation. The renal vasoconstriction as well as the tachycardia response (in atropinised animals) to defence area stimulation were thus significantly attenuated by a simultaneous afferent input from the cardiac receptors. With regard to the muscle vessels defence area stimulation and the ventricular receptor reflex mechanism, by means of an active cholinergic vasodilatation and sympathetic inhibition, respectively, worked synergistically to produce a virtual maximal vasodilatation. Like the arterial baroreceptors, the ventricular receptors thus modify the response pattern during a defence reaction in such a way that a high cardiac output and a maximal muscle blood flow is obtained, while excessive blood pressure elevations and work loads on the heart are prevented by the still efficiently operating reflex vasodilatory influences from the two receptor areas.

Cited studies were supported by grants from the Swedish Medical Research Counsil (SMF) No. 14X-644.

References

Blix, A.S., Wennergren, G. & Folkow, B. (1976). Cardiac receptors in ducks – a link between vasoconstriction and bradycardia during diving. *Acta physiol. scand.*, **97**, 13–19.
Little, R., Wennergren, G. & Öberg, B. (1975). Aspects of the central integration of arterial baroreceptor and cardiac ventricular receptor reflexes in the cat. *Acta physiol. scand.*, **93**, 85–96.

Mancia, G., Romero, J.C. & Shepherd, J.T. (1975). Continuous inhibition of renin release in dogs by vagally innervated receptors in the cardiopulmonary region. *Circulation Res., 36,* 529–35.

Öberg, B. & Thorén, P. (1973). Circulatory responses to stimulation of left ventricular receptors in the cat. *Acta physiol. scand., 88,* 8–22.

Thorén, P. (1972). Left ventricular receptors activated by severe asphyxia and by coronary artery occlusion. *Acta physiol. scand., 85,* 455–63.

Thorén, P. (1973). Reflex bradycardia elicited from left ventricular receptors during acute severe hypoxia in cats. *Acta physiol. scand., 87,* 103–12.

Thorén, P. (1974). Characteristics of left ventricular receptors with non-medullated vagal afferents. *Circulation, 50* (suppl. iii), 145.

Wennergren, G., Henriksson, B.-Å., Weiss, L.-G. & Öberg, B. (1976). Effects of stimulation of non-medullated cardiac afferents on renal water and sodium excretion. *Acta physiol. scand., 97,* 261–3.

Wennergren, G., Lisander, B. & Öberg, B. (1976). Interaction between the hypothalamic defence reaction and cardiac ventricular receptor reflexes. *Acta physiol. scand., 96,* 532–47.

Wennergren, G., Little, R. & Öberg, B. (1976). Studies on the central integration of excitatory chemoreceptor influences and inhibitory baroreceptor and cardiac receptor influences. *Acta physiol. scand., 96,* 1–18.

Discussion

(a) It was pointed out from the floor that there has recently been some evidence that rather severe haemorrhage in the awake animal does not lead to vasoconstriction but to vasodilatation; yet if the same animal is anaesthetised, haemorrhage results in a severe vasoconstriction. It was questioned whether studying these receptors in the anaesthetised animal shows what is really happening under more normal situations in unanaesthetised animals. Öberg replied that it was difficult to make more detailed studies on cardiac receptors in the unanaesthetised animal but there was evidence that nicotine applied to the epicardium had the same effect of bradycardia whether applied in the anaesthetised or the unanaesthetised dog. But of course anaesthesia can always change the reflex effects in one way or another.

(b) Hilton said there was some evidence that the chemoreceptor reflex, for instance, produces dilatation in skeletal muscle in unanaesthetised animals, not constriction. His group had recently been using a steroid anaesthetic in cats which seems to leave hypothalamic function intact and in addition, with that anaesthetic, chemoreceptor stimulation produces vasodilatation in skeletal muscle and not vasoconstriction. Öberg replied suggesting that the chemoreceptor mechanism is rather special in that it may relay to the hypothalamus, so responses from the chemoreceptor

reflex may change when the animal is decerebrate, but he said he did not know whether this was true also for the ventricular receptor mechanism.

(c) It was questioned from the floor whether the interpretation was correct that all the responses described resulted directly from stimulation of afferent fibres from the ventricle. It was pointed out that the extreme bradycardia must result in atrial dilatation. The blood must back up through the pulmonary vascular bed into the pulmonary artery as well so that the pressures through the low-pressure system, at least as far as the right ventricle, are raised, yet the particular pattern of efferent discharge, with a dilatation in the kidney rather more than the limb, was attributed to ventricular receptor stimulation. It was said that it was already known that stimulation of atrial receptors caused a decreased activity in efferent nerves to the kidney and that renal dilatation goes along with this. The additional evidence which has been given by Öberg is that when a shunt was produced there were changes in renal blood flow, but this shunt would also put pressure up in the left atrium. It was asked how was it possible to attribute that pattern of responses to ventricular receptors rather than all the other receptors that were being stimulated at the same time? In the discussion that followed the question was raised whether the renal vasodilatation was secondary to the bradycardia. Öberg said that the effects were undoubtedly mediated by non-medullated afferent fibres, including possibly non-medullated afferents from the atrium, and he pointed out that he had no evidence at all of vasodilatation in the kidney in response to stimulation of medullated fibres.

From the floor it was suggested that from the evidence so far presented in this symposium it was unlikely that the changes in pressure shown here would be sufficient to stimulate atrial receptors discharging to non-medullated fibres.

It was pointed out that in the experiments in which the diuresis, natriuresis and the renal blood flow conductance was followed the cardiac afferents were stimulated after giving atropine and the same increase in renal blood flow conductance was observed. This effect is therefore not secondary to the bradycardia.

14. Left ventricular reflexes during coronary arteriography

A. G. GARCIA-AGUADO & F. PEREZ-GOMEZ

Ventricular reflexes have been investigated in anaesthetised animals, mainly with open chests and using techniques such as ligature of the coronary arteries, injection of mercury into the root of the aorta, clamping of the aorta, etc. According to the results of these investigations the receptors are located throughout the whole left ventricle and they may be linked to type C fibres.

We studied these reflexes during coronary arteriography performed for diagnostic purposes. The 200 unselected cases investigated in our department were divided into groups according to their clinical diagnosis. The main groups being formed by patients with coronary artery disease, valvular heart disease, congestive cardiomyopathy, hypertrophic cardio-myopathy and normal cases. The greatest degree of bradycardia (longest P–P interval) achieved in each coronary injection was evaluated and correlated with the anatomical distribution and integrity of the coronary tree. In each case the origin of the sinus node artery and of the A–V node artery was determined. Another group of cases with normal coronary circulation was also studied.

We decided which artery provided the major part in the irrigation of the inferior wall of the left ventricle (so-called dominant artery) according to the following criteria: (a) it was considered that the right coronary artery was dominant in cases with classic right dominance without obstruction and also in those with left dominance but with blockage of the circumflex artery so that the blood supply came from the right coronary artery through collateral circulation. We were aware that in these cases, even when the right coronary artery was dominant, an important part of the irrigation to the inferior wall was provided by branches of the left coronary artery. (b) Similar criteria were used to diagnose the blood supply from the

293

left coronary artery to the inferior wall (left dominance). *(c)* It was considered that there were equal contributions from the two sides when there were two posterior descending arteries coming from each side or when there were bilateral obstructions and the blood supply to the inferior wall was provided by collaterals from either coronary.

Table 14.1 shows the decrease of sinus rate in different groups of patients. The decrease was greater in normal cases and in those with hypertrophic obstructive cardiomyopathy, and smaller during left coronary arteriography in those cases with poor left ventricular function (patients with congestive cardiomyopathy and valvular heart disease).

Table 14.1. *Mean decrease of heart rate (beats/min) caused by coronary arteriography*

	Right coronary injection	Left coronary injection
Hypertrophic obstructed cardiomyopathy	26.6	25.5
Normal	25.2	25.4
Ischaemic heart disease	15.7	18.4
Congested cardiomyopathy	18.2	10.5
Valvular heart disease	12.0	8.4

Fig. 14.1 shows the decrease in heart rate during injection into the dominant artery plotted against the decrease caused by injection into the non-dominant artery. The decrease of rate was closely related to the opacification of the artery which mainly irrigates the inferior wall of the left ventricle ($P < 0.001$; $t = 2.757$). The correlation was much better when we took only the cases with a left dominant artery and compared the slowing of sinus rate during injection into that artery with that obtained during injection into the non-dominent artery ($t = 4.196$).

In the group of cases with a dominant right coronary artery the correlation was poorer, although still significant ($P = 0.05$), but in these cases there is also irrigation to the inferior wall from the non-dominant artery (branches of the circumflex artery, posterolateral, marginals and sometimes from the anterior descending artery when it goes around the apex towards the inferior wall).

To clarify this point we studied a group of 25 cases taken from about 250

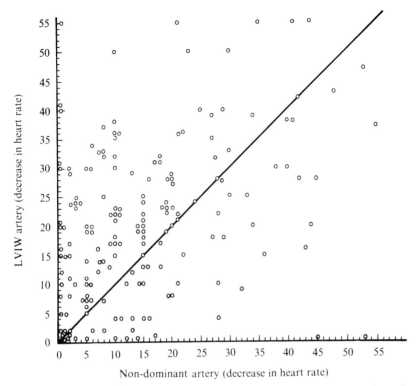

Fig. 14.1. Decrease of sinus rate during injection of contrast medium into the dominant artery (left ventricular inferior wall artery, LVIW artery; *y* axis) versus injection into the non-dominant artery (*x* axis). ($P=0.001$, $t=2.757$.)

cases with normal coronary arteries in whom the right coronary artery was clearly dominant, the size of the circumflex artery being small and the posterolateral artery distributed through the lateral wall of the ventricle and not through the postero-inferior wall (seen in the left lateral projection). In this group (Fig. 14.2*a*) the decrease of sinus rate was clearly related to the injection into the dominant right coronary artery ($P < 0.005$; $t=3.51$).

To evaluate the role of the A–V node artery we studied a group of 26 cases with equilibrated circulation which had the same blood supply from each side to the inferior wall of the left ventricle, normal coronary arteries and the A–V node artery originating from the right coronary artery (some were taken from the 200 cases previously described). The decrease of sinus rate (Fig. 14.2*b*) was similar during right and left coronary injections

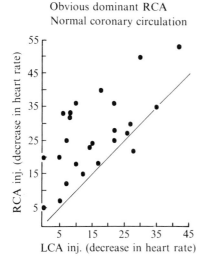

Obvious dominant RCA
Normal coronary circulation

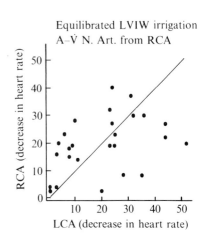

Equilibrated LVIW irrigation
A–V̇ N. Art. from RCA

Fig. 14.2. *(a)* Decrease of sinus rate during right coronary injection (RCA inj.) plotted against injection into left coronary artery (LCA inj.) ($P < 0.005$, $t = 3.51$, $n = 25$.) *(b)* Similar data to *(a)* in cases with equilibrated left ventricular inferior wall irrigation (LVIW irr.) ($t = -0.33$, $P = 0.75$; $r = 0.405$, $P = 0.023$). A–V N.Art. = A–V node artery.

($P = 0.75$), illustrating that the A–V node artery and the rest of the arteries do not influence the degree of bradycardia.

The main difference between this group (Fig. 14.2*b*) and the one previously described (Fig. 14.2*a*) is the pathway of the posterolateral artery which travels laterally in the group considered to have a clear dominant right coronary artery and posteriorly in the other group. The different behaviour of these groups during coronary arteriography also supports the hypothesis that the postero-inferior wall of the left ventricle is the origin of the reflexes caused by coronary arteriography.

According to several correlations we have made the degree of bradycardia was not related to the origin of the sinus node artery.

The incidence of sinus arrest longer than 2 s was also closely related to the injection of contrast medium into the dominant artery. Of 12 episodes, eight were caused by injection into the dominant artery and in two the circulation was such that there was the same blood supply from each side. In the other two sinus arrest appeared during injection into a non-dominant right coronary artery, but with large circumflex and posterolateral arteries. In only three cases did the sinus arrest appear during the injection into the artery which gave origin to the sinus node artery.

In summary. Ventricular reflexes have been studied in experimental animals, the response being bradycardia and peripheral vasodilatation. We studied the origin of these reflexes during coronary arteriography in human beings, using the degree of bradycardia as a measure of the response. According to our results the receptors are mainly located in the inferior wall of the left ventricle and they are not related to the areas supplied by the sinus node and A–V node arteries.

Several clinical findings appear to support our results: (1) the behaviour of an anterior cardiac infarction differs from that of an inferior infarction. An anterior infarction is usually followed by a sympathetic stimulation which can be assumed from the reports of a high degree of catecholamine discharge and good results of the treatment with β-blockers either in the acute phase or in long trials; we found a much higher degree of sinus rate than in inferior infarction and when hypotension appeared in the course of an anterior infarction it was normally associated with tachycardia and high peripheral vascular resistances as a result of a normal baroreceptor response. Later in this volume it will be shown that a partial occlusion of the anterior descending artery leads to sympathetic stimulation (see p. 353). (2) An inferior cardiac infarction tends to be followed by a parasympathetic discharge as judged from a tendency to observe bradycardia and lower catecholamine discharge. The trials with β-blocker treatment were found to be discouraging and when hypotension appears it is normally followed by bradycardia, vomiting and low peripheral vascular resistance as a result of an abolition of the normal baroreceptor response.

There is probably enough evidence to say that the ventricular reflexes may function differently according to their localisation and that the parasympathetic autonomic system plays as important a role in the initial evolution of an inferior cardiac infarction as it does in experimental ventricular reflexes.

Separation of arterial and vagal reflexes during changes in blood volume

V. S. BISHOP, D. F. PETERSON & H. O. STINNETT

from Department of Pharmacology, Health Science Center, University of Texas
San Antonio, Texas 78284, USA

Receptors in the cardiopulmonary region subserved by vagal afferent fibres have been shown to exert a restraint on the sympathetic adrenergic outflow. In addition it has been reported that the baroreflex sensitivity is reduced by volume loading. This study was designed to investigate the influence of acute volume expansion on the reflex heart rate and blood pressure responses to aortic nerve stimulation. In 24 anaesthetised rabbits stepwise increases in mean atrial pressure did not result in a significant change in heart rate or mean arterial pressure. However, increases in mean right atrial pressure resulted in a progressive diminution of the reflex heart rate and mean arterial pressure responses to aortic nerve stimulation with the threshold occurring at mean right arterial pressure increases of 0.1 kPa. When mean right arterial pressure was increased to 1.3 kPa the peak reflex and mean arterial pressure was attenuated by 44% and 52%, respectively. This attenuation of the reflex heart rate and mean arterial pressure responses to aortic nerve stimulation was eliminated by carotid sinus and aortic baroreceptor denervation. However, the attenuation was unaltered by bilateral vagotomy before arterial baroreceptor denervation. Bilateral vagotomy after arterial baroreceptor denervation did not alter the magnitude of reflex responses. These results suggest that the arterial baroreceptors can detect subtle changes in arterial pressure during acute volume loading, resulting in a diminished sympathetic efferent activity. Accordingly, the effects of sympathetic inhibition during subsequent aortic nerve stimulation would be less, causing smaller reflex changes in both heart rate and mean arterial pressure. The diminished sympathetic efferent activity may also explain the apparent lack of involvement of vagal afferents under these experimental conditions.

Electrophysiology and reflexes of 'sympathetic' afferents

15. Mechanisms of excitation of cardiac 'sympathetic' afferents

Y. UCHIDA

The heart and great vessels are innervated by baroreceptors and chemore-ceptors having afferent pathways in the vagus nerves. Recently, innervation of the heart and great vessels by receptors having afferent pathways in the cardiac sympathetic nerves has also been demonstrated (Brown, 1967; Ueda, Uchida & Kamisaka, 1969; Takenaka, 1970; Brown & Malliani, 1971; Uchida, Kamisaka & Ueda, 1971; Malliani, Recordati & Schwartz, 1973; Uchida, Kamisaka, Murao & Ueda, 1974; Uchida & Murao, 1974b; Uchida, 1975a). These afferent fibres in the cardiac sympathetic nerves have been called afferent sympathetic nerve fibres in order to differentiate them from those in the vagus nerves (Brown, 1967; Uchida *et al.,* 1974). Several workers have demonstrated that the afferent sympathetic nerve fibres from the heart and aorta have an action as a vasopressor (Malliani, Schwartz & Zanchetti, 1969; Brown & Malliani, 1971; Peterson & Brown, 1971; Uchida *et al.,* 1971; Lioy *et al.,* 1974). Also, their action as a nociceptor has been demonstrated in animals (Sutton & Lueth, 1930; Gutzman, Braun & Lim, 1962; Brown, 1967; Uchida *et al.,* 1971). In patients with coronary artery diseases angina pain can be eliminated by transection of the stellate ganglia through which the afferent sympathetic nerve fibres pass (White & Blard, 1948). It has been generally thought that these afferent fibres are excited during myocardial ischaemia and elicit pain in man and pseudo-affective response in animals. However, little is known about the mechanisms and modality of their excitation during myocardial ischaemia. In this chapter my report focuses mainly on the mechanisms and modality of excitation of afferent sympathetic nerve fibres of left ventricular origin.

Classification of afferent cardiac sympathetic nerve fibres
Conduction velocity of the afferent cardiac sympathetic nerve fibres was

measured in cats by Brown (1967) and in dogs by Uchida *et al.* (1974). Brown stimulated the thoracic communicating rami and recorded compound action potentials in the cardiac nerves and observed compound action potentials that had conduction velocities less than 25 m/s. He classified this action potential as A_δ fibres according to the method described by Erlanger & Gasser (1937). He divided this compound action potential into two subgroups; A_{δ_1} and A_{δ_2}. Conduction velocities of A_{δ_1} ranged from 15 to 25 m/s and those of A_{δ_2} from 9 to 14 m/s. Uchida *et al.* (1974) stimulated the middle cardiac nerve or ansa subclavia in dogs and recorded action potentials evoked from the thoracic communicating rami.

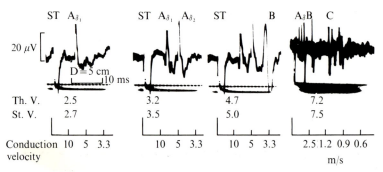

Fig. 15.1. Electrically evoked action potentials in the cardiac nerves. Stimulating electrodes: cut central end of the middle cardiac nerve. Recording electrodes: cut peripheral end of the third thoracic communicating ramus of the left side. Th.V. = threshold voltage for each group of fibres. St.V. = stimulation voltage. First infection (ST) in each photograph is stimulation artifact. From the left: A_{δ_1}; A_{δ_1} and A_{δ_2}; A_{δ_1}, A_{δ_2} and antidromic B potentials. In the last group (labelled C) action potentials of afferent C fibres are included. Conduction velocities are at the bottom of each photograph.

The action potentials contained three distinctly separated elevations. The conduction velocities for the fastest elevation ranged from 8 to 26 m/s. This elevation was classified as A_δ. This elevation was further divided into A_{δ_1} and A_{δ_2}. Following B elevation, this author found a more slowly propagating diphasic elevation with a conduction velocity less than 2 m/s, the C elevation. Conduction velocities for C ranged from 0.3 to 2 m/s (Fig. 15.1). In addition to A_δ and C potentials, Uchida also recorded the action potentials of the fibres whose conduction velocities dropped between 2 and 4.5 m/s (Uchida *et al.,* 1974). However, classification of this group of fibres is still controversial (Paintal, 1954).

Mechanosensitivity of afferent cardiac sympathetic nerve fibres

Mechanical stimulation of the cardiac surface in dogs can cause excitation of a certain group of both A_δ and C fibres (Fig. 15.2). Elevation of left ventricular pressure can also cause excitation of those which respond to mechanical stimulation of the cardiac surface (Ueda *et al.*, 1969; Uchida *et al.*, 1974). In addition to these mechanosensitive fibres the A_δ and C fibres which did not respond to mechanical stimuli but to myocardial ischaemia and/or chemical stimuli were found in dogs and cats (Takenaka, 1970; Brown & Malliani, 1971; Uchida & Murao, 1974*a, c,* 1975*a*). However, it remains to be elucidated whether the difference in mechanosensitivity is caused by the difference in character of the nerve endings or the difference in the environments in which the endings are embedded.

In dogs the majority of mechanosensitive A_δ and C fibres of ventricular origin fires synchronously to ventricular wall motion in response to tapping

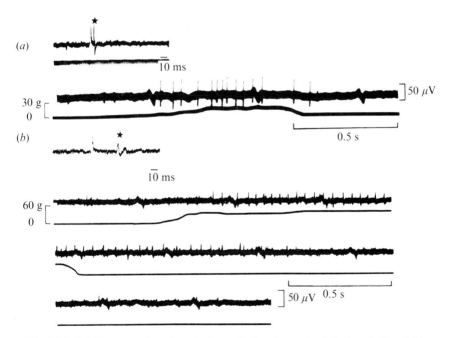

Fig. 15.2. *(a)* Upper tracing: electrically evoked action potential of an A_δ fibre (★). Lower tracing: action potentials evoked in same nerve filament by tapping anterior interventricular groove. Tapping is shown by strain gauge arch recording of pressure in g on receptive field. *(b)* Top tracing: electrically evoked action potential of a C fibre (★). Second to fourth tracings: action potentials evoked in same nerve filament by tapping anterior wall of left ventricle.

of the ventricular surface or elevation of ventricular pressure (Ueda *et al.,* 1969; Uchida *et al.,* 1974). Spontaneous discharge has been observed in both A_δ and C fibres in dogs (Uchida *et al.,* 1974; Uchida & Murao, 1974*b*, 1975*a*). In general, the A_δ fibres fire regularly in close relationship to each cardiac cycle while the C fibres fire irregularly and independently (Fig. 15.3.). Respiratory modulation of firing has frequently been observed in the fibres, especially of low pressure chamber origin (Uchida & Murao, 1974*b*, 1975*a*). Ventricular premature beats also modify the spontaneous discharge. In general, discharge frequency increases during the diastole of the premature beat and/or during post-extrasystolic contraction, indicating that augmented distension and/or contraction of the ventricular wall are responsible for the increased discharge frequency (Uchida *et al.,* 1974; Uchida & Murao, 1975*a*).

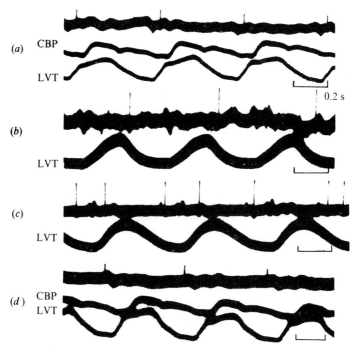

Fig. 15.3. Spontaneous firing of A_δ fibres and its relation to cardiac cycle. CBP = peripheral coronary blood pressure measured from a catheter which was introduced in retrograde fashion into a small peripheral branch of the coronary artery. LVT = left ventricular tension in g force measured from a force-displacement strain gauge arch sewn to the left ventricular wall adjacent to the receptive fields. Upward motion of LVT indicates left ventricular contraction.

Distribution of the receptive fields of afferent cardiac sympathetic nerve fibres

Distribution of the receptive fields has been examined in dogs by Uchida and his colleagues by tapping the cardiac surfaces and great vessels (Ueda *et al.*, 1969; Uchida & Murao, 1974*b*; Uchida, 1975*a, b*). They observed the receptive fields of both A_δ and C fibres in all four chambers of the heart, caval veins, pulmonary artery, aorta and pleurae (Fig. 15.4).

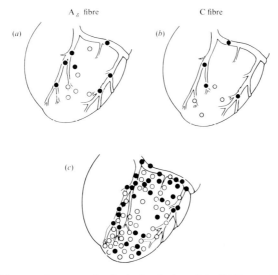

Fig. 15.4. Fields receptive to mechanical stimulation. *(a), (b)* Receptive fields of A_δ and C fibres. *(c)* Receptive fields of fibres whose conduction velocity was not measured. ○, Spontaneously discharging fibres; ●, fibres silent except when stimulated mechanically.

Responses of afferent cardiac sympathetic nerve fibres to chemical stimuli

Distilled water and solutions of sodium chloride in concentrations below 900 $\mu g \cdot cm^{-3}$ and over 18 $mg \cdot cm^{-3}$ excite the afferent fibres of left ventricular origin (Uchida & Murao, 1974*a*). Potassium chloride in a concentration of 100 $\mu g \cdot cm^{-3}$ or over also caused excitation of the fibres when it was dripped to the cardiac surface (Uchida & Murao, 1974*a*). This concentration was the same as that required for activation of the carotid chemoreceptors (Eyzaguirre & Koyano, 1965) and it is also close to the concentration achieved during myocardial ischaemia (Haddy & Scott,

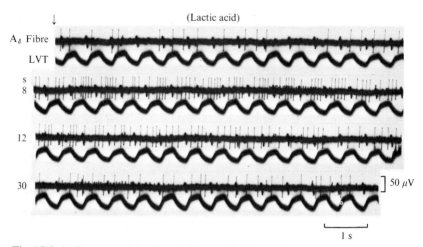

Fig. 15.5. Action potentials of an A_δ fibre evoked by application of 750 $\mu g \cdot cm^{-3}$ of lactic acid to the anterior wall of the left ventricle. LVT = left ventricular tension.

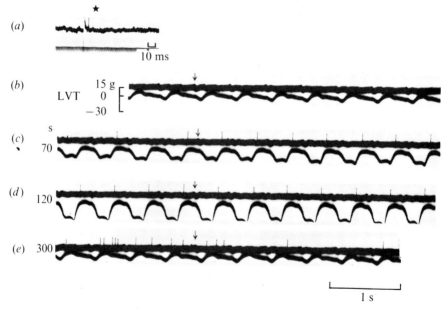

Fig. 15.6. *(a)* Electrically evoked action potential of an A_δ fibre (★). Conduction time for response to mechanical stimulation = 30 ms. *(b)–(d)* Control; 70 and 120 s after coronary occlusion; and *(e)* 300 s after release of coronary occlusion. ↓ Indicates QRS of ECG. Downward deflection of left ventricular tension (LVT) following QRS indicates systolic bulge of the left ventricle.

1971). Bradykinin in a concentration of 50 ng·cm^{-3} and lactic acid in a concentration of 7.5 μg·cm^{-3} also caused excitation of the afferent fibres (Uchida & Murao, 1974a, d) (Fig. 15.5). The concentrations of these agents increases much more during myocardial ischaemia (Conn, Wood & Morales, 1959).

Excitation of afferent cardiac sympathetic nerve fibres during coronary occlusion

Excitation of afferent cardiac sympathetic nerve fibres during coronary occlusion has been demonstrated in cats (Brown, 1967) and in dogs (Uchida & Murao, 1974c). In general, the A$_\delta$ fibres fired in synchrony with

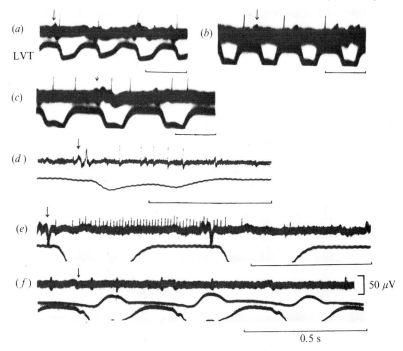

Fig. 15.7. Relations of afferent discharge to cardiac cycle during coronary occlusion. *(a)* An A$_\delta$ fibre. Conduction time = 25 ms. Excitation is therefore at beginning of passive stretching of ischaemic left ventricular wall (systolic bulge). *(b)* An A$_\delta$ fibre. Conduction time = 54 ms. Shortening of ischaemic left ventricular wall is responsible for discharge. *(c)* An A$_\delta$ fibre. Conduction time = 35 ms. *(d)* An A$_\delta$ fibre. Conduction time = 35 ms. *(e)* An A$_\delta$ fibre which did not respond to mechanical stimuli. Continuous discharge from beginning of a cardiac beat to next beat. This discharge occurred at each three cardiac beats. *(f)* A C fibre. Discharge was irregular. Arrows indicate QRS of ECG. LVT = left ventricular tension.

Table 15.1. *Relationship between cardiac cycle and discharge of afferent fibres during coronary occlusion*

| | Firing patterns | | | | | |
	A	B	C	D	E	Miscellaneous
A_δ fibres	2	3	1	2	—	—
C fibres	—	—	—	5	—	—
Unclassified fibres	4	7	3	9	9	5

passive stretching of the ventricular wall during systole, the systolic bulge, while the C fibres fired irregularly and independently (Figs 15.6, 15.7 and Table 15.1). The difference in the relationship between discharge and ventricular wall motion indicates that unphysiological motion of the ischaemic ventricular wall played an important role in the excitation of the A_δ fibres but it was not responsible for excitation of the C fibres.

In order to gain insight into the underlying mechanism for excitation of the afferent fibres during coronary occlusion the effects of intravenous injections of several agents on afferent discharge were examined. Trasylol, which blocks the formation of bradykinin, could not suppress the afferent discharge induced by coronary occlusion. On the other hand sodium bicarbonate, a buffer of acids, effectively suppressed the discharge of the C fibres but not the discharge of the A_δ fibres (Uchida & Murao, 1975a). This fact indicates that acidosis or H^+ played an important role in excitation, induced by myocardial ischaemia, of the C fibres. Acetylsalicylic acid suppressed excitation of both groups of fibres. This agent antagonises with bradykinin and blocks the synthesis of prostaglandins. However, it remains unclear whether this agent suppressed excitation through antagonising with bradykinin or through suppression of prostaglandin synthesis (Uchida & Murao, 1975a).

Nitroglycerine and propranolol are the most widely used anti-anginal agents. Since a certain group of afferent cardiac sympathetic nerve fibres participate in nociception of the heart the effects of these agents on excitation of the afferent fibres during coronary occlusion were examined. Intravenous injections of nitroglycerine in a dose of 30 $\mu g \cdot kg^{-1}$ or over effectively suppressed excitation of both A_δ and C fibres. Suppression of the discharge occurred with improvement of left ventricular wall motion. The duration of the beneficial effect, however, was up to 4 min (Uchida & Murao, 1974e). Intravenous injection of propranolol in a dose of 500

$\mu g \cdot kg^{-1}$ also suppressed excitation of both groups of fibres. Since the number of the action potentials per second was reduced while the number of action potentials per cardiac beat was unchanged propranolol may have suppressed excitation through reduction in heart rate (Uchida & Murao, 1974*f*).

Effect of partial constriction of coronary artery on afferent cardiac sympathetic nerve fibres

Excitation of both A_δ and C fibres was produced not only by coronary occlusion but also by partial constriction of the coronary artery. When mean peripheral coronary blood pressure was reduced below 50% of the

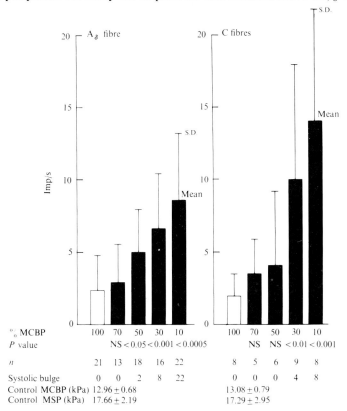

% MCBP	100	70	50	30	10		100	70	50	30	10
P value		NS	<0.05	<0.001	<0.0005			NS	NS	<0.01	<0.001
n	21	13	18	16	22		8	5	6	9	8
Systolic bulge	0	0	2	8	22		0	0	0	4	8
Control MCBP (kPa)	12.96 ± 0.68						13.08 ± 0.79				
Control MSP (kPa)	17.66 ± 2.19						17.29 ± 2.95				

Fig. 15.8. Relationship between discharge frequency of afferent fibres and severity of coronary artery constriction. Severity of coronary artery constriction was expressed as percentage of mean peripheral coronary blood pressure (% MCBP). MSP = mean systolic blood pressure; n = number of fibre preparations; NS = not significant; s.d. standard deviation.

control value by constriction discharge frequency of both A_δ and C fibres increased. Since 50% mean peripheral coronary blood pressure corresponded to a decrease of 18% in coronary blood flow in another series of experiments (Uchida, 1977) it was considered that the afferent cardiac sympathetic nerve fibres can respond to mild myocardial ischaemia (Fig. 15.8).

Roles of extracardiac factors in the regulation of afferent cardiac sympathetic nerve fibres

In clinical conditions the inciting factors of anginal attacks are still not well known. Therefore it is important to clarify the factors which may lead to excitation of the afferent cardiac sympathetic nerve fibres. In our study mean peripheral coronary blood pressure was reduced stepwise by constriction and a brief electrical stimulation was applied to the efferent cardiac sympathetic nerves. When mean peripheral coronary blood pressure was lowered below 50% of the control value a brief electrical stimulation of the efferent cardiac sympathetic nerves resulted in further excitation of the afferent fibres of left ventricular origin. Excitation of the fibres occurred after cessation of stimulation and continued for up to several minutes. In the same condition the stimulation resulted in a decrease of blood flow in the partially constricted coronary artery and the development of systolic bulge of the left ventricle. This decrease in coronary blood flow was eliminated by pretreatment with phentolamine. The result indicates that contraction of the constricted coronary artery caused by excitation of α-adrenergic receptors, and an increased work load and energy requirement caused by excitation of β_1-adrenergic receptors in the myocardium resulted in more severe myocardial ischaemia and excitation of the afferent fibres (Uchida & Murao, 1975b). The same phenomenon was also produced by a brief electrical stimulation of the cervical vagus nerves (Uchida et al., 1975b).

Mechanical stimulation of the carotid sinus also modified activity of the afferent fibres when it was performed during partial constriction of the coronary artery. Before coronary constriction, elevation of carotid sinus pressure to 26.7 kPa could not produce an obvious influence on the activity of the afferent fibres. However, the same stimulation suppressed afferent fibre activity when it was applied during partial constriction of the coronary artery (Fig. 15.9). This fact indicates that carotid sinus stimulation may be beneficial for the suppression of anginal pain in man and the pseudo-affective response in animals caused by myocardial ischaemia.

Cyclical excitation of afferent cardiac sympathetic nerve fibres

Slow spontaneous cyclical reductions in blood pressure and flow of the partially constricted coronary artery frequently occurs. This phenomenon is independent of the changes in systemic blood pressure and heart rate and more frequently occurs in female dogs (Uchida & Ueda, 1973; Uchida & Murao, 1974*g*; Uchida, Yoshimoto & Murao, 1975*a*; Uchida, 1977). It is also accompanied by elevation of the ST segment of the electrocardiogram and systolic bulge of the left ventricle when it occurs at low pressure or flow levels (Uchida *et al.*, 1975*a*), which closely resembles Prinzmetal's variant form of angina pectoris since cyclical ST elevation and vasospasm of the coronary artery are frequently observed in both (Prinzmetal *et al.*, 1959; Dhurendahl *et al.*, 1972; Murao, Harumi & Mashima, 1974; Uchida,

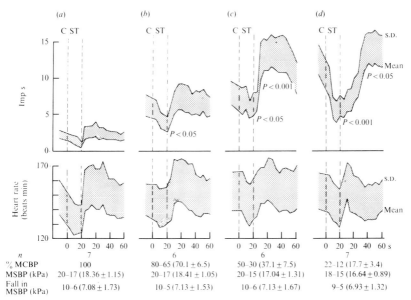

Fig. 15.9. Effect of elevation of the right carotid sinus pressure to 26.7 kPa on discharge frequency, heart rate and systemic blood pressure before and during constriction of coronary artery. Severity of coronary artery constriction was expressed as % MCBP. MSBP = mean systemic blood pressure before carotid stimulation. % MCBP = percentage mean peripheral coronary blood pressure before carotid stimulation. *(a)* Before coronary constriction; *(b)–(c)* during partial constriction; *(d)* during occlusion. C = Discharge frequency and heart rate before carotid stimulation; ST = carotid stimulation for 20 s. Time in s at the bottom of each panel indicates the time from cessation of stimulation.

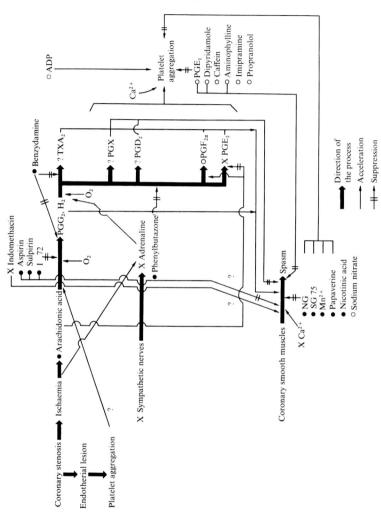

Fig. 15.10. Mechanisms for coronary artery spasm induced by partial constriction. ●, The agents which eliminated the cyclical reduction in blood flow of partially constricted coronary artery; ○, the agents which did not affect the cyclical reduction; ×, the agents which augmented the cyclical reduction; ?, the agents whose action on the cyclical reduction was not examined. PG = prostaglandin.

Inoue, Yoshimoto & Murao, 1975). This phenomenon was suppressed by intravenous injection of aspirin (25 mg·kg^{-1}), sulpirin (100 mg·kg^{-1}), I 72 (2.5 mg·kg^{-1}), benzydamine (1 mg·kg^{-1}) and phenylbutazone (1 mg·kg^{-1}), which suppress prostaglandin synthesis. Among these agents, I 72 has no action on platelet aggregation, and no anti-inflammatory action. This phenomenon was also suppressed by the injections of nitroglycerine (60 μg·kg^{-1}), SG 75 (100 μg·kg^{-1}), Mn^{2+} (18 mg·kg^{-1}), papaverine (500 μg·kg^{-1}) and nicotinic acid (20 mg·kg^{-1}). Among these agents, nicotinic acid has no action on platelet aggregation. On the other hand, prostaglandin E$_1$, dipyridamole, caffeine and imipramine, which suppress platelet aggregation, and imidazole and 1-methyl-imidazole, which inhibit synthesis of thomboxane A$_2$, could not suppress this phenomenon (Uchida, 1977; Yoshimoto, Uchida & Murao, 1976). The cyclical reduction in pressure and flow which was suppressed with phenylbutazone and benzydamine, was reproduced with prostaglandin E$_2$ but not with prostaglandin F$_{2\alpha}$. In the preparations in which the cyclical reduction did not appear spontaneously, the injections of adrenaline in a

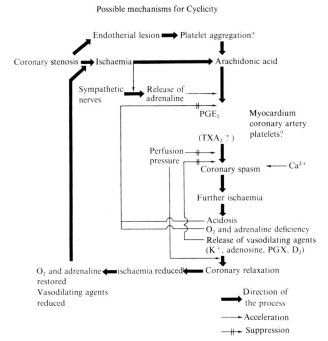

Fig. 15.11. Mechanisms for cyclicity of the changes in coronary blood flow of partially constricted coronary artery. PG = prostaglandin.

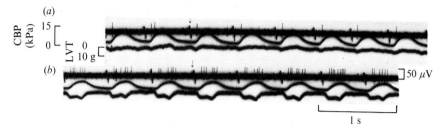

Fig. 15.12. Cyclical excitation of an A_δ fibre in synchrony with the cyclical reduction in blood pressure of partially constricted coronary artery *(a)* during spontaneous rise in blood pressure of partially constricted coronary artery; *(b)* during spontaneous fall in blood pressure of partially constricted coronary artery. ↓ Indicates QRS of ECG. CBP = coronary blood pressure; LVT = left ventricular tension.

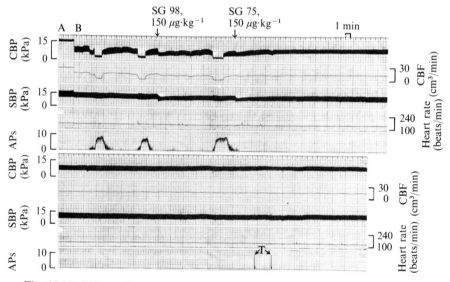

Fig. 15.13. Effects of two coronary vasodilating agents on cyclical reduction in blood pressure and flow of partially constricted artery and cyclical excitation of an A_δ fibre. Upper half is followed by lower half. A: before coronary artery constriction. B: during coronary artery constriction. From top channel in both upper and lower halves: peripheral coronary blood pressure (CPB), coronary blood flow (CBF), systemic blood pressure (SBP), heart rate and integrated action potentials (APs) of an A_δ fibre. The agents were injected intravenously at ↓. T = tapping receptive field of the fibre.

dose of 1 $\mu g \cdot kg^{-1}$ or electrical stimulation of the left stellate ganglion induced the cyclical reduction. On the other hand, noradrenaline, isoprenaline, methoxamine, thrombine, Cu-EDTA and ADP could not induce the cyclical reduction. These findings indicate that (1) ischaemia-induced synthesis of prostaglandins, especially prostaglandin E_2 which acts as a constrictor of the coronary smooth muscles, leads to vasospasm of the partially constricted coronary artery; (2) adrenaline acts as an accelerator of prostaglandin synthesis; and (3) that platelet aggregation itself does not play an important role in this phenomenon (Figs 15.10 and 15.11).

During the course of cyclical reduction in blood pressure and flow of the partially constricted coronary artery, excitation of the afferent sympathetic nerve fibres also occurred. Excitation of the fibres always occurred during the reduction in coronary blood pressure and flow as shown in Figs 15.12 and 15.13 and was suppressed by the intravenous injections of nitroglycerine ($60 \mu g \cdot kg^{-1}$), SG 75 ($100 \mu g \cdot kg^{-1}$), Mn^{2+} ($18 \, mg \cdot kg^{-1}$), aspirin ($25 \, mg \cdot kg^{-1}$), sulpirin ($100 \, mg \cdot kg^{-1}$), phenylbutazone ($1 \, mg \cdot kg^{-1}$) and benzydamine ($1 \, mg \cdot kg^{-1}$).

References

Brown, A.M. (1967). Excitation of afferent cardiac sympathetic nerve fibres during myocardial ischemia. *J. Physiol.*, **190**, 34–53.

Brown, A.M. & Malliani, A. (1971). Spinal sympathetic reflexes initiated by coronary receptors, *J. Physiol.*, **212**, 685–93.

Conn, H.L., Jr, Wood, J.C. & Morales, G.S. (1959). Rate of change in myocardial glycogen and lactic acid following arrest of coronary circulation. *Circulation Res.*, **7**, 721–7.

Dhurendahl, R.W., Watt, D.L., Silver, M.D., Trimble, A.S. & Adelman, A.G. (1972). Prinzmetal's variant form of angina with arteriographic evidence of coronary artery spasm. *Am. J. Cardiol.*, **30**, 902–4.

Erlanger, J. & Gasser, H.S. (1937). *Electrical signs of nervous activity.* University of Pennsylvania Press: Philadelphia.

Eyzaguirre, C. & Koyano, H. (1965). Effects of pharmacological agents on chemoreceptor discharges. *J. Physiol.*, **178**, 410–37.

Gutzman, F., Braun, C. & Lim, R.K. (1962). Visceral pain and pseudoaffective response to intra-arterial injection of bradykinin and other agents. *Arch. int. Pharmacodyn. Thér.*, **136**, 353–84.

Haddy, F.J. & Scott, J.B. (1971). Bioassay and other evidence for participation of chemical factors in local regulation of blood flow. *Circulation Res.*, **28** (suppl. I), 86–92.

Lioy, F., Malliani, A., Pagani, N., Recordati, G. & Schwartz, P.J. (1974). Reflex hemodynamic responses initiated from the thoracic aorta. *Circulation Res.*, **34**, 78–84.

Malliani, A., Recordati, G. & Schwartz, P.J. (1973). Nervous activity of afferent cardiac sympathetic nerve fibres with atrial and ventricular endings. *J. Physiol.*, **229**, 457–69.

Malliani, A., Schwartz, P.J. & Zanchetti, A. (1969). A sympathetic reflex elicited by experimental coronary occlusion. *Am. J. Physiol.*, **217**, 703–9.

Murao, S., Harumi, K. & Mashima, S. (1974). Clinical study on the mechanisms of Prinzmetal's variant angina. *J. Jap. Soc. internal med.*, **63**, 751.

Paintal, A.S. (1954). A comparison of nervous impulses of mammalian non-medullated nerve fibres with those of the smallest diameter medullated fibres. *J. Physiol.*, **126**, 255–70.

Peterson, D.F. & Brown A.M. (1971). Pressor reflexes produced by stimulation of afferent fibres in the cardiac sympathetic nerves of the cat. *Circulation Res.*, **28**, 605–10.

Prinzmetal, M., Kennamer, R., Merliss, R., Wada, T. & Bor, N. (1959). Angina pectoris. 1. A variant form of angina pectoris. *Am. J. Med.*, **27**, 375–88.

Scott, J.B. & Radawski, D. (1971). Role of hyperosmolarity in the genesis of active and reactive hyperemia. *Circulation Res.*, **28** (suppl. I), 26–32.

Sutton, D.C. & Lueth, H.C. (1930). Experimental pain production on excitation of the heart and great vessels. *Archs intern. Med.*, **45**, 827–67.

Takenaka, N. (1970). Afferent discharges in the left cardiac sympathetic nerve induced by asphyxia in the cat. *Jap. Heart J.*, **11**, 550–8.

Uchida, Y. (1975*a*). Afferent sympathetic nerve fibers with mechanoreceptors in the right heart. *Am. J. Physiol.*, **228**, 223–30.

Uchida, Y. (1975*b*). Afferent aortic nerve fibers with their pathways in cardiac sympathetic nerves. *Am. J. Physiol.*, **228**, 990–5.

Uchida, Y. (1977). Neural control of abnormal coronary circulation. In *Kawaguchi Lake symposium on coronary circulation*, ed. K. Nakamura, pp. 20–40. Japan Boeringersohn Ltd: Japan.

Uchida, Y., Inoue, K., Yoshimoto, N. & Murao, S. (1975). Angiographic demonstration of experimental coronary artery spasm. *J. Jap. Coll. Angiol.*, **15**, 100.

Uchida, Y., Kamisaka, K., Murao, S. & Ueda, H. (1974). Mechanosensitivity of afferent cardiac sympathetic nerve fibers. *Am. J. Physiol.*, **226**, 1088–93.

Uchida, Y., Kamisaka, K. & Ueda, H. (1971). Anginal pain: excitation of the afferent cardiac sympathetic nerve. *Jap. Circulation J.*, **35**, 147–61.

Uchida, Y. & Murao, S. (1974*a*). Potassium-induced excitation of afferent cardiac sympathetic nerve fibers. *Am. J. Physiol.*, **226**, 603–7.

Uchida, Y. & Murao, S. (1974*b*). Afferent sympathetic nerve fibers originating in left atrial wall. *Am. J. Physiol.*, **227**, 753–7.

Uchida, Y. & Murao, S. (1974*c*). Excitation of afferent cardiac sympathetic nerve fibers during coronary occlusion. *Am. J. Physiol.*, **226**, 1094–9.

Uchida, Y. & Murao, S. (1974*d*). Bradykinin-induced excitation of afferent cardiac sympathetic nerve fibers. *Jap. Heart J.*, **15**, 84–91.

Uchida, Y. & Murao, S. (1974*e*). Nitroglycerine-induced suppression of excitation of afferent cardiac sympathetic nerves during myocardial ischemia. *Jap. Heart J.*, **15**, 154–65.

Uchida, Y. & Murao, S. (1974*f*). Effect of propranolol on excitation of afferent sympathetic nerve fibers during myocardial ischemia. *Jap. Heart J.*, **15**, 280–8.

Uchida, Y. & Murao, S. (1974*g*). Cyclic changes in peripheral blood pressure of partially constricted coronary artery. *J. Jap. Coll. Angiol.*, **14**, 383.

Uchida, Y. & Murao, S. (1975*a*). Acid-induced excitation of afferent cardiac sympathetic nerve fibres. *Am. J. Physiol.*, **228**, 27–33.

Uchida, Y. & Murao, S. (1975*b*). Sustained decrease in coronary blood flow and excitation of cardiac sensory fibers following sympathetic stimulation. *Jap. Heart J.,* **16,** 265–79.

Uchida, Y. & Ueda, H. (1973). Experimental studies on anginal pain. *Annual Report of Japan Cardiovascular Foundation.*

Uchida, Y., Yoshimoto, N. & Murao, S. (1975*a*). Cyclic fluctuations in coronary blood pressure and flow induced by coronary artery constriction. *Jap. Heart J.,* **16,** 454–64.

Uchida, Y., Yoshimoto, N. & Murao, S. (1975*b*). Excitation of afferent cardiac sympathetic nerve fibers induced by vagal stimulation. *Jap. Heart J.,* **16,** 548–63.

Ueda, H., Uchida, Y. & Kamisaka, K. (1969). Distribution and responses of the cardiac sympathetic receptors to mechanically induced circulatory changes. *Jap. Heart J.,* **10,** 70–81.

White, J.C. & Blard, E.F. (1948). Surgical relief of severe angina pectoris; methods employed and results in 83 patients. *Medicine,* **27,** 1–7.

Yoshimoto, N., Uchida, Y. & Murao, S. (1976). Effects of anti-anginal agents and Ca^{2+} on minute rhythm of the partially constricted coronary artery. *Jap. Circulation J.,* **40,** 85.

Discussion

(a) It was suggested that the application of chemicals to the epicardial surface probably affects the properties of the fibres rather than the endings. However, Brown pointed out that from a functional rather than a biophysical point of view it may not matter.

(b) There was also some concern about the method of localising the nerve endings. Strong probing might be misleading and it was thought to be necessary to open the heart and probe carefully because it was felt to be important to study only fibres which had a known origin.

(c) It was thought to be too early to correlate results from animals with those from patients with angina. For example, injection of contrast media resulted in arterial spasm and could not be regarded as a specific stimulus.

16. Afferent cardiovascular sympathetic nerve fibres and their function in the neural regulation of the circulation

A. MALLIANI

The classic conception of the organisation of the neural control of circulation implies an all-important function of the supraspinal centres. Accordingly, the only afferent nervous activity from cardiovascular receptors that is considered essential to this regulatory function is the one conveyed to the medulla, while the role of the afferent nerve fibres projecting from the cardiovascular system to the spinal cord has been thought to be restricted to the transmission of pain (Sutton & Lueth, 1930; White, 1957), i.e. to signalling abnormal events. In brief, although for many years sympathetic reflexes have been known to exist in spinal animals (Sherrington, 1906), they have not been attributed a fundamental role comparable to that recognised for somatic spinal reflexes.

The general endeavour of our research during these last years has been, instead, to discover the possible similarities existing in the spinal organisation of the somatic and visceral nervous system.

The first attempt made was to explore the possibility of eliciting reflexes from the heart through an afferent sympathetic limb. We found that a transient occlusion of a coronary artery could alter the firing frequency, most often toward excitation, of preganglionic sympathetic fibres in animals with either an intact central nervous system or with a cervical spinal section and bilateral vagotomy: a sympatho-sympathetic spinal reflex was thus described (Malliani, Schwartz & Zanchetti, 1969). Furthermore, as the single fibres were isolated from a sympathetic ramus communicans (T3), a great many fibres of which participate in the efferent innervation of the heart, we hypothised the existence of spinal cardio-cardiac reflexes.

This finding raised the more general problem about the further possibility that natural haemodynamic events could also elicit sympathetic reflexes in spinal animals.

We demonstrated that arterial blood pressure rises, sometimes of very moderate magnitude (2.0–2.7 kPa), could reflexly modify the preganglionic sympathetic impulse activity in spinal vagotomised animals: reflex responses were either excitatory or inhibitory and were consistent for each individual fibre (Malliani, Pagani, Recordati & Schwartz, 1971).

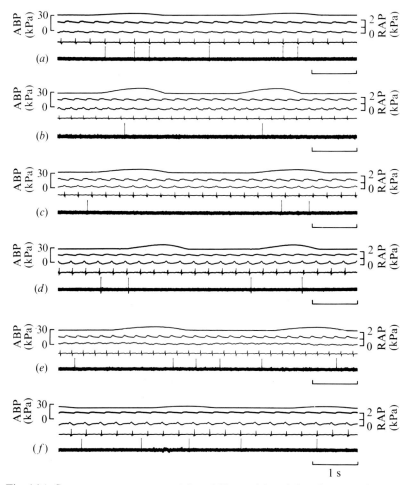

Fig. 16.1. Spontaneous nervous activity of fibres with atrial endings. Each record shows the activity of a different fibre. Fibres illustrated in *(a)*, *(d)*, and *(e)* had their endings in the right atrium; fibres in *(b)*, *(c)*, and *(f)* had their endings in the left atrium. Tracings, from top to bottom, represent the endotracheal pressure (inflations upwards), the arterial blood pressure (ABP), the right atrial pressure (RAP), the ECG and the nervous recording. (From Malliani, Recordati & Schwartz, 1973*b*, unpublished observations.)

The following obvious step for our experiments had to be a search for cardiovascular receptors and their afferent sympathetic fibres which mediated these reflexes. We began by investigating the properties of afferent sympathetic fibres with cardiac sensory endings. In a first study (Malliani & Brown, 1970; Brown & Malliani, 1971) we selected fibres which were excited during increases in pressure and flow in the perfused left coronary artery; in addition they were excited during interruptions of left coronary perfusion leading to ischaemia. Those fibres had an irregular or no spontaneous impulse activity.

In a subsequent study (Malliani, Recordati & Schwartz, 1973*b*) we purposely searched for afferent cardiac sympathetic fibres which displayed a spontaneous impulse activity: it occurred in phase with normal cardiac

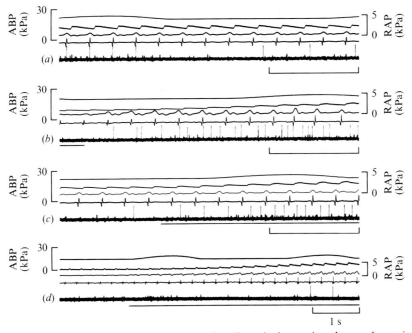

Fig. 16.2. Activity of a fibre with left atrial endings during various haemodynamic stimuli. *(a)* Spontaneous activity; *(b)* effects of releasing a constriction of the inferior vena cava (end of constriction marked by the end of a bar below the left side of the record); *(c)* constriction of the thoracic aorta (marked by a bar); *(d)* re-injection of blood (marked by a bar) performed after a bleeding of 40 cm³ performed in about 4 min. Tracings, from top to bottom, represent the endo-tracheal pressure (inflations upwards), the arterial blood pressure (ABP), the right atrial pressure (RAP), the ECG and the nervous recording. (From Malliani, Recordati & Schwartz, 1973*b*, unpublished observations.)

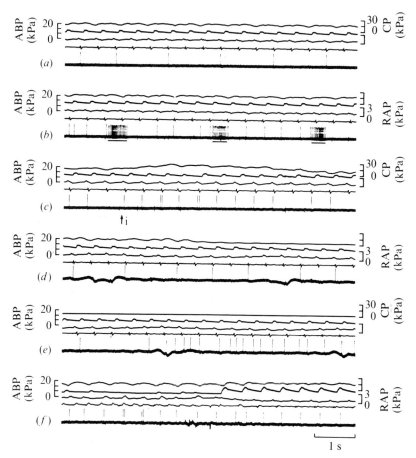

Fig. 16.3. Activity of a fibre with left atrial endings during various haemodynamic stimuli and during interruption of left coronary artery perfusion. (a) Spontaneous activity; (b) mechanical probing of an area of the left atrium performed on the beating heart; (c) fast injection (i) of 2 cm³ of saline into the inflow coronary tubing; (d) interruption of left coronary artery perfusion (indicated by the decrease of coronary inflow pressure); (e) starts 52 s after the end of the preceding record. Ten seconds after the end of (e) the coronary perfusion is instituted again; however, the ventricles do not contract efficiently (beginning of (f)). (f) The heart recovers during the second half of the record. Tracings from top to bottom, represent the coronary inflow pressure (CP), the arterial blood pressure (ABP), the right atrial pressure (RAP), the ECG and the nervous recording. (From Malliani, Recordati & Schwartz, 1973, unpublished observations.)

events and its frequency was related to the mechanics of the cardiac chamber in which the receptive endings were located.

As examples Figs 16.1–16.3 show the impulse activity of various afferent sympathetic fibres with atrial endings. In Fig. 16.1, the spontaneous activity of six different atrial receptors is displayed: we found fibres discharging in phase with atrial systole (*a* wave of atrial pressure, Fig. 16.1*a*), with bulging of the atrio-ventricular valves (*c* wave, Fig. 16.1*b*) and with atrial filling (*v* wave, Fig. 16.1*c*). For some fibres a particular phasic pattern of spontaneous discharge was constant for periods of about 1–2 h. However, a particular pattern could also change spontaneously into another pattern, sometimes with continuous oscillations (Fig. 16.1*d*, activity in phase with *c* and *a* waves; Fig. 16.1*e*, activity in phase with *c* and *v* waves; Fig. 16.1*f*, activity in phase with *v* and *a* waves).

The spontaneous discharge of fibres with atrial endings was often influenced by ventilation of the animal (Fig. 16.2*a*). The superimposed respiratory rhythm was mostly in phase with inflations of the lungs and did not depend on friction between heart and lungs as it also occurred when the lungs were lifted from the heart. We attributed the respiratory rhythm to the concomitant rises in atrial pressure which are known to occur during inflations of the lungs with positive pressure (Opdyke *et al.*, 1948).

Fig. 16.2. illustrates the effects of various haemodynamic events on the activity of a left atrial receptor. The spontaneous activity was in phase with either atrial systole (first four impulses occurring during inflation of the lungs, in Fig. 16.2*a*) or with the initial part of atrial filling and lengthening of atrial diameters (Recordati, Lombardi, Malliani & Brown, 1974) (last three impulses in Fig. 16.2*a*). The fibre was markedly excited after the release of an occlusion of the inferior vena cava (Fig. 16.2*b*) or during the constriction of the thoracic aorta (Fig. 16.2*c*). The impulse activity was abolished during haemorrhage (initial part of Fig. 16.2*d*).

To study the activity of cardiac sympathetic receptors during acute reductions in coronary flow we used a preparation in which the left coronary artery was perfused with an extra-corporeal pump as described by Brown (1968). With this technique it is possible to interrupt left coronary perfusion without those local distortions which accompany the occlusion of a vessel by a snare and which would directly affect the discharge of mechanoreceptors.

In Fig. 16.3 a left atrial receptor, spontaneously active during atrial systole (Fig. 16.3*a*), excited by local mechanical probing (Fig. 16.3*b*) and by fast injection of 2 cm^3 of saline into the inflow coronary tubing (Fig. 16.3*c*), was studied during a prolonged interruption of left coronary flow (beginning in Fig. 16.3*d*). After about 58 s a clear excitation was present (Fig. 16.3*e*) together with systemic hypotension and increased right atrial pressure. Resumption of coronary perfusion rapidly restored cardiac function (last part of Fig. 16.3*f*). The mean latency for the excitation of six fibres (three left atrial, two right ventricular and one left ventricular) similarly studied during interruption of left coronary flow, was 79 s (Malliani *et al.*, 1973*b*). The discharge of each of the six fibres was clearly reduced or abolished during an acute bleeding, although this event is accompanied by a marked decrease in coronary blood flow (Granata, Huvos, Pasquè & Gregg, 1969). It is likely that the crucial difference between the two situations is represented by the size of the heart which is progressively increased during interruption of coronary flow but is reduced during acute bleeding. Typical cardiac vagal receptors of medullated nerve fibres, studied in similar conditions, showed very similar patterns of responses (Recordati *et al.*, 1971).

Therefore, sympathetic and vagal cardiac endings, which differ in their spontaneous activity, are both mainly sensitive to mechanical stimuli. The conduction velocity of these afferent sympathetic cardiac nerve fibres, when determined, was always in the range characteristic for medullated fibres (Malliani *et al.*, 1973*b*).

The afferent neural channel connecting the cardiovascular system to the spinal cord is thus tonically active in relation to specific and normal as well as abnormal haemodynamic events. Through this afferent channel spinal reflexes can be elicited. In electrophysiological experiments both excitatory and inhibitory reflex responses can be recorded (Malliani *et al.*, 1969, 1971). In a subsequent series of experiments (Pagani *et al.*, 1974), specifically designed to analyse excitatory and inhibitory components of spinal reflexes, we tested the activity of preganglionic fibres isolated from a sympathetic outflow mainly directed to the heart during various haemo-dynamic and mechanical events. When the stimuli were applied to the heart, thus eliciting a possible cardio-cardiac reflex, an excitatory response was the rule. Conversely, when the stimuli affected cardiac and vascular receptors (rises in arterial blood pressure) simultaneously the discharge of the same fibre was either increased or decreased, each response being reflex in nature and consistent for each individual fibre. A special cannula was then devised to stretch the walls of the thoracic aorta without interfering with aortic blood flow: it was thus possible to restrict a physical stimulus to the aortic sympathetic sensory endings. Fig. 16.4 shows a reflex inhibition

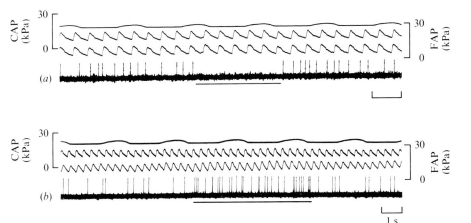

Fig. 16.4. Effects of stretching (marked by a bar) the aortic walls (as described in the text) on the discharge of two different (*(a)* and *(b)*) sympathetic preganglionic neurons. Tracings represent, from top to bottom, the endotracheal pressure (inflations upwards), the carotid and the femoral arterial blood pressure (CAP and FAP) and the nervous recording. (From Pagani, Schwartz, Banks, Lombardi & Malliani, 1974.)

(Fig. 16.4*a*) and excitation (Fig. 16.4*b*) of the impulse activity of two different preganglionic neurons during a similar aortic stretch (Pagani *et al.*, 1974).

In the course of the experiments we were impressed with the promptness of such reflex responses obtained in animals with an acute spinal section, as well as with the moderate magnitude of stretch which had to be applied.

We have no explanation yet for some general principles which seem to be suggested by our experiments and concern the excitatory and inhibitory components of spinal reflexes. While it is obvious that receptors located in various cardiovascular sites exert different reflex effects on the same sympathetic outflow, as a working hypothesis we may also speculate about a possible prevalence of excitatory mechanisms in segmental cardio-cardiac reflexes, while more complexity would characterise those spinal reflexes involving numerous segments.

To conclude on the electrophysiological reflex responses induced by stimulation of afferent sympathetic fibres, I have to mention that the impulse activity of efferent cardiac vagal fibres isolated from a cardiac nerve at its junction with the heart was inhibited by the stimulation of afferent cardiac sympathetic fibres (Schwartz *et al.*, 1973). This cardio-cardiac sympatho-vagal reflex represents a total overturning of the classic vago-sympathetic circuitry. The same stimulus, furthermore, excited the efferent sympathetic fibres running into the same nerve: it is obvious that this can result in a synergistic excitatory effect on cardiac functions. Both sympatho-vagal and sympatho-sympathetic cardio-cardiac reflexes were found to interact with baroreceptive mechanisms (Schwartz *et al.*, 1973). Fig. 16.5 shows a schematic representation of these electrophysiological reflex responses.

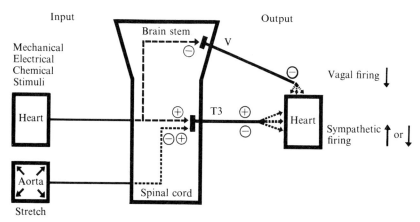

Fig. 16.5. Electrical responses to stimulation of afferent sympathetic fibres (see text). (From Malliani *et al.*, 1975*b*, by courtesy of *Brain Research*.)

In checking the functional significance of these electrophysiological results, after Peterson & Brown (1971) had proved that the stimulation of afferent cardiac sympathetic fibres could induce a reflex increase in arterial blood pressure, we found that a similar stimulation could evoke a reflex increase in myocardial contractility (Malliani, Peterson, Bishop & Brown, 1972) and heart rate (Malliani, Parks, Tuckett & Brown, 1973*a*): incidentally, these experiments definitely demonstrated the existence of cardio-cardiac sympatho-sympathetic reflexes. More recently we also found that stimulation of afferent cardiac sympathetic fibres modifies the aortic diastolic pressure–diameter relationship, determining a lower diameter for any given pressure, i.e. an increased aortic smooth muscle tone (Pagani, Schwartz, Bishop & Malliani, 1975).

However, it remained to be demonstrated that cardiovascular reflex responses could be elicited by mechanical stimuli affecting cardiovascular sympathetic sensory endings. To this purpose, we used the same aortic

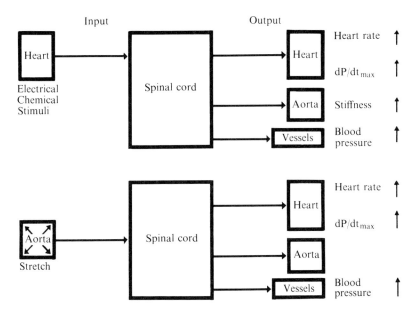

Fig. 16.6. Haemodynamic responses to stimulation of afferent sympathetic fibres (see text). (From Malliani *et al.*,1975*b*, by courtesy of *Brain Research*.)

cannula previously described (Pagani *et al.*, 1974). In vagotomised cats with both carotid arteries occluded, the mechanical stretch of the thoracic aorta was found to induce a reflex excitation of sympathetic activity increasing heart rate, myocardial contractility and arterial blood pressure (Lioy *et al.*, 1974). Fig. 16.6 represents a scheme of these haemodynamic reflex responses.

On the basis of our present understanding of blood pressure regulation (Fig. 16.7*a*) stretch of a reflexogenic area (R), simulating a rise in arterial blood pressure, produces a reflex decrease in systemic blood pressure through an inhibition of the sympathetic (S) discharge affecting vascular (V) resistances and thus blood pressure (BP). In this closed loop a negative feedback assures a tonic control of the system regulating the dynamic equilibrium of its operating point (Franz, 1974).

Our experiments, in which supraspinal inhibitory mechanisms were not operative, have revealed spinal excitatory reflexes that seem to exhibit positive feedback characteristics. Thus a stimulus likely to simulate the effects on aortic walls of an increase in mean aortic pressure (Fig. 16.7*b*) produces a further rise in systemic arterial pressure by increasing the sympathetic activity. We suggest that these spinal neural mechanisms, in addition to the possible increase of the central command (C in Fig. 16.7), may contribute to the tonic maintenance of pathophysiological states such as arterial hypertension (Malliani *et al.,1975a*). However, the co-existence of inhibitory components, as revealed by the electrophysiological analysis of spinal sympathetic reflexes, should also be considered (and, accordingly, is indicated by the broken line in Fig. 16.7). Furthermore, it has been

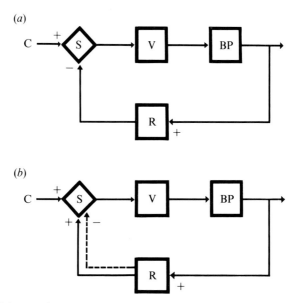

Fig. 16.7. Scheme of suggested mechanisms underlying nervous control of blood pressure regulation (see text). *(a)* Our present understanding of blood pressure regulation; *(b)* the feedback system where a stimulus likely to simulate the effects on aortic walls of an increase in mean aortic pressure produces a further rise in systemic arterial pressure by increasing sympathetic activity. R = reflexogenic area; S = sympathetic discharge; V = vascular resistances; BP = blood pressure; C = central command. (From Malliani *et al.*, 1975a, by courtesy of the Medical Research Society and of the Biochemical Society.)

proven by Lioy & Szeto (1975) that cardiovascular reflex responses initiated by the stretch of the thoracic aorta are also evident when supraspinal inhibitory mechanisms are operative. Thus the modulation of spinal sympathetic reflexes may be rather complex with respect to fundamental properties, such as the range of operation, gain, stability and interactions with other regulatory mechanisms.

On similar grounds, cardio-cardiac sympathetic excitatory reflexes are likely to participate in the pathogenesis of cardiac arrhythmias associated with myocardial ischaemia (Malliani *et al.*, 1969; Schwartz, Foreman, Stone & Brown, 1976). Furthermore, cardio-vascular changes occurring during particular physiological states such as exercise may partly depend upon reflex mechanisms mediated by sympathetic afferents.

Recently we have expanded the investigation into the functional properties of sympathetic vascular receptors: we have selected a high- and low-pressure area.

We (Pagani, 1975; Malliani & Pagani, 1976) recorded the activity of 35 single afferent sympathetic fibres with aortic endings: 29 fibres were myelinated and six were unmyelinated. Their spontaneous discharge was characterised by single action potentials in phase with pressure pulses (however, the discharge of unmyelinated fibres was often more erratic): the impulse activity was increased during increases in aortic pressure and, conversely, decreased during decreases in aortic pressure. All of these fibres, both myelinated and unmyelinated, did not appear to be sensitive to changes in blood gases or pH as their discharge was abolished by bleeding the animal to death and as they were not excited by asphyxia, provided that this stimulus was not accompanied by an arterial pressure rise. This is in contrast with another report (Uchida, 1975) claiming an increased activity of unmyelinated aortic afferent sympathetic fibres during asphyxia. In the attempt of interpreting such a discrepancy we have to point out that in our experiments it was always ascertained that nervous impulses elicited by electrical or mechanical stimuli as well as those occurring spontaneously originated from the same fibre, by analysing their configuration on photographic records with expanded time bases. We considered only the constancy of amplitude and shape of action potentials to be a safe criterion, especially in absence of other impulses. Moreover, one should be aware of the fact that recordings obtained from the cut peripheral end of a sympathetic ramus communicans may include not only afferent but also postganglionic efferent nerve activity; the latter would be obviously excited during asphyxia.

Fig. 16.8 shows an example of the mechanoreceptive properties of an aortic sympathetic unmyelinated afferent nerve fibre. The fibre, the conduction velocity of which was about 1 m/s (Fig. 16.8g), was well excited *in vivo* by aortic pressure rises (Fig. 16.8b), but with an irregular rhythm, caused by cardiac arrhythmias. After the animal had been killed a large, saline-filled latex balloon was introduced into the aorta through the opened left ventricle. The pressure used to distend the balloon was used as an approximate index of aortic wall stretch, as we observed that this pressure

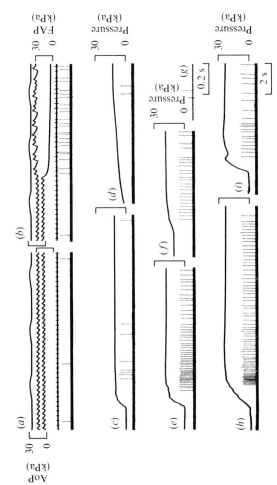

Fig. 16.8. Activity of an afferent sympathetic unmyelinated nerve fibre (Group C) with one receptive field located in the distal third of the aortic arch. (a) Control; (b) occlusion of the descending aorta; (c)–(f), (h) and (i) effects of stretching the aortic wall by distending a latex balloon located in the distal part of the aortic arch (see text); (g) electrical stimulation of the left inferior cardiac nerve activating the fibre. Approximate length of the fibre, 5 cm. Calculated conduction velocity 1 m/s. In (a) and (b) tracings, from top to bottom are the endotracheal pressure, the aortic pressure (AoP), the femoral arterial pressure (FAP), the ECG and the nervous recording. In (c)–(f), (h) and (i), top tracing: pressure applied to the distending balloon; bottom tracing: nervous recording. (From Malliani & Pagani, 1976, by courtesy of the Physiological Society.)

was approximately linearly related to the external aortic dimension measured with ultrasound techniques (Pagani *et al.*, 1975).

When the balloon was inflated (Fig. 16.8*c* and following parts of the figure) the highest impulse frequencies were attained during the rising phase of the pressor stimulus, followed thereafter by an adapted discharge (Fig. 16.8*e* and *h*). A dynamic component of the stimulus was also proven by reaching the same level of absolute pressure at different speeds (Fig. 16.8*c* and *d*) or from different initial pressures (Fig. 16.8*e* and *f*). Finally, a response suggesting receptor fatigue and/or mechanical alterations of the vascular wall was observed: in Fig. 16.8*h* a pressure rise produced a much higher activation of the impulse activity than that represented in Fig. 16.8*i* where the same pressure step was applied after the stimulus had been sustained for 100 s and just released for a few seconds. Post-mortem probing indicated that 11 out of the 35 aortic sympathetic afferents had several and distinct receptive fields: they were usually located in nearby aortic areas or, in addition, in other proximal portions of the arterial tree or in the adjacent pleura and connective tissue (Pagani, 1975; Malliani & Pagani, 1976). Thus these afferents seem particularly apt to signal pulsatile aortic and arterial stretches. They may also be part of a nervous pathway through which pressor reflexes, with positive feedback characteristics, can be elicited (Malliani *et al.*, 1975*a*).

It should be realised that it was crucial to localise the sensory endings of each fibre after the animal had been killed. For instance, Fig. 16.9 shows the activity of three different afferent sympathetic fibres which were erroneously considered to be aortic at the preliminary probing performed on the pulsating aorta. The fibre illustrated in Fig. 16.9*c* had a spontaneous discharge in phase with aortic pulses; all three fibres were well excited by the constriction of the aorta (Fig. 16.9). When the probing was repeated post-mortem, the sensory fields of the fibres turned out to be located as follows: in the periaortic pleura for the fibre of Fig. 16.9*a* and in the periaortic connective tissue for the fibres of Fig. 16.9*b* and *c*. Findings of this type may account for many discrepancies furnished by the literature and strongly indicate that a sound localisation of the receptive fields requires that not more than one single fibre is studied during each experiment.

As a low-pressure area we selected the pulmonary veins. As long ago as 1941, Nonidez had reported that no receptor endings were present in the pulmonary veins of previously sympathectomised cats and he suggested that the afferent nerve fibres might run in the sympathetic nerves. We

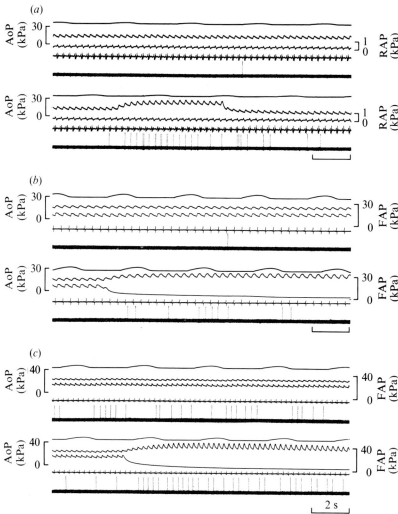

Fig. 16.9. Activity of three different afferent sympathetic myelinated nerve fibres with receptive fields located outside of the arterial tree. *(a)* Fibre with one receptive field in the periaortic pleura. Conduction velocity 4.6 m/s. Upper records, control; lower records, occlusion of the descending aorta (indicated by the rise in aortic pressure). Second tracing, aortic pressure (AoP); third tracing, right atrial pressure (RAP). *(b)* and *(c)* Fibres with their receptive field in the periaortic connective tissue. Conduction velocity 20 m/s and 14 m/s, respectively. Upper records, control; lower records, occlusion of the descending aorta (indicated by diverging blood pressure traces). The second tracing represents the aortic and the third tracing the femoral arterial pressure (FAP). (From Malliani & Pagani, 1975, unpublished observations.)

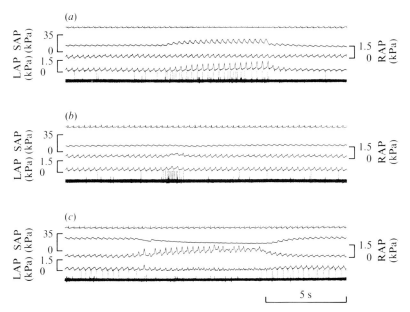

Fig. 16.10. Nervous activity of a single afferent sympathetic myelinated nerve fibre innervating the left middle pulmonary vein. In each display recordings are, from top to bottom: ECG, systemic arterial blood pressure (SAP), right atrial pressure (RAP), left atrial pressure (LAP) and nervous activity. *(a)* Constriction of the descending aorta; *(b)* forced short inflation of the lungs, corresponding to the increased nervous activity and to the small oscillation in right atrial pressure; *(c)* constriction of the pulmonary artery. (From Lombardi, Malliani & Pagani, 1976, unpublished observations.)

recorded the activity of 19 sympathetic afferent fibres which all had one single receptive field localised to one pulmonary vein and extending to the atrial junction (Lombardi, Malliani & Pagani, 1976). They discharged spontaneously, mainly in phase with atrial systole (Fig. 16.10). Their activity was increased during elevations of left atrial pressure (Fig. 16.10*a*) and decreased during reductions in left atrial pressure (Fig. 16.10*c*); they were also markedly excited by forced inflations of the lungs (Fig. 16.10*b*). Thus with electrophysiological techniques we confirmed Nonidez's data.

These receptors seem particularly suited for signalling pulmonary congestion: they may, therefore, contribute, through excitatory sympathetic reflexes, to the reflex tachycardia and hypertension which often accompanies this pathophysiological state. The functional importance of low-pressure vascular sympathetic receptors has been recently demonstrated by the finding that unanaesthetised and atropinised chronic spinal cats respond to intravenous infusions of saline with a consistent tachycardia which is abolished by deafferentiation (Bishop *et al.,* 1976). Similar findings have been obtained in anaesthetised dogs by Gupta (1975). Thus sympathetic afferents can mediate the phenomenon first described by Bainbridge (1915) and indisputably attributed, if reflex in nature, to vagal afferents.

A few general points are worth discussing. It is evident that present knowledge of cardiovascular sympathetic afferents indicates that their functional properties are very complex. Some of these fibres seem to have no regular spontaneous impulse activity (Ueda, Uchida & Kamisaka, 1969; Malliani & Brown, 1970; Brown & Malliani, 1971), whereas others display a spontaneous discharge in phase with specific haemodynamic events (Malliani *et al.,* 1973*b*; Nishi, Sakanashi & Takenaka, 1974; Pagani, 1975; Lombardi *et al.,* 1976; Malliani & Pagani, 1976). It is difficult to reconcile this latter observation with the classical view of an exclusive nociceptive function of cardiovascular sympathetic afferents (White, 1957) and indeed this suggests their participation in the tonic reflex control of circulation (Malliani *et al.,* 1975*b*). Some of these sympathetic afferents have a single and very restricted receptive field (Malliani *et al.,* 1973*b*; Nishi *et al.,* 1974; Lombardi *et al.,* 1976) whereas others have either a large or more than one receptive field (Malliani *et al.,* 1973*b*; Coleridge, Kidd, Coleridge & Banzett, 1975; Pagani, 1975; Malliani & Pagani, 1976).

In a preliminary report Coleridge *et al.* (1975) have recently emphasised that some multi-terminal sympathetic afferents innervate several organs of widely different function. In relation to these findings these authors concluded that sympathetic afferents are unlikely to provide the central nervous system with specific information. However, it is my opinion that it is not that simple to decide what may be specific for the central nervous system. Perhaps their statement may hold true for those particular fibres the impulse activity of which is only indirectly influenced by haemodynamic events and which most often innervate large portions of the pleura and the connective tissue surrounding the various thoracic organs (Holmes & Torrance, 1959).

Another debated question concerns the nature of the stimulus to which afferent sympathetic unmyelinated fibres are responsive. While for some of them the mechanoreceptive properties appear clearly proven (Fig. 16.8), it is more difficult to interpret the action of chemicals (Uchida & Murao,

1975) and therefore to accept the conclusion that myocardial ischaemia excites unmyelinated fibres and thereby produces pain, through chemical stimuli (Uchida & Murao, 1975).

We are currently investigating whether ventricular unmyelinated afferent sympathetic fibres may be excited by haemorrhage (as reported for vagal ventricular afferents by Öberg & Thorén (1972)): should this be the case, their chemoreceptive properties would be defined on much better grounds.

A last point, to conclude. We have recently explained our reasons for using the term 'sympathetic afferents' (Malliani *et al.*, 1975*b*). This is not at all a problem of terminology but a point essential to the concept that the sympathetic input to the spinal cord may be as important to the regulation of the sympathetic output as the somatic spinal afferents are to the regulation of the somatic nervous outflow.

'The reflex-arc is the unit mechanism of the nervous system when that system is regarded in its integrative function' (Sherrington, 1906): I am convinced that this also holds true for the visceral nervous system. Without questioning the 'dominance of the brain.'

References

Bainbridge, F.A. (1915). The influence of venous filling upon the rate of the heart. *J. Physiol.*, **50**, 65–84.

Bishop, V.S., Lombardi, F., Malliani, A., Pagani, M. & Recordati, G. (1976). Reflex sympathetic tachycardia during intravenous infusions in chronic spinal cats. *Am. J. Physiol.*, **230**, 25–9.

Brown, A.M. (1968). Motor innervation of the coronary arteries of the cat. *J. Physiol.*, **198**, 311–28.

Brown, A.M. & Malliani, A. (1971). Spinal sympathetic reflexes initiated by coronary receptors. *J. Physiol.*, **212**, 685–705.

Coleridge, H.M., Kidd, C., Coleridge, J.C.G. & Banzett, R.B. (1975). Multi-terminal sympathetic afferent fibers supplying the thoracic organs of cats and dogs. *Physiologist, Wash.*, **18**, 173.

Franz, G.N. (1974). On blood pressure control. *Physiologist, Wash.*, **17**, 73–86.

Granata, L., Huvos, A., Pasquè, A. & Gregg, D.E. (1969). Left coronary hemodynamics during hemorrhagic hypotension and shock. *Am. J. Physiol.*, **216**, 1583–9.

Gupta, P.D. (1975). Spinal autonomic afferents in elicitation of tachycardia in volume infusion in the dog. *Am. J. Physiol.*, **229**, 303–8.

Holmes, R. & Torrance, R.W. (1959). Afferent fibres of the stellate ganglion. *Q. Jl exp. Physiol.*, **44**, 271–81.

Lioy, F., Malliani, A., Pagani, M., Recordati, G. & Schwartz, P.J. (1974). Reflex hemodynamic responses initiated from the thoracic aorta. *Circulation Res.*, **34**, 78–84.

Lioy, F. & Szeto, P.M. (1975). Baroreceptor and chemoreceptor influence on the cardiovascular responses initiated by stretch of the thoracic aorta. *Physiologist, Wash.*, **18**, 293.

Lombardi, F., Malliani, A. & Pagani, M. (1976). Nervous activity of afferent sympathetic fibers innervating the pulmonary veins. *Brain Res.*, **113**, 197–200.

Malliani, A. & Brown, A.M. (1970). Reflexes arising from coronary receptors. *Brain Res.*, **24**, 352–5.

Malliani, A., Lombardi, F., Pagani, M., Recordati, G. & Schwartz, P.J. (1975a). Spinal sympathetic reflexes in the cat and the pathogenesis of arterial hypertension. *Clin. Sci. mol. Med.*, **48**, 259s–60s.

Malliani, A., Lombardi, F., Pagani, M., Recordati, G. & Schwartz, P.J. (1975b). Spinal cardiovascular reflexes. *Brain Res.*, **87**, 239–46.

Malliani, A. & Pagani, M. (1976). Afferent sympathetic nerve fibres with aortic endings. *J. Physiol.*, **263**, 157–69.

Malliani, A., Pagani, M., Recordati, G. & Schwartz, P.J. (1971). Spinal sympathetic reflexes elicited by increases in arterial blood pressure. *Am. J. Physiol.*, **220**, 128–34.

Malliani, A., Parks, M., Tuckett, R.P. & Brown, A.M. (1973a). Reflex increases in heart rate elicited by stimulation of afferent cardiac sympathetic nerve fibres in the cat. *Circulation Res.*, **32**, 9–14.

Malliani, A., Peterson, D.F., Bishop, V.S. & Brown, A.M. (1972). Spinal sympathetic cardiocardiac reflexes. *Circulation Res.*, **30**, 158–66.

Malliani, A., Recordati, G. & Schwartz, P.J. (1973b). Nervous activity of afferent cardiac sympathetic fibres with atrial and ventricular endings. *J. Physiol.*, **229**, 457–69.

Malliani, A., Schwartz, P.J. & Zanchetti, A. (1969). A sympathetic reflex elicited by experimental coronary occlusion. *Am. J. Physiol.*, **217**, 703–9.

Nishi, K., Sakanashi, M. & Takenaka, F. (1974). Afferent fibres from pulmonary arterial baroreceptors in the left cardiac sympathetic nerve of the cat. *J. Physiol.*, **240**, 53–66.

Nonidez, J.F. (1941). Studies on the innervation of the heart. II. Afferent nerve endings in the large arteries and veins. *Am. J. Anat.*, **68**, 151–89.

Öberg, B. & Thorén, P. (1972). Increased activity in left ventricular receptors during hemorrhage or occlusion of caval veins in the cat. A possible cause of the vaso-vagal reaction. *Acta physiol. scand.*, **85**, 164–73.

Opdyke, D.F., Duomarco, J., Dillon, W.H., Schreiber, H., Little, R.C. & Seely, R.D. (1948). Study of simultaneous right and left atrial pressure pulses under normal and experimentally altered conditions. *Am. J. Physiol.*, **154**, 258–72.

Pagani, M. (1975). Afferent sympathetic nerve fibres with aortic endings. *J. Physiol.*, **252**, 45P–6P.

Pagani, M., Schwartz, P.J., Banks, R., Lombardi, F. & Malliani, A. (1974). Reflex responses of sympathetic preganglionic neurones initiated by different cardiovascular receptors in spinal animals. *Brain Res.*, **68**, 215–25.

Pagani, M., Schwartz, P.J., Bishop, V.S. & Malliani, A. (1975). Reflex sympathetic changes in aortic diastolic pressure–diameter relationship. *Am. J. Physiol.*, **229**, 286–90.

Peterson, D.F. & Brown, A.M. (1971). Pressor reflexes produced by stimulation of afferent cardiac nerve fibers in the cardiac sympathetic nerves of the cat. *Circulation Res.*, **28**, 605–610.

Recordati, G., Lombardi, F., Malliani, A. & Brown, A.M. (1974). Instantaneous dimensional changes of the right atrium of the cat. *J. appl. Physiol.*, **36**, 686–92.

Recordati, G., Schwartz, P.J., Pagani, M., Malliani, A. & Brown, A.M. (1971). Activation of cardiac vagal receptors during myocardial ischemia. *Experientia,* **27,** 1423–4.

Schwartz, P.J., Foreman, R.D., Stone, H.L. & Brown, A.M. (1976). Effect of dorsal root section on the arrhythmias associated with coronary occlusion. *Am. J. Physiol.,* **231,** 923–8.

Schwartz, P.J., Pagani, M., Lombardi, F., Malliani, A. & Brown, A.M. (1973). A cardiocardiac sympathovagal reflex in the cat. *Circulation Res.,* **32,** 215–20.

Sherrington, C.S. (1906). *The integrative action of the nervous system.* Yale University Press: Connecticut.

Sutton, D.C. & Lueth, H.C. (1930). Experimental production of pain on excitation of the heart and great vessels. *Archs intern. med.,* **45,** 827–67.

Uchida, Y. (1975). Afferent aortic nerve fibers with their pathways in cardiac sympathetic nerves. *Am. J. Physiol.,* **228,** 990–5.

Uchida, Y. & Murao, S. (1975). Acid-induced excitation of afferent cardiac sympathetic nerve fibers. *Am. J. Physiol.,* **228,** 27–33.

Ueda, H., Uchida, Y. & Kamisaka, K. (1969). Distribution and responses of the cardiac sympathetic receptors to mechanically induced circulatory changes. *Jap. Heart J.,* **10,** 70–81.

White, J.C. (1957). Cardiac pain. Anatomic pathways and physiologic mechanisms. *Circulation,* **16,** 644–55.

Discussion

(a) Weaver described some experiments done on cats with the IXth and Xth cranial nerves cut at the jugular foremen to remove carotid sinus and aortic depressor afferents. She found that if afferent fibres in the inferior cardiac nerve were stimulated with high frequencies or high intensities, similar excitation responses to those described by Malliani were obtained. At lower intensities of stimulation there were marked depressor responses and inhibition of renal nerve activity. She also found that if she expanded the blood volume of these animals there was inhibition of renal nerve activity which did not occur after cutting the dorsal roots T1 to T5. She suggested that this was evidence for contribution to volume homeostasis by sympathetic afferents.

(b) Folkow remarked that normally negative feedback mechanisms predominate and that positive feedback mechanisms were potentially dangerous. However, he suggested that during exercise, positive feedback, e.g. from an increase in heart rate, might give a further boost to the autonomic nervous system.

(c) In answer to a question, Malliani said that the chronic spinal preparations were made by transecting at C7–C8 to allow breathing. Connections with spinal centres were later prevented by cutting dorsal roots.

(d) Malliani noted that most of the receptors in the heart that he had recorded from had only one 'point' source. In a few cases there were two in adjacent areas. In the pulmonary veins there were only single sources for these very sensitive fibres. In the aorta both myelinated fibres and C fibres could have multiple sources. For most of the fibres with multiple fields the sensitive spots were in adjacent areas – aorta and intercostal artery or adjacent connective tissues. The physiological role of these receptors has not yet been worked out.

(e) Malliani gave details of the method used to locate endings from C fibres: when the action potential was observed in a prospective 'C' fibre from the ventricle he injected saline, or caused asphyxia or haemorrhage. He then opened the pericardium and probed the heart. Then the animal was killed, the heart opened and the receptive field for one single fibre located. The conduction velocity was measured.

It was suggested from the floor that these methods would result in mechanosensitive fibres being selected; chemosensitive C fibres from the heart signalling, e.g., cardiac pain would not be found. Malliani explained that he studied all strands with haemorrhage. He did not use veratridine because this was not a physiological stimulus to chemoreceptors. Only mechanoreceptors had, so far, been detected.

(f) Since responses from sympathetic afferents were shown to be overridden by the arterial baroreflex, the question was raised as to whether the converse might be true, that the baroreflex response might be less during stimulation of sympathetic afferents. However, Malliani had no evidence on this.

17. Intrinsic characteristics of baroreceptors

A. M. BROWN

Cardiovascular baroreceptors monitor the mechanical status of the circulation, i.e. its volume and complex pressures. It is only natural, therefore, that the effects of these stimuli upon baroreceptor discharge have received the lion's share of attention from investigators interested in control of the circulation. As in the case of any sensory receptor, however, the receptor's membrane properties would also be expected to influence its discharge. Since this aspect of baroreceptor function has been largely ignored, my collaborators, Drs Saum, Tuley, Kunze and Ayachi and I decided to investigate what we refer to as the intrinsic or membrane properties of baroreceptors. Our studies have involved mainly aortic baroreceptors and to a lesser extent, atrial receptors, but it is likely that our findings apply to all baroreceptors and to mechanoreceptors in general. We found that baroreceptors have an electrogenic sodium pump that contributes to their pressure thresholds and their supra-threshold pressure–discharge curves (Saum, Brown & Tuley, 1976). The electrogenic pump is responsible for post-excitatory depression following restoration of pressure from a higher level (Landgren, 1952; Saum $et\ al.$, 1976), and limits the steady-state discharge achieved by increased pressures (Brown $et\ al.$, 1976). The reflex baroreceptor effects of the cardiac glycosides are attributable to blockage of this electrogenic sodium pump. Baroreceptors are also sensitive to changes in extracellular sodium (Na^+_0) and potassium (K^+_0) (Saum, Ayachi & Brown, 1978). In fact, reductions in Na^+_0 of as little as 5% increase threshold and reduce sensitivity producing a reflex increase in arterial blood pressure and urine flow (Kunze, Saum & Brown, 1977, 1978). The response of the receptors is interpreted quantitatively using a model based on conventional membrane theory (Saum $et\ al.$, 1978). The ionic sensitivity of baroreceptors may, therefore, be significantly

339

involved in prompt regulation of extracellular fluid volume, and it is possible that their intrinsic properties are as essential as their mechano-sensitive properties for bodily function. In the remainder of this article the methods we have used, the principle results and the main conclusions will be presented in greater detail.

Receptor experiments

We have used *in vitro* preparations from the rat. The main reasons for this are that experimental control is better in the *in vitro* situation, and the rat is a relatively cheap, plentiful animal in which many interesting models for human pathophysiology, such as hypertension, are readily available. The two preparations we have developed are an aortic arch–aortic nerve preparation and an atrial–vagal nerve preparation. The former has been completely described (Brown *et al.*, 1976) but the latter has not. Briefly, the heart is exposed and the cervical and thoracic vagi are dissected free and reflected backwards on to it. The hila of the lungs are ligated and the heart is removed. A double-barrelled concentric cannula is inserted into the left atrial appendage for perfusion and pressure stimulation. In both preparations pressure stimuli are delivered via a shaker which is an electro-dynamic transducer that faithfully and easily reproduces a wide variety of pressure stimuli, such as steps, ramps, sine waves and broad band noise (Fig. 17.1). Nervous discharge is recorded with platinum–iridium wires connected to a low noise, AC coupled pre-amplifier and pressures are recorded via metal cannulae connected to strain gauge manometers. Data are recorded on FM tape and subsequently analysed using a PDP 11/40 computer. The interspike interval and blood pressure are digitised every 70 μs and converted to a Fortran binary file which is stored on a disc pack or on magnetic tape. The reciprocal of the interspike interval is used to plot instantaneous frequency. Although this is a point process, the instan-taneous frequency reflects, as we shall show, the underlying receptor potential which is a graded function of the applied pressure. Therefore, the instantaneous frequency may also be thought of as a continuous function.

Receptors connected to myelinated fibres have a regular discharge but this is not always true when the fibres are unmyelinated (Thorén, Saum & Brown, 1977). The regularly discharging receptors may be considered to be unbranched and our conception of such a receptor is shown in Fig. 17.2. The experiments reported presently concern only regularly discharging receptors.

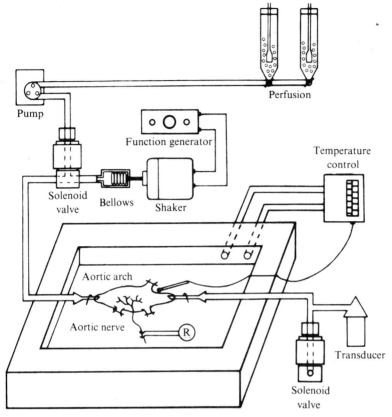

Fig. 17.1. Experimental arrangement. R refers to the recording electrodes. In atrial preparations the isolated atria and attached vagal nerve replace the aortic arch and aortic nerve.

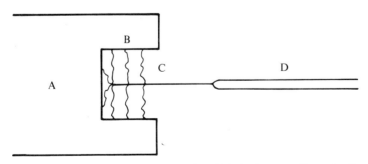

Fig. 17.2. Idealised baroreceptor. A, vessel wall; B is the wavy lines coupling the receptor to A; C is the receptor and junction with axon D. The junction is also the spike-initiating zone. This model applies to regularly discharging baroreceptors.

Electrogenic sodium pump

When an aortic baroreceptor or an atrial receptor is discharging in the steady state, a subsequent temporary step of increased pressure is followed by a period of silence when the previous lower pressure level is restored. This is referred to as post-excitatory depression (PED) and is illustrated for an aortic baroreceptor in Fig. 17.3 and for an atrial receptor in Fig. 17.4. It was first described in baroreceptors by Landgren (1952), and is a well-recognised property of mechanoreceptors (Ottoson & Shepherd, 1971). Since the baroreceptor ending is not recorded directly it is necessary to demonstrate that this effect upon axonal discharge does result from changes in the ending. As records (a_1) and (a_2) in Fig. 17.3 indicate the axon is still capable of conducting an action potential during PED.

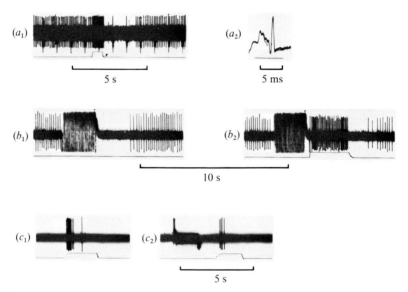

Fig. 17.3. Post-excitatory depression (PED) in a rat baroreceptor. (a_1). Top tracing is the unit discharge. Bottom tracing is pressure. A steady-state discharge was elicited by a pressure step to 16.0 kPa. A subsequent step to 21.3 kPa causes an increased discharge. During restoration of pressure the receptor is silent due to PED and then recovers. The first two impulses during PED were elicited by electrical stimulation of the more distal nerve trunk. (a_2) One of the conducted action potentials referred to above shown with a faster time-scale. (b_1) PED elicited by a burst of antidromic action potentials. Steady-state discharge was induced by a pressure step. (b_2) Pressure step during electrically evoked PED can still elicit a discharge indicating that the endings are not refractory. (c_1) Pressure-elicited discharge which is reduced by prior antidromic stimulation in (c_2).

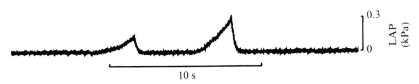

Fig. 17.4. Post-excitatory depression (PED) in a left atrial receptor. Steady-state discharge was elicited by a 0.2 kPa pressure step in the left atrium. Two superimposed ramps elicited bursts of action potentials followed by PED. LAP = left atrial pressure.

Moreover, PED can be elicited by antidromic stimulation at frequencies greater than the prevailing steady-state discharge (Fig. 17.3b_1). But, as record *(b_2)* shows, the endings are not refractory during PED for a further increase in pressure can still elicit a burst of impulses. The inhibitory effect of antidromic action potentials on the mechanical response is shown in another way in records *(c_1)* and *(c_2)* of Fig. 17.3. The antidromic action potentials can also be applied at the end of a pressure-evoked burst of impulses and they will prolong PED further. This indicates that the antidromic impulses can be conducted at least as far as the spike-initiating zone (Fig. 17.2) so that the cause of PED can be attributed to changes in the spike-initiating zone or nerve ending. From a functional point of view, any further distinction in the site of origin of PED is not crucial.

It should be noted that adaptation of the discharge elicited by a pressure step is not affected by pump blockage (Fig. 17.7*b*) and we have attributed this process to visco-elastic processes which couple the receptor to the vessel wall (Fig. 17.2) (Saum *et al.*, 1976).

The duration of PED is related to the number of impulses preceding it as illustrated in Fig. 17.5*a*. The relationship is a quantitative one which is described by a rectangular hyperbola (Fig. 17.5*b*) (Saum *et al.*, 1976). It is also influenced by the control steady-state discharge and by the impulse

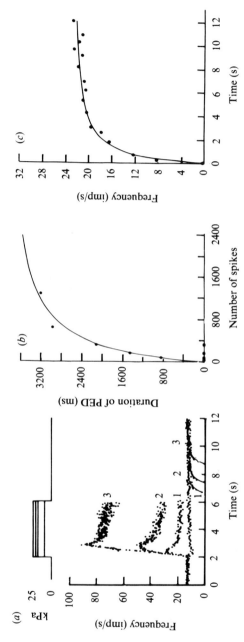

Fig. 17.5. Quantitative aspects of post-excitatory depression (PED). (a) Instantaneous frequencies during three increasing pressure steps. Each dot represents the reciprocal of the interspike interval between successive action potentials. The duration of PED is related to the frequency at constant durations as shown in (b). (c) Recovery from PED.

frequency during the step stimulus (Saum *et al.*, 1976). The recovery from PED is shown in Fig. 17.5*c* and the time for half recovery ranges from 0.5 to 2 s so that a single cardiac cycle can be readily spanned by this phenomenon.

The cause of PED can, in part, be inferred from the hyperbolic relationship between the number of impulses eliciting it and its duration (Fig. 17.5*b*). The number of impulses may be considered as a sodium load and the PED as the hyperpolarisation resulting from increased activation of a membrane electrogenic sodium pump by the increase in intracellular sodium. Enzymatic reactions generate such curves. Our interpretation is strengthened by several pharmacological experiments. Thus, cardiac glycosides which block the electrogenic sodium pump (Sokolove & Cooke, 1971) abolish PED (Fig. 17.6*a*); the same is true of zero potassium solutions (Fig. 17.6*b*). Another pump blocker lithium is also effective. These drugs at the doses used did not have any action upon the steady-state pressure–volume curves of the excised vascular segments used in these experiments. Redistribution of ions is unlikely to be involved in PED since accumulation of extracellular potassium during increased baroreceptor activity should lead to an increased discharge when the pressure is restored from a higher level.

The electrogenic sodium pump also operates at rest. When it is blocked by the pharmacological agents described above the threshold pressure for steady-state baroreceptor discharge is lowered and the pressure–steady-state discharge curve is steeper (Fig. 17.7*a*). These effects are immediate and are attributed mainly to pump blockage but some ionic redistribution could also be involved.

Effects of Na^+_0 and K^+_0 on baroreceptor discharge

The experiments on PED and the electrogenic sodium pump indicated to us that ions might affect baroreceptor function. It is well known that the function of other mechanoreceptors is greatly affected by changes in Na^+_0 and K^+_0 (Katz, 1950; Edwards, Terzuolo & Washizu, 1963; Gray & Sato, 1953). However, except for one study by Matsuura (1973), who reported that lowering Na^+_0 decreased the electrotonic potential elicited from common carotoid artery baroreceptors by probing the vessel wall with a glass rod, there is no information on this point.

We find that lowering Na^+_0 by 12.5–50% increased theshold and

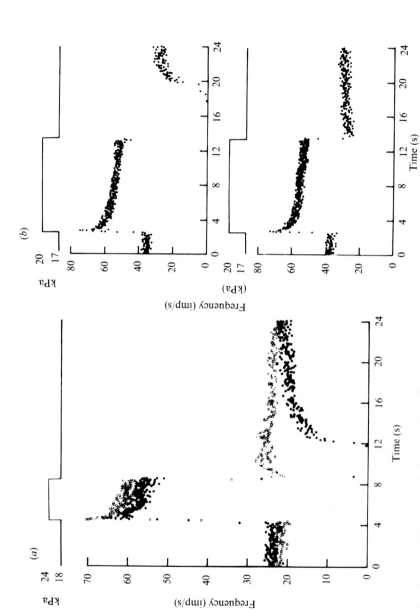

Fig. 17.6. Effects of blockage of electrogenic sodium pump upon post-excitatory depression by (a) ouabain. Filled circles are control response; and (b) zero potassium.

Top record is control and bottom record is after 2 min in zero potassium Krebs–Henseleit.

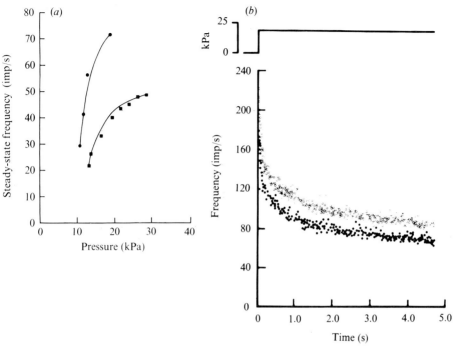

Fig. 17.7. Actions of electrogenic sodium pump on *(a)* steady-state discharge, and *(b)* adaptation. *(a)* Steady-state pressure–discharge curves for a single baroreceptor in control solution (rectangles) and following 1 min of pump blockage with ouabain (filled circles). *(b)* Discharge response to a pressure step. The transient peak discharge adapts to a steady-state level. Filled circles before and open circles after ouabain. The time course of adaptation is unaffected by the drug.

reduced baroreceptor sensitivity to supra-threshold pressures (Fig. 17.8) (Saum *et al.*, 1978). This occurs in atrial receptors as well. Smaller changes in Na^+_0 are also effective but the results are not stastistically significant because of the variability of receptor response, and the small number of receptors sampled. We shall demonstrate in a subsequent part of this report, however, that changes in Na^+_0 of as little as 5% can elicit baroreceptor reflexes through an action upon the receptors.

Increasing K^+_0 had the opposite effects to reducing Na^+_0. Thus, threshold is lowered and sensitivity is enhanced (Fig. 17.9; Saum *et al.*, 1978). Reducing K^+_0 appears to have the opposite effects to increasing K^+_0, but we have not done enough experiments to establish this definitely.

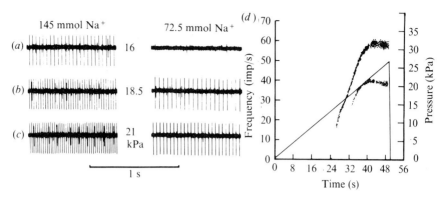

Fig. 17.8. Effects of lowering extracellular Na^+_0 on baroreceptor discharge. *(a)–(c)* Steady-state unit discharge elicited by the corresponding pressure steps in control and $0.5 Na^+_0$. *(d)* Continuous records of frequency response to slow ramps in control solution (filled circles) and $0.5 Na^+_0$ (open circles).

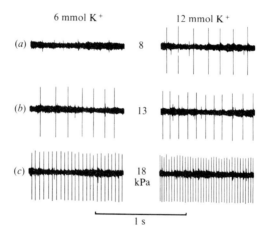

Fig. 17.9. *(a)–(c)* Effects of increasing K^+_0 on steady-state baroreceptor discharge at the pressure indicated.

The small changes in sodium and potassium do not affect the vascular pressure–volume curves so that the actions are upon the receptors.

These results can be interpreted qualitatively from conventional membrane theory. In an attempt to be more quantitative and also to predict other ionic effects, we have developed a model baroreceptor membrane which is based upon the equivalent electrical circuit shown in Fig. 17.10. It

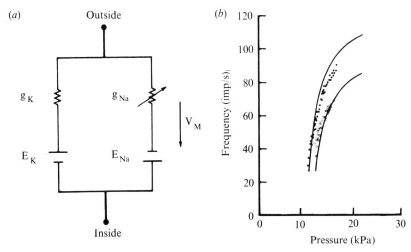

Fig. 17.10. Baroreceptor model for predicting ionic sensitivity. *(a)* Equivalent electrical circuit; g_K, g_{Na}, E_K and E_{Na} are potassium conductance, sodium conductance, potassium battery and sodium battery, respectively. *(b)* Steady-state pressure–discharge curves for a single baroreceptor in control solution (filled circles) and $0.5 \, Na^+_0$ (open circles). The continuous lines are predicted from the circuit equations for the model represented by *(a)*.

is composed of a sodium battery, E_{Na}, a sodium conductance, g_{Na}, a potassium battery, E_K, and a potassium conductance, g_K. The chloride ion concentration is ignored since the results are not dependent upon its presence. At rest the potassium conductance predominates but increasing pressures, P, increase g_{Na}. The baroreceptor membrane depolarises and gives rise to the generator current which flows to the spike-initiating zone. The steady-state nervous discharge, fss, is a linear function of this current and the final expression relating it to pressure, Na^+_0 and K^+_0, is:

$$fss = A_2(E_{Na} - E_K)\{A_1(P - Pss)/[1 + A_1(P - Pss)]\} \qquad (17.1)$$

where E_{Na} and E_K are the sodium and potassium equilibrium potentials, respectively, Pss is the pressure threshold for steady-state discharge, A_1 is a constant relating the sodium–potassium conductance ratio to pressure and A_2 is a constant relating fss to generator current (Saum *et al.*, 1978). These constants are obtained from a baroreceptor pressure–discharge curve in control solution. Changes in Na^+_0, K^+_0 or pressure then generate all other curves. As Fig. 17.10*b* shows, the theory accurately predicts the experimental data. The model holds for regularly discharging baroreceptors and does not distinguish between ionic effects upon the receptor or its spike-initiating zone. From a functional point of view such distinctions are unnecessary.

Reflex experiments

Function of the electrogenic sodium pump

The reflex role of the electrogenic pump was evaluated by blocking it. Injection of ouabain ($250 \mu g$ in $0.1 \ cm^3$ of heparine saline) into the common carotid artery of acutely anaesthetised rats produces a transient hypotension followed by an increase in arterial pressure (Ayachi, Saum & Brown, 1978). Similar diphasic pressure responses have been reported elsewhere (Quest & Gillis, 1971). The initial hypotension is mediated by the carotid sinus nerve since it is abolished by sectioning the nerve. This result is predicted from the threshold-lowering and sensitivity-increasing actions of cardiac glycosides upon the membrane electrogenic sodium pump of baroreceptors (Saum et al., 1976).

The potential hypotensive actions of cardiac glycosides were examined in spontaneously hypertensive rats (SHRs). We found that the chronic administration of $4-5 \ mg \cdot kg^{-1}$ digitoxin by gastric tube for one week to 23 SHR rats and to 11 comparable normotensive rats produced a significant reduction in arterial blood pressure in the SHRs. The pressure was measured by the tail cuff method. The results suggest that electrogenic sodium pump activity is greater in SHRs although the experiments on single unit baroreceptor discharge did not bear this out (Saum et al., 1976). This aspect of baroreceptor function, particularly as it may apply to resetting in hypertensive animals, requires further investigation.

Sodium sensitivity of baroreceptors

We have found that changes in Na^+_0 of as little as 5% can produce substantial reflex baroreceptor increases in arterial blood pressure and urine flow (Kunze et al., 1977, 1978). The experiments were performed upon acutely anaesthetised cats in which one carotid sinus was isolated and perfused and the other sinus nerve, both cervical vagi and the depressor nerves were cut. The results are shown in Fig. 17.11. We have not checked the actions of Na^+_0 on atrial reflexes but we suspect that similar results will be obtained. Changes in Na^+_0 of this magnitude are known to occur physiologically (Bartter et al., 1951). Prompt regulation of these changes can be effected via the baroreceptors and atrial receptors so that extracellular volume may be controlled on a fast time-scale, as well as the better known slower mechanism that involves renal output and oral intake of salt and water.

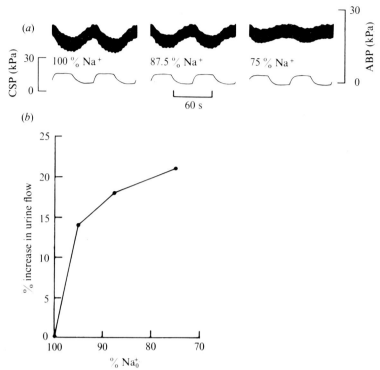

Fig. 17.11. Reflex effects of lowering Na^+_0 in an isolated carotid sinus on arterial blood pressure *(a)* and urine flow *(b)*. Acutely anaesthetised cat with bilateral cervical vagotomy and the other carotid sinus nerve cut. Top tracing in *(a)* is arterial blood pressure (ABP) and the bottom tracing is carotid sinus pressure (CSP). Lowering sinus Na^+_0 increases mean arterial pressure and reduces sinus pressure induced oscillations of arterial pressure. Changes in Na^+_0 of as little as 5% had detectable effects on blood pressure and urine flow.

Conclusion

It is truly remarkable that despite years of investigation of baroreceptor function, their intrinsic or membrane properties have been ignored. The present results suggest that the ionic sensitivity of baroreceptors may be as important as their well-known and extensively studied mechanosensitivity. By virtue of their connections in the central nervous system and the reflex responses they elicit, they provide a mechanism for rapid regulation of extracellular sodium and body fluid volume.

This work was supported by DHEW grants HL-16657 and HL-18798.

References

Ayachi, S., Saum, W.R. & Brown A.M. (1978). The hypotensive effects of cardiac glycosides in normotensive and spontaneously hypertensive rats. *Circulation Res.* (In press.)

Bartter, F.C., Albright, F., Forbes, A.P., Leaf, A., Dempsey, E. & Carroll, E. (1951). The effects of adrenocorticotropic hormone and cortisone in the adrenogenital syndrome associated with congenital adrenal hyperplasia: an attempt to explain and correct its disordered hormonal pattern. *J. clin. Invest.*, **30**(3), 237–51.

Brown, A.M., Saum, W.R. & Tuley, F.H. (1976). A comparison of aortic baroreceptor discharge in normotensive and spontaneously hypertensive rats. *Circulation Res.*, **39**, 488–96.

Edwards, C., Terzuolo, C.A. & Washizu, T. (1963). The effect of changes of the ionic environment upon an isolated crustacean sensory neuron. *J. Neurophysiol.*, **26**, 948–57.

Gray, J.A.B. & Sato, M. (1953). Properties of the receptor potential in Pacinian corpuscles. *J. Physiol.*, **122**, 610–36.

Katz, B. (1950). Depolarization of sensory terminals and the initiation of impulses in the muscle spindle. *J. Physiol.*, **111**, 261–83.

Kunze, D.L., Saum, W.R. & Brown, A.M. (1977). Sodium sensitivity of baroreceptors mediates reflex changes of blood pressure and urine flow. *Nature*, **267**, 75–8.

Kunze, D.L., Saum, W.R. & Brown, A.M. (1978). Sodium sensitivity of baroreceptors: reflex effects on blood pressure and fluid volume. *Circulation Res.* (In press.)

Landgren, S. (1952). On the excitation mechanism of the carotid baroreceptors. *Acta physiol. scand.*, **26**, 1–34.

Matsuura, S. (1973). Depolarization of sensory nerve endings and impulse initiation in common carotid baroreceptors. *J. Physiol.*, **235**, 31–56.

Ottoson, D. & Shepherd, G.M. (1971). Transducer characteristics of the muscle spindle as revealed by its receptor potential. *Acta physiol. scand.*, **82**, 545–54.

Quest, J.A. & Gillis, R.A. (1971). Carotid sinus reflex changes produced by digitalis. *J. Pharmac. exp. Ther.*, **177**(3), 650–61.

Saum, W.R., Ayachi, S. & Brown, A.M. (1978). Actions of sodium and potassium ions on baroreceptors of normotensive and spontaneously hypertensive rats. *Circulation Res.*, **41**, 768–74.

Saum, W.R., Brown, A.M. & Tuley, F.H. (1976). An electrogenic sodium pump and baroreceptor function in normotensive and spontaneously hypertensive rats. *Circulation Res.*, **39**, 497–505.

Sokolove, P.G. & Cooke, I.M. (1971). Inhibition of impulse activity in a sensory neuron by an electrogenic pump. *J. gen. Physiol.*, **57**, 125–63.

Thorén, P., Saum, W.R. & Brown, A.M. (1977). Characteristics of rat aortic baroreceptors with nonmedullated afferent nerve fibers. *Circulation Res.*, **40**, 231–7.

Discussion

(a) Discussion initially centred around the mechanisms responsible for post-excitatory depression. Brown stated that post-excitatory depression always occurred and suggested that it was a property of nerve endings and

was related to the high surface area to volume ratio which favoured sodium loading. It was caused by a different mechanism from that responsible for adaptation because, unlike adaptation, it was influenced by ouabain, lithium, cooling and reducing the potassium concentration to zero; it was unaffected by alterations in calcium ion concentration. Brown did not know whether post-excitatory depression was a property of the receptor membrane or the spike-initiating zone but considered that from a functional point of view this was not important.

(b) The other major point of discussion concerned adaptation and post-excitatory depression in atrial receptors during balloon distension. Brown maintained that atrial 'type B' receptors adapted sufficiently so that, within seconds of balloon inflation, their discharge would have adapted out. He suggested that atrial muscle contraction was necessary for sustained discharge and that receptors mediating reflex responses would be those in series with the muscle. This view was supported by Recordati who noted that there was a very high frequency of discharge when the atrium contracts isovolumically and he pointed out that, in the region of the balloons at the vein–atrial junctions, the atrium effectively contracts isovolumically. This was challenged by other participants and it was noted that atrial receptors continued to discharge as long as the balloons were distended, i.e. for several minutes. Also when atrial pressure increased there was an increase in impulse frequency per cardiac cycle. Arndt remarked that a steady-state discharge from atrial receptors had been observed by Paintal and Chapman as well as himself.

The role of afferent cardiac sympathetic fibres in arrhythmia production during coronary occlusion

A. M. BROWN, R. D. FOREMAN, P. J. SCHWARTZ &
H. L. STONE

from Department of Physiology & Biophysics, University of Texas Medical Branch,
Galveston, Texas 77550, and

Istituto Ricerche Cardiovascolari, University of Milan, Via F. Sforza 35, 20122 Milan,
Italy

Coronary artery occlusion is often associated with cardiac arrhythmias which are reduced by bilateral ablation of the stellate ganglia (Harris, Estandia & Tillotson, 1951) and also by unilateral left stellate ganglion blockade by cold (Schwartz, Stone & Brown, 1977). We have previously shown that coronary artery occlusion excites afferent cardiac sympathetic fibres with their sensory endings in both atria and ventricles (Malliani, Recordati & Schwartz, 1973) and elicits a reflex increase in sympathetic activity present in intact and spinal animals (Malliani Schwartz & Zanchetti, 1969). We have also shown that excitation of afferent cardiac sympathetic fibres increases cardiac sympathetic efferent activity and reduces cardiac vagal efferent activity (Schwartz et al., 1973).

The possibility that arrhythmias associated with brief coronary artery occlusions may partially depend upon afferent cardiac sympathetic fibres was investigated. In 37 vagotomised, chloralose-anaesthetised dogs and cats a laminectomy was performed to expose the dorsal roots C8–T5. The left anterior descending and/or the circumflex coronary arteries were occluded for up to 90 s and the number of ectopic beats occurring during occlusion and at release were counted. Arrhythmias were consistently elicited for at least three consecutive trials in 12 animals. At this point, in six dogs and two cats, dorsal roots were sectioned (interrupting the major sensory input from the heart while leaving intact the efferent pathway) and three identical occlusions were performed. Dorsal root section significantly reduced the arrhythmias in all the animals (mean decrease: $62 \pm 19\%$). Four dogs served as controls and the second group of coronary artery occlusions was performed without dorsal root section: arrhythmias never decreased, ruling out number of occlusions or time as causative factors.

We suggest that afferent cardiac sympathetic fibres contribute to arrhythmia production during coronary artery occlusions and that these arrhythmias depend therefore in part upon a cardio-cardiac sympathetic reflex.

References

Harris, A.S., Estandia, A. & Tillotson, R.F. (1951). Ventricular ectopic rhythms and ventricular fibrillation following cardiac sympathectomy and coronary occlusion. *Am. J. Physiol.*, **165**, 505–12.

Malliani, A., Recordati, G. & Schwartz, P.J. (1973). Nervous activity of afferent cardiac sympathetic fibres with atrial and ventricular endings. *J. Physiol.*, **229**, 457–69.

Malliani, A., Schwartz, P.J. & Zanchetti, A. (1969). A sympathetic reflex elicited by experimental coronary occlusion. *Am. J. Physiol.*, **217**, 703–9.

Schwartz, P.J., Pagani, M., Lombardi, F., Malliani, A. & Brown, A.M. (1973). A cardio-cardiac sympathovagal reflex in the cat. *Circulation Res.*, **32**, 215–20.

Schwartz, P.J., Stone, H.L. & Brown, A.M. (1977). Effects of unilateral stellate ganglion blockade on the arrhythmias associated with coronary occlusion. *Am. Heart J.*, **92**, 589–99.

Autonomic afferents at T1 in elicitation of volume-induced tachycardia in the dog

P. D. GUPTA & M. SINGH

from the Vallabhbhai Patel Chest Institute, University of Delhi, India

Intravenous infusion of blood elicited tachycardia in anaesthetised dogs with cardiac β-receptor blockade (dogs with vagal efferents intact but cardiac sympathetic efferents blocked) and bradycardia in dogs with combined cardiac β-receptor blockade and rhizotomy at T1 (dogs with vagal efferents intact but autonomic afferents at T1 plus cardiac sympathetic efferents blocked). The difference in heart rate response between these two preparations suggests the existence of a cardio-acceleratory reflex with its afferent pathway at T1, and its efferent pathway in vagus nerves. Furthermore, heart rate response in dogs in which spinal roots were sectioned selectively at T1 indicates that these afferents predominate on the right side and their entry at T1 may be via both dorsal and ventral roots. Also, in dogs spinalised at T1 or T7, or dogs rhizotomised at individual segments at C8 and T2 to T7, infusion evoked heart rate response which was not significantly different from the response in control dogs, indicating that afferents at these segments make an insignificant contribution to infusion-induced tachycardia.

Some characteristics of cardiac receptors with fibres in sympathetic rami

D. BAKER, H. M. COLERIDGE, J. C. G. COLERIDGE & C. KIDD

from the Cardiovascular Research Institute, University of California and the Department of Cardiovascular Studies, University of Leeds, UK

We recorded impulses from 41 cardiac afferent fibres dissected from the 2nd, 3rd and 4th left thoracic sympathetic rami communicantes of anaesthetised dogs and cats. Control activity was characteristically sparse. A single spike occurred irregularly but in fairly constant relation to the ECG. When pressure was increased in the cardiac chambers firing increased to 1–3 impulses per cardiac cycle. Occlusion of the thoracic aorta commonly caused a sharp initial increase in discharge in ventricular fibres which subsided after a few seconds, firing often reverting to the control level while ventricular pressure was still elevated. Atrial endings usually had a more sustained response to an increase in intracardiac pressure. Generally, both atrial and ventricular endings required a high pressure to stimulate them, but occasionally atrial endings were stimulated when atrial pressure rose by no more than 0.4–0.5 kPa. Receptive fields were defined with a fine probe or bristle. In marked contrast to their modest response to an increase in intracardiac pressure fibres fired at a high frequency when their terminals were touched lightly. Of 41 cardiac fibres 17 had a single terminal apiece (located on either the atrium or ventricle), 11 had 2–4 terminals on a single cardiac chamber, and 13 had 2–7 terminals widely scattered over the heart, great vessels and pericardium.

Efferent and central nervous mechanisms

18. Cardiac vagal efferent activity

E. NEIL

Cardiac vagal efferent impulses should ideally be recorded intra-thoracically from slips of vagal branches coursing to the heart itself. Such records were first obtained by Rijlant (1936*a, b*). Marguth, Raule & Schaefer (1951), Green (1959), Kunze (1972) and Kordy (1973) further developed such studies.

Unfortunately, the technique involves surgical procedures which combine to minimise the information obtained about the activity of cardiac vagal units in animals in 'physiological condition'. Moreover, most of the experiments listed above were performed on anaesthetised cats – a species which manifests little evidence of tonic cardiac vagal restraint even when subjected to minimal surgical interference. Jewett (1964) and Iriuchijima & Kumada (1963, 1964) pioneered the technique of sampling single- or few-fibre efferent units from the cervical vagus in dogs, which showed impulse activity consonant with that expected of cardiac vagal fibres. We have extended such studies as reported in a series of communications (Kordy, 1973; Kordy, Neil & Palmer, 1974; Neil, 1975; Kordy, Neil & Palmer, 1975) and have demonstrated some of our main findings to the Physiological Society (Neil & Palmer, 1975).

The dog

Dogs premedicated with morphine (2 mg·kg^{-1} body weight) are anaesthetised by the intravenous administration of a-chloralose (60 mg·kg^{-1}) and urethane (600 mg·kg^{-1}) dissolved in saline. After the insertion of an endotracheal tube or after tracheostomy the animals are allowed to breathe spontaneously. End-tidal carbon dioxide, the diaphragmatic electromyogram and a thermistor device attached to the tracheal tube are monitored simultaneously with systemic blood pressure (recorded from the abdominal aorta or femoral artery), the ECG and the heart rate. Arterial pCO_2, pO_2 and pH are determined approximately half hourly.

One or both carotid sinuses are prepared so that blood flow from the carotid artery occurs only via the external carotid artery and the carotid body veins. The pressure in the carotid bifurcation is recorded via a catheter in the lingual artery. On clipping the common carotid and external carotid arteries, the carotid bifurcation is 'tight' and the pressure in the sinus and carotid body segment can be varied by connecting the lingual catheter system to a pressure reservoir.

When free carotid blood flow is permitted, the dogs usually show marked sinus arrhythmia. The depth of anaesthesia is important in this respect. If the level of anaesthesia becomes light, sinus arrhythmia disappears and the heart rate quickens. The loss of vagal tone in such conditions is probably caused by a central influence of nociceptive afferents, which is inhibitory to the baroreceptor cardiac vagal reflex as noted by Iriuchijima & Kumada (1964).

One cervical vagus is dissected clear between the level of the superior laryngeal nerve and that of the thyroid artery. The vagal sheath is incised and its blood vessels ligated. Vagal strands are separated in continuity and tested for the presence of presumed cardiac efferents as follows:

(1) Electrical stimulation of their peripheral end causes bradycardia.
(2) The central end shows impulse activity which occurs during expiration (and is accompanied by bradycardia) and which falls silent during inspiration.
(3) Efferent cardiac vagal activity is provoked or exacerbated by a rise of pressure in the isolated innervated carotid bifurcation.

The cardiac vagal efferents intermingle with those of the recurrent laryngeal nerve. Murray (1957) showed that in cats this nerve (containing the large myelinated fibres destined to supply the abductors and adductors of the larynx) forms a discrete bundle at the perimeter of the cervical vagal trunk. Its cell bodies lie in the nucleus ambiguus. The cell bodies of the cardiac vagal efferents were previously thought to lie in the dorsal motor nucleus of x. Doubts were first raised about this by Szentagothai (1952), who found scant degeneration of cardiac vagal efferent branches after destroying the dorsal motor nucleus but obvious degenerative changes after lesions of the nucleus ambiguus. Calaresu & Pearce (1965b) could not induce bradycardia by selective stimulation of the dorsal motor nucleus and Kerr (1969) showed that vagal stimulation caused bradycardia in chronically prepared cats whose dorsal motor nucleus had been destroyed. The elegant experiments of McAllen & Spyer (1975, 1976) show unequivocally that the nucleus ambiguus is indeed the site of the cardiac vagal cell bodies and I have asked our Chairman, Professor Hilton, to expand on their work (done in his laboratory) in the discussion.

This apparent digression leads me to an important practical point. When splitting the tiny strands, proved to contain cardiac efferents, the rhythmic activity of the large 'abductor' fibres (Green & Neil, 1955), which discharge with inspiration, is a useful guide to the closeness of the small myelinated fibres of the cardiac efferents. Moreover, the isolation of a strand which contains a single 'inspiratory' fibre as well as the cardiac efferent yields evidence of the natural rhythm of the reticular complex known as the respiratory centre. This rhythm still occurs during artificial respiration of animals given muscle relaxants – providing the pump ventilation is not excessive – and enables us to dissociate between the influence of the *central* respiratory rhythm on the pattern of vagal efferent activity and that of peripheral influences, such as those of the Hering–Breuer afferents, which are determined by the rate and stroke of the pump.

Fig. 18.1 shows that vagal cardiac efferent activity occurs during natural expiration and does not occur during natural inspiration (signalled by the discharge of the laryngeal abductor fibre). Moreover, it shows that when the pressure in one carotid sinus is raised in expiration (upper record) the increase in cardiac vagal efferent activity and the accompanying bradycardia occurs promptly. When sinus pressure is raised in inspiration, no bradycardia and no increase in vagal activity occurs until the 'inspiratory' neuron has ceased to discharge. Each record shows that even during sinus hypertension every inspiration is attended by tachycardia and is characterised by a disappearance of cardiac vagal activity.

Fig. 18.2 indicates strikingly the importance of the carotid baroreceptors in determining cardiac vagal discharge. In the upper part of the figure both carotid sinuses were exposed to the systemic arterial blood pressure; sinus arrhythmia is obvious and the phasing of cardiac vagal activity clear. When one common carotid artery is clipped and the carotid sinus pressure in the other bifurcation is lowered to zero there is no vagal discharge and no sinus arrhythmia (lower record). Sinus hypertension induces vigorous vagal discharge and cardiac slowing, showing the usual respiratory phasing.

Chemoreceptor stimulation alone often causes no bradycardia because of the reflex hyperpnoea which it induces (Daly & Scott, 1963). However, if this hyperpnoea is inhibited by stimulation of the superior laryngeal nerve

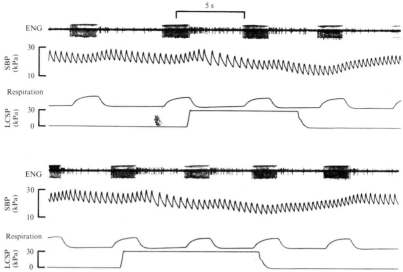

Fig. 18.1. Dog, 12.4 kg. Morphine–chloralose–urethane anaesthesia. Left carotid sinus isolated as described in text. Right carotid circulation free. Spontaneous respiration. Upper record and lower record each shows from above downward: electroneurogram (ENG) of a strand of the cervical vagus, systemic blood pressure (SBP), respiration (inspiration upwards) and left carotid sinus pressure (LCSP). Note that the electroneurogram shows the activity of a single inspiratory neuron and one cardiac vagal fibre which discharges only during expiration. In the upper record the left carotid sinus pressure is raised from zero to 30 kPa at the beginning of expiration; cardiac vagal discharge occurs promptly and is accompanied by bradycardia. During sinus hypertension, each spontaneous inspiration is accompanied by tachycardia and a disappearance of cardiac vagal discharge. In the lower record, sinus pressure is raised to 30 kPa during inspiration. No bradycardia and no cardiac vagal discharge occurs until expiration supervenes. The time interval at the top of the record signals 5 s.

then an injection of sodium cyanide into the carotid body provokes asystole (Fig. 18.3), as reported by Kordy *et al.* (1974, 1975). Such results are similar to those of Angell-James & Daly (1972), who showed that nasal reflexes which induce reflex apnoea similarly unmask the reflex bradycardia provoked by chemoreceptor stimulation.

When dogs breathing spontaneously are paralysed by succinyl choline the diaphragmatic electromyogram and the tidal air movements (recorded either by end-tidal carbon dioxide variations or by thermistor) disappear leaving the central rhythm of the laryngeal abductor neuron identifying the activity of the respiratory centre. It is then clear that the impulse bursts of the cardiac vagus occur between the natural discharges of the inspiratory

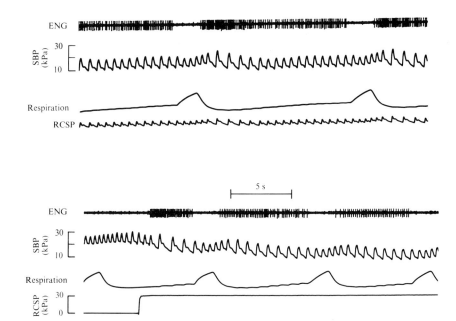

Fig. 18.2. Dog, 17.3 kg. Morphine–chloralose–urethane anaesthesia, spontaneous respiration. Right carotid sinus prepared for isolation as required. Both the upper and lower records show from above downwards; electroneurogram (ENG) of a single cardiac vagal fibre, systemic blood pressure (SBP), respiration (inspiration upwards) and pressure recorded (via the right lingual artery) in the right carotid bifurcation (RCSP).

The upper record shows that when both carotid bifurcations are supplied by systemic arterial blood vagal discharge is brisk during expiration but vanishes during inspiration; inspiration is accompanied by tachycardia. The lower record shows the situation when the left common carotid artery is clamped and the right carotid sinus is isolated. Initially the pressure in the right carotid sinus is zero. No vagal discharge is seen; tachycardia is obvious. When the pressure in the right carotid sinus is raised during expiration, vagal activity and bradycardia supervene. Both vagal discharge and bradycardia are abolished during each inspiratory phase. The time interval shows 5 s.

neuron and do not alternate with the rhythm of the pump, providing that the tidal ventilation volume is not excessive. Similarly, sinus inflation induces reflex vagal discharges with a latency depending on the phase of the natural inspiratory firing and not on the phase of pump ventilation (Fig. 18.4).

Flaxedil (gallamine triethiodide) has been extensively used as a muscle relaxant with the intention of suppressing secondary cardiac chronotropic

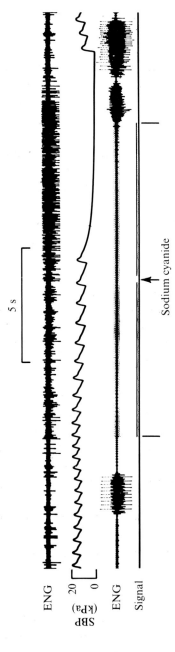

Fig. 18.3. Dog, 15.0 kg. Morphine–chloralose–urethane anaesthesia, spontaneous respiration. Left external carotid artery and left occipital artery tied. Left lingual artery catheterised. Records show from above downwards: electroneurogram (ENG) of cardiac vagus, systemic blood pressure (SBP), diaphragmatic electromyogram (EMG) and signal marker. Time interval 5 s. Between the vertical lines the central end of the right superior laryngeal nerve is stimulated (10 V, 25/s, 0.5 ms) causing inhibition of respiration and some cardiac slowing. At the arrow (shown as a break in the signal line) 50 μg of sodium cyanide in 0.5 cm^3 saline is injected via the lingual catheter. Its arrival at the carotid body promotes massive discharge of the cardiac vagal efferent fibres and cardiac asystole. On ceasing stimulation of the superior laryngeal nerve inspiration supervenes followed shortly by recovery from cardiac asystole.

reflexes induced by chest movements (e.g. Öberg & Thorén, 1972). It is, however, vagolytic – it abolishes the effects of peripheral vagal efferents (Fig. 18.5).

To summarise: in the dog under morphine–chloralose–urethane anaesthesia, cardiac vagal efferent impulses can be recorded electroneurographically and can be related to the well-known phenomenon of sinus arrhythmia. Cardiac efferent impulses are dependent on baroreceptor afferent input from the sino-aortic areas, as was shown from measurements of heart rate by Koch (1931) before and after sino-vagal deafferentation. The carotid sinus baroreceptors seem much the more important source of this afferent input for if both isolated sinus segments are exposed to low perfusion pressure vagal efferent activity is sparse or non-existent even though both vagi, containing all the aortic, all the atrio-ventricular and all the pulmonary vascular and juxtacapillary afferents are intact. This statement is not meant to imply that these multitudinous afferents *cannot* provoke reflex cardiac vagal slowing – for they can – but only to stress that in dogs in good physiological condition the carotid sinus baroreceptors play the major role in reflex cardiac vagal restraint. Further analysis of the part played by the aortic (or other thoracic cardiovascular) receptors in the intact animal is difficult. The aortic nerve is rarely separate in the dog and when it is, it is a mixed nerve always containing chemoreceptors as well as baroreceptors; other afferents are usually present and often thoracic sympathetic fibres accompany them. The mere identification of the aortic nerve by its joining the superior laryngeal root as described by Cyon (see Koch, 1931, for references) is no longer permissible. Identification can only be certain by recording the electroneurogram and demonstrating baroreceptor activity. Even then the mixed character of the nerve renders the results of stimulation of its central end almost protean. In the rabbit there is no chemoreceptor component; the effect of central stimulation of the so-called aortic nerve led Cyon and Ludwig to term it the 'nervus depressor'. The dog and the cat, unlike the rabbit, show variable responses to central stimulation of the central end of the aortic nerve; it seems prudent to dispense with the description 'aortic depressor' for the nerve in the cat and the dog.

The cat

The anaesthetised cat usually has a heart rate of 180 beats/min or more, perhaps suggesting that its tonic vagal restraint is poor. Even when both carotid sinus preparations are inflated bradycardia may be brief and

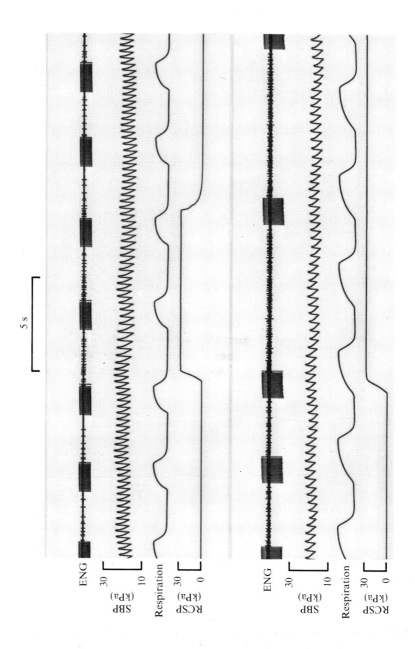

transient although reflex systemic hypotension is indeed brisk and impressive. Thus, the analysis of unitary activity in cardiac vagal fibres is difficult in the cat, although to her credit Kunze (1972) was successful. Some, who possibly should be nameless, have argued that there is a relative scarcity of cardiac vagal efferents in the cat. It is true that Agostoni, Chinnock, Daly & Murray (1957) found only 500 efferent fibres in the cardiac vagal branches of the cat. Nevertheless, it is a simple class demonstration to show that stimulation of the peripheral end of the cervical vagus will stop the heart of the cat. It follows that the feeble degree of tonic vagal restraint shown by the anaesthetised cat is more likely due to a *central* inhibition of the baroreceptor reflexes.

Lopes & Palmer (1975) examined the role of diencephalic structures in modifying the reflex response of the heart rate to stimulation of the central end of the sinus nerve. In the 'intact' cat sinus nerve stimulation causes only mild bradycardia; this response is moderately increased by simultaneous stimulation of the central end of the superior laryngeal nerve. However, when the central connections between the diencephalon and mesencephalon are bilaterally severed at the level of the mamillary bodies employing a special knife (Lopes & Neto, 1973) (which effects a semi-circular cut at the anteroposterior plane A 7.0 to a depth of from 0 to -5 in the vertical plane), the situation is altered. The lesion itself causes only a minor and transient fall of blood pressure but lowers the heart rate by some 25%. After

---

Fig. 18.4. Dog, 16.3 kg. Morphine–chloralose–urethane anaesthesia. Right carotid bifurcation prepared for isolation and is temporarily isolated in each of these records. Left carotid bifurcation undisturbed and receives its natural blood supply. In both records, from above downwards: electroneurogram (ENG) of cervical vagus efferent strand (containing one 'abductor' neuron and one cardiac vagal efferent), systemic blood pressure (SBP), respiration (inspiration upwards) and right carotid sinus pressure (RCSP). Time interval is 5 s. In the upper record, cardiac vagal activity occurs sparsely and only in expiration. When the right carotid sinus pressure is raised from 0, cardiac discharge occurs promptly but is still inhibited during each phase of activity of the inspiratory neuron. The lower record shows the results 10 min after the i.v. administration of succinylcholine (5 mg in 1 cm³ saline) which, paralysing respiration, required artificial respiration provided by a pump (stroke volume 160 cm³, rate 14/ min). The same vagal strand shows rhythmic firing of the 'abductor' neuron, interspersed with discharge of the cardiac vagal fibre. Cardiac vagal impulse activity is unrelated to the rhythm of the pump but is locked to that of the central discharge. On raising the right carotid sinus pressure – initially 0 – cardiac impulse traffic is increased but still bears no rhythmic relation to the chest movements caused by the pump.

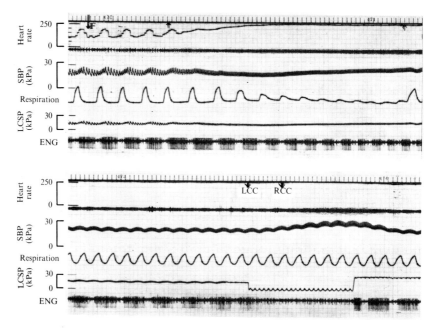

Fig. 18.5. Dog, 14.6 kg. Morphine–chloralose–urethane anaesthesia, initially
breathing spontaneously. The left carotid bifurcation was prepared either for
natural blood flow through the common carotid artery or for carotid sinus-body
perfusion. The right external carotid and occipital arteries were ligated. Each record
shows from above downwards: time in seconds, heart rate, systemic blood pressure
(SBP), respiration, left carotid sinus blood pressure (LCSP) (recorded via a lingual
catheter) and the electroneurogram (ENG) of a cervical vagal strand. The upper
record shows the animal initially breathing spontaneously and manifesting sinus
arrhythmia, accompanied by rhythmic bursts of cardiac vagal discharge occurring
during expiration. At the arrow (F) flaxedil infusion is commenced and is suspended
at the second arrow. Note that the heart rate accelerates prior to any failure of
natural respiration reaching a value of more than 250/min. Natural breathing fails
but the rhythmic discharge of the cardiac vagal unit continues. Pump respiration is
commenced at the third arrow. The lower record (taken about 1.25 min later) shows
the heart rate to be above 250/min (for the recording is identical with that of the
seconds time-scale) and the central rhythmic vagal discharge continuing without
relation to the rhythm of pump ventilation. In this lower record two arrows show
clamping of first the left and then the right carotid arteries, respectively (LCC and
RCC). Note that the pressure in the left carotid artery is now submitted to a
pulsatile stimulus slightly in excess of 0 and that the cardiac vagal unit falls silent.
On raising the left carotid sinus pressure to 30 kPa during the inspiratory phase of
pump ventilation, cardiac vagal discharge reappears immediately and then shows a
rhythm unrelated to that of the pump, but virtually identical with that of the
'respiratory centre'.

the lesion, sinus nerve stimulation now causes a much more striking bradycardia. When the central end of the superior laryngeal nerve is stimulated simultaneously the heart rate falls from 150 to 30 beats/min (Fig. 18.6).

These results are consonant with Hilton's (1963) findings that stimulation of the hypothalamic defence area (Abrahams, Hilton & Zbrozyna, 1960) inhibits the baroreceptor cardiac vagal reflex.

Lopes & Palmer (1976a) have not only confirmed the results of Hilton (1963) but have additionally shown that stimulation of the hypothalamic defence area will still inhibit the reflex vagal bradycardia induced by sinus nerve stimulation even in cats in which the spinal cord has been sectioned at either the level of C1 or that of C6. It is well known that hypothalamic defence area stimulation induces powerful discharge of the cardiac sympathetic efferents in the 'intact' cat. However, such an exacerbation of cardiac sympathetic activity cannot occur after section of the cervical spinal cord – yet hypothalamic stimulation still inhibits the reflex vagal bradycardia which might be expected on sinus nerve stimulation. Further analysis reveals that this inhibitory effect of hypothalamic stimulation on cardiac vagal bradycardia is exerted by the arousal or increased activity of hypothalamic neurons which activate 'inspiratory' units of the medullary respiratory reticular complex. Thus, cats with spinal sections at the level of C6, which still breathe spontaneously, show inspiratory spasm of the diaphragm in response to hypothalamic stimulation. Cats with spinal section at the level of C1 can no longer breathe spontaneously and require pump ventilation. However, recordings of 'abductor' laryngeal fibres show that the central rhythm of respiration may continue and that stimulation of the hypothalamic defence area still provokes sustained firing of these inspiratory units. This increased response of inspiratory medullary neurons to hypothalamic stimulation can be almost completely prevented by simultaneous excitation of the central end of the superior laryngeal nerve. When the response of the heart rate to sinus nerve stimulation is examined against the background of combined stimulation of the defence area plus that of the superior laryngeal nerve, then the spinalised cat shows striking bradycardia.

Lopes & Palmer (1976a) conclude that the inhibitory effect of the hypothalamus upon the reflex cardiac vagal bradycardia induced by sinus nerve stimulation is caused partly by sympathetic discharge aroused from the hypothalamus and partly by the excitation of medullary inspiratory neurons, which block or 'gate' the effect of the baroreceptor input on the

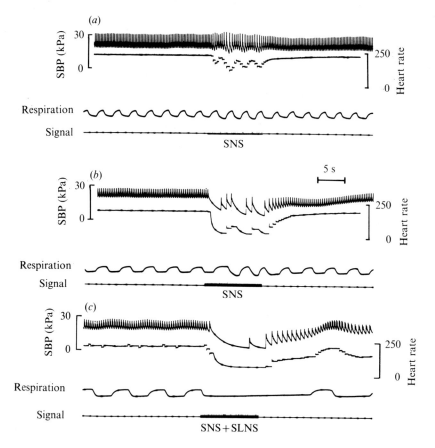

Fig. 18.6. Cat, 3.8 kg. Chloralose–urethane anaesthesia, spontaneous respiration. Records show from above downwards: systemic blood pressure (SBP), heart rate, respiration (inspiration upwards) and signal marker. Time interval 5 s. *(a)* Shows the bradycardia and sinus arrhythmia evoked by stimulation of the central end of the sinus nerve (15 V, 25/s, 0.5 ms) in the 'intact' animal before making the hypothalamic lesion described in the text. *(b)* Shows that sinus nerve stimulation provokes a much more pronounced bradycardia following section of the hypothalamic connections with the rhombencephalon. *(c)* Reveals that when the sinus nerve and the central end of the superior laryngeal nerve are stimulated together the inhibition of respiration is accompanied by a profound bradycardia. (Note: before the hypothalamic lesion the stimulation of the superior laryngeal nerve only moderately increased the response of the heart rate to sinus nerve stimulation (SNS).) SLNS = superior laryngeal nerve stimulation

cardiac vagal centre. They have extended their concept of this respiratory gating mechanism recently (Lopes & Palmer, 1976*b*).

To summarise: the influence of the discharge of central respiratory neurons should not be ignored in experiments which purport to examine cardiac vagal reflex responses. The administration of muscle relaxants to intact or thoracotomised animals, necessitating thereby the use of artificial respiration, in no way excludes the possibility that afferent inputs still exercise their effect upon central respiratory rhythms.

References

Abrahams, V.C., Hilton, S.M. & Zbrozyna, A. (1960). Active muscle vasodilation produced by stimulation of the brain stem: its significance in the defence reaction. *J. Physiol.,* **154,** 491–513.

Agostoni, E., Chinnock, J.E., Daly, M. De B. & Murray, J.G. (1957). Functional and histological studies of the vagus nerve and its branches to the heart, lungs and abdominal viscera in the cat. *J. Physiol.,* **135,** 182–205.

Angell-James, Jennifer, E. & Daly, M. De B. (1972). Reflex respiratory and cardiovascular effects of stimulation of receptors in the nose of the dog. *J. Physiol.,* **220,** 673–96.

Calaresu, F.R. & Pearce, J.W. (1965*a*). Electrical activity of efferent vagal fibres and dorsal nucleus of the vagus during reflex bradycardia in the cat. *J. Physiol.,* **176,** 228–40.

Calaresu, F.R. & Pearce, J.W. (1965*b*). Effects on heart rate of electrical stimulation of medullary vagal structures in the cat. *J. Physiol.,* **176,** 241–51.

Daly, M. De B. & Scott, Mary J. (1963). The cardiovascular responses to stimulation of the carotid body chemoreceptors in the dog. *J. Physiol.,* **165,** 179–97.

Green, J.H. (1959). Cardiac vagal efferent activity in the cat. *J. Physiol.,* **149,** 47P.

Green, J.H. & Neil, E. (1955). The respiratory function of the laryngeal muscles. *J. Physiol.,* **129,** 134–41.

Hilton, S.M. (1963). Inhibition of baroreceptors on hypothalamic stimulation. *J. Physiol.,* **165,** 56–7P.

Iriuchijima, J. & Kumada, M. (1963). Efferent cardiac vagal discharge of the dog in response to electrical stimulation of sensory nerve. *Jap. J. Physiol.,* **13,** 599–706.

Iriuchijima, J. & Kumada, M. (1964). Activity of single vagal efferent cardiac fibres in the dog. *Jap. J. Physiol.,* **14,** 479–97.

Jewett, D.L. (1964). Activity of single efferent fibres in the cervical vagus of the dog with special reference to possible cardio-inhibitory fibres. *J. Physiol.,* **175,** 321–57.

Kerr, F.W.L. (1969). Preserved vagal visceromotor function following destruction of the dorsal motor nucleus. *J. Physiol.,* **202,** 755–69.

Koch, E. (1931). *Die reflektorische Selbsteuerung des Kreislaufes Sreinkopf.* Steinkopff: Leipzig.

Kordy, M.T. (1973). An electroneurographic study of cardiac vagal reflexes. Ph.D. thesis, University of London.

Kordy, M.T., Neil, E. & Palmer, J.F. (1974). The combined influence of chemoreceptor and superior laryngeal nerve stimulation on cardiac vagal discharge. In *Proc. XXVIth Int. Congr. Physiol.*, vol. XI, p. 62, Abstract 185. Secretary General of the XXVIth Congress: India.

Kordy, M.T., Neil, E. & Palmer, J.F. (1975). The influence of laryngeal afferent stimulation on cardiac vagal responses to carotid chemoreceptor excitation. *J. Physiol.*, **247**, 24–5P.

Kunze, D.L. (1972). Reflex discharge patterns of cardiac vagal efferent fibres. *J. Physiol.*, **222**, 1–15.

Lopes, O.U. & Neto, J.C. (1973). The effect of hypothalamic lesions on the carotid occlusion reflex. *J. Physiol.*, **232**, 37P.

Lopes, O.U. & Palmer, J.F. (1975). Tonic diencephalic inhibition of the cardiac vagal centre in the cat. *J. Physiol.*, **254**, 62–3P.

Lopes, O.U. & Palmer, J.F. (1976a). Hypothalamic inhibition of vagal component of the sinus nerve cardiac reflex. *J. Physiol.*, **260**, 45–6P.

Lopes, O.U. & Palmer, J.F. (1976b). Proposed respiratory 'gating' mechanism for cardiac slowing. *Nature (Lond.)*, **264**, 454–6.

Marguth, H., Raule, W. & Schaefer, H. (1951). Aktionströme in zentrifugalen Herznerven. *Pflügers Arch. ges. Physiol.*, **254**, 224–45.

McAllen, R.M. & Spyer, K.M. (1975). The origin of cardiac vagal efferent neurones in the medulla of the cat. *J. Physiol.*, **244**, 82P.

McAllen, R.M. & Spyer, K.M. (1976). The location of cardiac vagal preganglionic motorneurones in the medulla of the cat. *J. Physiol.*, **258**, 187–204.

Murray, J.G. (1957). Innervation of the intrinsic muscles of the cat's larynx by the recurrent laryngeal nerve: a unimodal nerve. *J. Physiol.*, **135**, 206–12.

Neil, E. (1975). Integration of cardiovascular reflexes. In *Proc. IIIrd Congr. Turkish Physiol.*, pp. 23–4. Sermet: Istanbul.

Neil, E. & Palmer, J.F. (1975). Effects of spontaneous respiration on the latency of reflex cardiac chronotropic responses to baroreceptor stimulation. *J. Physiol.*, **247**, 16P.

Öberg, B. & Thorén, P. (1972). Increased activity in left ventricle during haemorrhage or occlusion of caval veins in the cat; a possible cause of the vago-vagal reaction. *Acta physiol. scand.*, **85**, 164–73.

Rijlant, P. (1936a). Le contrôle extrinsique de la fréquence du battement cardiaque. *C.R. Soc. Biol. Paris*, **123**, 99–101.

Rijlant, P. (1936b). L'arrhythmie cardiaque respiratoire. *C.R. Soc. Biol. Paris*, **123**, 997–1001.

Szentagothai, J. (1952). The general visceral efferent column of the brain stem. *Acta morph. hung.*, **2**, 313–28.

Discussion

(a) There was some discussion of the connections of the vagal cardiomotor neurons and Hilton referred to mapping work of Spyer and McAllen which indicated that baroreceptor pathways converged on the nucleus ambiguus. They had inserted micro-electrodes and mapped the region from which they could evoke antidromic activity from stimulating the cardiac branch of the right vagus. The area was in a region containing so-called inspiratory neurons. The units were normally silent but could be activated by micro-iontophoresis. They also showed that the effectiveness of brief baroreceptor or chemoreceptor pulses depended on their timing in relation to the respiratory cycle.

(b) In answer to a comment Neil remarked that although Hering–Breuer afferents discharged during normal inspiration they appeared to be unimportant in controlling respiration in unanaesthetised man.

(c) Gootman had noted that when she had stimulated the pressor area the sympathetic discharge was greater during inspiration and less during expiration.

(d) Freyschuss indicated that he had studied sinus arrhythmia in conscious man breathing normally and had noted that during inspiration there were increases in heart rate and stroke volume. When the subjects were placed in a ventilator and instructed to breathe to follow the ventilator, sinus arrhythmia reversed. He offered no explanation for this observation. It was remarked that there are fewer vagal efferent fibres to the heart in the cat compared with the dog and that higher baroreceptor pressures were required in the cat to stimulate the efferent vagal discharge. However, it was pointed out that because there are fewer fibres in the cat it does not imply that the effect is less.

19. Central neurons activated by cardiac receptors

C. KIDD

There exists an extensive body of knowledge on receptors in the cardiac chambers and great vessels which have myelinated and non-myelinated afferent fibres in the vagal nerves (see the following chapters in this volume) and there is also detailed knowledge of the reflex responses evoked when these receptors are activated. However, when the literature is searched with the particular brief of the central nervous connections of cardiac receptors in mind (the functional anatomy of the central pathways, particularly the location of interneurons and the central nervous pathways to the efferent neurons of the vagal and sympathetic efferent systems) then the information is very sparse. In this account the existing evidence specifically on the 'cardiac' pathways will be considered; where useful, comment will be made about carotid sinus and aortic pathways but this evidence will not be considered in detail. The existing evidence on central connections within the brain stem will be considered under several headings: histological evidence for the distribution of vagal afferent fibres; the distribution and characteristics of the populations of neurons excited by selective stimulation of a discrete group of cardiac receptors – the left atrial receptors.

Anatomical and degeneration studies of vagal afferent fibres

Since the rootlets of the IXth and Xth cranial nerves are closely approximated about their entrance into the medulla most authors consider the effects of their division together. Those studies, using the Marchi, Nauta and Fink–Heimer techniques, which were based on the effects of division of all vagal and glossopharyngeal rootlets in a variety of animals (mouse, rat, guinea pig, opposum, cat, dog and monkey), are unanimous in concluding that the major site of termination for afferent fibres in the IXth and Xth nerves is the nucleus tractus solitarius (Allen, 1923; Ingram & Dawkins, 1945; Åström, 1953; Urabe, Yamazaki & Araki, 1960; Kerr, 1962; Cottle, 1964; Kimmel, 1965; Torvik, 1965; Rhotan, O'Leary & Ferguson, 1966; Culberson & Kimmel, 1972). The nucleus consists of a dorso-medial section composed of small, densely packed cells and a ventrolateral part which is composed of more scattered small and medium cells with some large cells. It is clear that afferent fibres enter the medulla via the rootlets, then travel medially through the intervening tissue to make a ventral approach towards the ipsilateral tractus solitarius and thence pass caudally giving off collaterals to the shell-like nucleus tractus solitarius. The most dense aggregations of degenerating terminals are in the intermediate and caudal areas of the nucleus just above and below the obex. A further zone of terminations is noted in the commissural nucleus, immediately caudal to the obex (Fig. 19.1a). However, since all IXth and Xth rootlets are divided the origins of afferent fibres concerned are likely to be widely distributed throughout the thorax and abdominal viscera.

A more selective approach in which only a few rootlets were divided at one time was adopted by Cottle (1964). The basis for this were electrophysiological studies on the rootlets of the IXth and Xth nerves (Bonvallet & Sigg, 1958; von Baumgarten & Arenda, 1959) which showed that fibres with a 'cardiac rhythm' were found predominantly in the IXth nerve rootlets and the most rostral vagal rootlets. Cottle therefore separately divided small groups of rootlets and examined the brain stem for resultant degeneration using the Nauta technique. She concluded that the most rostral part of the nucleus tractus solitarius received only IXth nerve fibres, the intermediate and caudal regions received fibres from both nerves, while degenerating fibres in the most caudal part were wholly vagal in origin. In the intermediate zone, the areas of most dense termination were the dorso-medial region and a large cell region in the ventrolateral region of the nucleus; more caudally there were some degenerating fibres in the commissural nucleus (Fig. 19.1b). Cottle concluded that since the most rostral rootlets contained fibres with a 'cardiac rhythm' the cardiovascular afferents were likely to be in this intermediate area of the nucleus tractus solitarius. However, several points should be borne in mind. Few animals were involved and the histological techniques for obtaining degenerating fibres are notoriously fickle and there are no adequate methods of confirming whether all or only a proportion of degenerating fibres are stained by this technique; Fink & Heimer (1967) indicate that probably all degenerating fibres are not stained but are unable to give quantitative data.

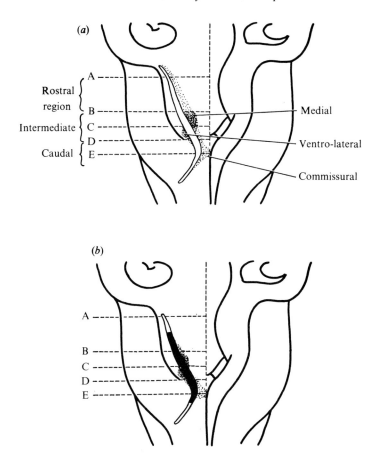

Fig. 19.1. Dorsal view of cat brain stem showing subdivisions of nucleus tractus solitarius. Dotted areas indicate degenerating fibres following division of all IXth and Xth rootlets *(a)* and a group of rostral Xth nerve rootlets *(b)*. From Cottle (1964)

The electrophysiological evidence from the rootlets is also equivocal. There is a significant proportion of 'respiratory' afferents in the middle and rostal rootlets and in no case was the origin of the receptors defined for the 'cardiac' afferents. Furthermore, no account was taken of fibres which did not have an overt cardiac and respiratory rhythm, e.g. many receptors with non-myelinated fibres do not have such patterns.

The application of the Falck and Hillarp technique for the demonstration of monoamines in combination with low-power electron microscopy has revealed further details of the neural structures within the dorso-medial area of the nucleus tractus solitarius. The neurons are small, 10–20 μm in diameter, and there is a rich network of myelinated and non-myelinated fibres. A high proportion of these fibres are catecholaminergic (Fuxe, 1965; Fuxe, Hokfeldt & Nilsson, 1965) but recent degeneration evidence shows that these fibres do not originate from the IXth and Xth nerves (Chiba & Doba, 1975, 1976). Electron microscopic examination of the medial area of the nucleus tractus solitarius of the cat reveals that synaptic contacts made by primary afferent fibres are largely axo-dendritic or axo-somatic in type; axo-axonic synapses were not observed. On the other hand, in the commissural area, while axo-dendritic and axo-somatic synapses predominate, a significant proportion of axo-axonic synaptic contracts could be demonstrated but these were not derived from primary afferent fibres (Chiba & Doba, 1976).

In summary, histological evidence shows that the major distribution of vagal afferent fibres is to the tractus solitarius and fibres terminate in the intermediate and caudal regions of the nucleus tractus solitarius; there are particular aggregations of vagal terminals in the dorso-medial area, a part of the ventrolateral area and the commissural area of the nucleus tractus solitarius. Within the dorso-medial and commissural areas there are rich axo-dendritic and axo-somatic synaptic linkages with the small second order neurons. The detail of the synaptic linkages in the ventrolateral area has not been defined. There is some evidence that IXth nerve afferents do not extend as far caudally as the Xth nerve fibres and in the intermediate zone there is evidence of a separation within the tractus solitarius with Xth nerve fibres more medial to IXth nerve afferents. Although the detailed topography of the termination of cardiac afferent fibres is equivocal as derived from the electrophysiology of vagal rootlets some, at least, are likely to be found in the intermediate and caudal areas of the nucleus tractus solitarius.

Carotid sinus and aortic afferent fibres

The major purpose of this account is to consider primarily the evidence which relates to central neuron connections of vagal cardiac afferents and I shall not therefore consider in detail the evidence on central projections of carotid sinus and aortic baroreceptors. However, in passing, it is important to note the regions of the brain stem shown to be involved. Thus, electrical stimulation of the carotid sinus and aortic nerve afferents have shown projections to the nucleus tractus solitarius, a parahypoglossal area immediately ventral to the hypoglossal nucleus and to neurons in the area of the nucleus ambiguus (e.g. Humphrey, 1967; Crill & Reis, 1968; Miura & Reis, 1969; Seller & Illert, 1969; Biscoe & Sampson, 1970*a, b;* Gabriel & Seller, 1970; Spyer & Wolstencroft, 1971; Lipski, McAllen & Spyer, 1975). There is some doubt about a postsynaptic projection of these nerves to the medial reticular area in the brain stem (see Lipski *et al.,* 1975). A number of studies have examined the effects of distension of the carotid sinus area on medullary neurons (Trzebski *et al.,* 1962; Humphrey, 1967; Koepchen *et al.,* 1967; Biscoe & Sampson, 1970*b;* Miura & Reis, 1972; Lipski *et al.,* 1975; Lipski & Trzebski, 1975) and although the precise detail of the location of neurons is not always given the major pattern is similar to that described above.

Electrical stimulation of vagal nerves

In a number of studies, electrical stimuli were applied to the cervical vagal nerves and evoked 'slow wave' potentials, with latencies compatible with activity in myelinated fibres, have been recorded from a number of areas of the brain stem; the major ones are the medial and commissural components of the nucleus tractus solitarius, the nucleus intercalatus and the dorsal motor nucleus of the vagus in the medial part of the brain stem while, more laterally, responses were also evoked in the region of the nucleus ambiguus (Harrison & Bruesch, 1945; Lam & Tyler, 1952; Anderson & Berry, 1956; Hellner & von Baumgarten, 1961; Porter, 1963).

This distribution is illustrated in Fig. 19.2*a*, which shows the distribution of 'slow wave' responses observed by Porter (1963) in the anaesthetised cat in response to stimulation of the cervical vagal nerves. Recordings from single neurons were also made from these areas and evidence was adduced to show that a proportion were likely to be interneurons.

These investigations were concerned with neurons activated by mye-linated fibres; however, most afferent fibres within the vagal nerves are non-myelinated (e.g. Agostini, Chinnock, Daly & Murray, 1957). Although there have been few studies of the distribution of neurons activated by non-myelinated fibres (Lam & Tyler, 1952; Fussey, Kidd & Whitwam, 1973), they indicate a similar distribution to that described above for myelinated fibres. Fig. 19.2*b* indicates the distribution of evoked responses in the dog medulla following electrical stimulation of the cervical vagal nerves (Fussey *et al.,* 1973). Activity with latencies appropriate to myelinated afferents is evoked within the dorso-medial areas of the nucleus

382   *Efferent and central mechanisms*

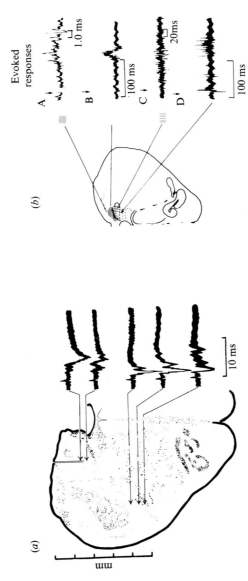

Fig. 19.2. (*a*) 'Slow wave' responses evoked in cat brain stem by electrical stimulation of the cervical vagal nerve (Porter 1963). (*b*) 'Slow wave' and multi-unit activity evoked from dog brain stem by stimulation of cervical vagal nerve. A, Short latency evoked 'slow' wave and spike activity; B, long latency evoked 'slow' wave and spike activity; C, evoked spike activity with medium and long latency; D, spike activity of medium and long latency from dorsal motor nucleus of vagus nerve (Fussey, Kidd & Whitwam, 1973).

tractus solitarius. Overlapping this distribution and extending ventrally is a region in which long-latency responses occur; measurements of conduction velocity in the vagal nerves showed that the excitatory fibres are non-myelinated (< 2.5 m/s). More deeply, the dorsal motor vagal nucleus was another site of evoked responses following excitation of non-mye-linated fibres. Some of this activity was shown to be excited synaptically and some was being evoked antidromically. Of course, the major difficulty in interpreting these responses arises not only from the difficulties in interpreting the nature of some of the different components within these responses but also, the diverse origins in the thorax and abdomen of the afferent fibres at this level render it impossible to assess their role in specific reflex pathways. Clearly, useful progress in defining brain stem pathways involved in 'cardiac' reflexes requires selective excitation of afferent fibres. Recently, we have made some attempts in this direction by restricting the electrical stimuli to a branch of the right thoracic vagal nerve (right cardiac nerve) which passes to the heart behind the azygos vein (Jarisch & Zotterman, 1948; Öberg & Thorén, 1972). Recordings of impulse activity from strands of this nerve show that afferent fibres originate mainly from the atria and ventricles although there are some fibres with receptors in the adjacent bronchi and pulmonary tissue (S. Donoghue, R. Fox & C. Kidd, unpublished observations).

A search was made for electrically evoked responses in anaesthetised cats and recordings were made from that part of the medulla extending from 2 mm caudal to 4 mm rostral to the obex. 'Slow wave' activity which had latencies appropriate to activation by myelinated and non-myelinated fibres was observed throughout the nucleus tractus solitarius and the dorsal motor nucleus of the vagus; responses with latencies appropriate to myelinated fibres were also observed in the region of the nucleus ambiguus. In short, the distribution of 'slow wave' evoked responses confirmed that observed with cervical vagal stimulation in the dog. However, in order to apply tests which allow an adequate differentiation of fibre and cellular recording sites it is essential to record activity from individual neurons and we therefore concentrated upon extracellular recordings from single neurons activated by non-myelinated fibres within the cardiac nerve (Donoghue, Fox, Kidd & Koley, 1977). Recordings were made from a total of 48 neurons and the conduction velocity of the fibres activating them ranged from 0.5 to 2.5 m/s; that is, the peripheral fibres were non-myelinated.

There are several electrophysiological tests which allow differentiation of presynaptic and postsynaptic sites for the recordings (e.g. Fussey, Kidd & Whitwam, 1970); however, these relate only to neurons activated by myelinated fibres. We therefore defined a number of criteria to separate the population into two classes, presynaptic and postsynaptic. Neurons were recorded as being excited synaptically if they responded with more than one spike to supramaximal vagal stimulation, they had large changes in latency when the stimulus intensity was increased from threshold to supramaximal levels and had a large variability of latency at threshold intensities. Conversely, neurons from which only one spike could be evoked, which had small changes in latency between threshold and supramaximal intensities and which exhibited relatively fixed latencies, were categorised as being presynaptic. Unfortunately at this time we do not have adequate information to allow a separation of this latter group (15 cells) into primary afferent fibres excited orthodromically or efferent cell bodies (or fibres) excited antidromically. A total of 22 neurons satisfied criteria which indicated that they were excited synaptically by cardiac vagal non-myelinated fibres but since the cardiac nerve was tied it was not possible to define further the nature of the receptors attached to these fibres. Tests on the remaining 11 neurons were insufficient to allow confident placing in either of the groups. Histological examination of the brain stem showed that the neurons were distributed in the dorso-medial area of the nucleus tractus solitarius, the area postrema and the dorsal motor nucleus of the vagus, with some distributed in the parahypoglossal area. Despite an extensive search through the brain stem these are the only areas where electrically evoked activity from non-myelinated fibres could be recorded. Although these are only preliminary observations on non-myelinated fibre projections several questions now arise. The dorso-medial area of the nucleus tractus solitarius is also the major site of termination for carotid sinus, aortic and vagal myelinated fibres and it will be important to define whether there is topographic separation of the inputs within the area of the nucleus tractus solitarius. What is the extent of the convergence onto an individual neuron of terminals of myelinated and non-myelinated afferent fibres? There are also a number of questions relating to the nature of the receptors and extent of possible inhibitory effects which remain to be defined. We hope to have some answers soon.

In conclusion, the results of electrical stimulation of vagal afferents provide only limited useful information on central nervous connections of cardiac afferents. However, they tend to suggest that the nucleus tractus solitarius and the area surrounding it are likely to be concerned but they reveal little evidence for the involvement of other areas, apart from the nucleus ambiguus. Considerable evidence now exists to show that cardio-motor efferent fibres have their cell bodies in this region (e.g. McAllen & Spyer, 1976) but details of the connections between this area and the nucleus tractus solitarius have still to be defined. Although there are a large number of investigations of sites within the medulla from which cardiac and vasomotor changes may be evoked by electrical stimulation there is no evidence which allows assessment of the relationship of these to reflex pathways involving cardiac receptors.

Physiological excitation of central neurons

Another approach to the problem of identifying central projections of cardiac receptors has been to search for brain stem neurons with spontaneous activity associated with the cardiac cycle (pulse phased) since many of the receptors in the heart with myelinated vagal afferent fibres have patterns of activity which are related to particular phases of the cardiac cycle. A number of investigations have described such neurons (e.g. Hellner & von Baumgarten, 1961; Smith & Pearce, 1961; Salmoiraghi, 1962; Humphrey, 1967; Fussey, Kidd & Whitwam, 1967; Koepchen *et al.,* 1967; Seller & Illert, 1969; Miura & Reis, 1972; Middleton, Woolsey, Burton & Rose, 1973). In a number of investigations the vagal nerves were divided and it can be assumed that the activity was derived from carotid sinus baroreceptors; in others, carotid sinus, vagal and aortic afferents were not differentiated and they are also of little help in defining central pathways of cardiac receptors.

Hellner & von Baumgarten (1961) and Middleton *et al.* (1973), in the cat, have reported neurons in the nucleus tractus solitarius with spontaneous bursts of impulses similar to those of type A or type B atrial receptors (Paintal, 1973). However, in neither study was unequivocal evidence produced to show that the recordings were from second-order neurons rather than primary afferent fibres; furthermore, experiments were not performed to confirm that the receptors involved were in the atria. This is important since there is adequate evidence to indicate that the location of receptors cannot be reliably interpreted solely from consideration of the patterns of discharge (e.g. Coleridge, Hemingway, Holmes & Linden, 1957). There are no studies in which the activity has been shown to originate, on the basis of independent evidence, from interneurons as

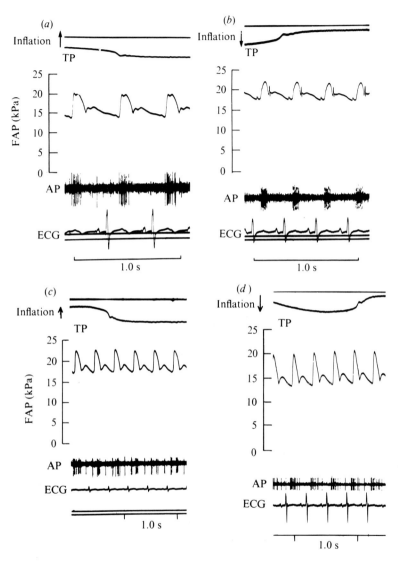

Fig. 19.3. Each panel shows spontaneous activity of an interneuron *(a, b, c, d)* from the nucleus tractus solitarius of the dog. In each case, electrical stimulation indicated that myelinated vagal afferent fibres were activating them. FAP = femoral aterial pressure; AP = action potentials; TP = tracheal pressure (Fussey, Kidd & Whitwarm, unpublished).

compared with fibres and in which the afferent fibres are present in the vagal nerves.

Some time ago at Leeds (Fussey *et al.*, 1967) we attempted to define, in the anaesthetised dog, the distribution within the brain stem of neurons with spontaneous activity associated with the cardiac cycle (pulse phased) and which were excited by vagal cardiac afferents. We used a series of criteria to separate activity recorded extracellularly with platinum/iridium micro-electrodes from primary afferent fibres from that elicited in interneurons (Fussey *et al.*, 1970), in response to electrical stimuli delivered to the intact vagal nerves. Briefly the criteria were: the maximum number of spikes evoked by a single shock, the shift of latency with increases in stimulus intensity, variability of latency, the maximum frequency at which the neuron would respond to repetitive stimulation and the duration of the action potential spike. In order to be included each neuron had to satisfy at least two criteria; most satisfied more. We were able to record from a total of 44 neurons, identified on this basis as interneurons and were able to exclude mechanical movements of the brain stem as a possible artifact. The range of latencies to electrical excitation of the cervical vagal nerves were 3.2–7.2 ms and when conduction time is calculated it is clear that the fibres exciting the neurons are myelinated; furthermore, when allowance is made for intermedullary conduction these latencies are compatible with the notion that few synapses intervene between the primary afferent fibres and the recording site and we suggest that the neurons lie 'early' in the brain stem pathways of the reflex pathways in which they are involved.

Examples of spontaneous activity of four such interneurons are shown in Fig. 19.3. Each cardiac cycle is associated with a burst of action potentials (10–40 imp/cycle) which begins approximately 60–110 ms after the R wave of the ECG. For a number of neurons the pattern of the burst was similar to that observed in aortic or pulmonary arterial baroreceptors; thus interspike intervals reached a minimum early in the burst and thereafter began to decline. In a few cases we were able to analyse the timing of the individual spikes within the cardiac cycle more closely and differences between the response of primary afferent fibres and interneurons became clear. Fig. 19.4 compares the post-R wave interval histograms for an interneuron and for an aortic arch baroreceptor. The successive peaks in Fig. 19.4*a* represent the timing of successive spikes during the burst and it can be seen that there is little variability in the early spikes. In Fig. 19.4*b*, which is the record from the interneuron, the variability of the first spike is much

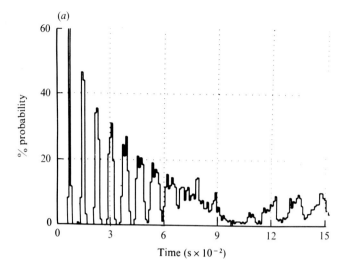

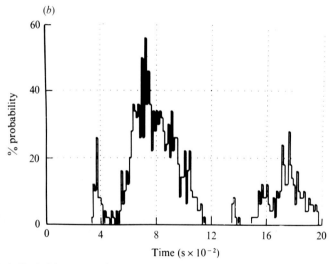

Fig. 19.4. Cycle histograms from aortic baroreceptor fibre *(a)* and an interneuron from the nucleus tractus solitarius *(b)*. The first spike in each of 200 cardiac cycles triggered an interval histogram programme. Abscissa = ratio of number of spikes/min to number of cycles (percentage probability). In *(a)* successive peaks represent the sequential occurrence of individual action potentials and indicate little variability for the early spikes. In *(b)* the wide variability in the timing of successive action potentials is reflected in the histogram (Fussey, Kidd & Whitwam, unpublished).

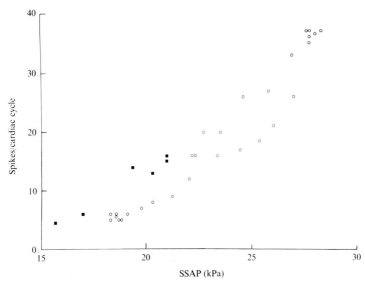

Fig. 19.5. Changes in activity of an interneuron induced by bleeding (■) and intravenous dextran infusion (○). SSAP = systolic systemic arterial pressure (Fussey Kidd & Whitwam, unpublished).

greater. Presumably, this variability is reflecting the convergence on to the neuron of activity in a number of primary afferent fibres.

In the case of some neurons the bursts of activity were broken into several components (Fig. 19.3c); in others, the relationship of the spikes to particular phases of the cardiac cycle was very inconstant and they appeared to be additionally influenced by changes in inflation pressures. In each case, when attempts were made to locate the receptors, using the established techniques of occluding vessels and mechanical stimulation of the cardiac chambers with a fine probe, the neurons were damaged as a result of the consequent oscillations in arterial and venous pressures. In some cases it was possible to alter the number of spikes in each cycle by intravenous infusion of small volumes of saline or bleeding. An example of the response of one neuron to the changes is shown in Fig. 19.5 and the activity is related to arterial pressure only to give an index of the change induced. Increases in pressure of 4.0–5.3 kPa were associated with an increase from 15 to 40 spikes per cycle. Of course those changes may be caused by alterations in a number of different parameters; however, as indicated above, any attempt to apply a more rigorously quantitative approach invariably resulted in the loss of the neuron.

Repeated attempts were made to demonstrate convergence from carotid sinus and vagal nerve afferents. In some experiments electrical stimuli were applied to the intact carotid sinus nerve, in others, an isolated carotid sinus preparation was perfused with arterial blood. In each preparation, every neuron with a pulse-phased activity was tested repeatedly by transient occlusion of the common carotid artery and particular attention was placed upon transient alterations in impulse firing immediately following occlusion and release of the carotid artery. Excitatory convergence was demonstrated for only three neurons. One example is shown in Fig. 19.6 where during a period of anodal blockade, which had been induced by repeated high-intensity stimulation and which had abolished the burst of action potentials associated with arterial pressure pulse, there was now a single spike in response to the oscillations in pressure within the carotid sinus. We were surprised to find such sparse evidence for convergence between carotid sinus and vagal afferent fibres.

Histological examination of the brain stem revealed that the pulse-phased neurons were located in very restricted regions extending from 1 mm caudal to 3 mm rostral to the obex. Approximately 30% of multi-unit recordings and all of the single neurons are represented in Fig. 19.7. The major distribution is the dorso-medial section of the nucleus of the tractus solitarius. An extensive search of the remaining brain stem, lateral and ventral, did not reveal further neurons with these character-istics. In conclusion, the investigation demonstrated that pulse-phased neurons excited by vagal afferent fibres are restricted to the nucleus of the tractus solitarius. This is the area which has previously been shown to be an area of termination of primary afferent fibres from the carotid sinus and aortic nerves and the region from which activity evoked by increases in pressure from carotid sinus baroreceptors has also been recorded. However, a high degree of convergence of the afferents from carotid sinus and the cardio-aortic region to the same neuron has not been demon-strated. Further studies are required to define whether this reflects the particular conditions of these experiments in relation to anaesthesia, etc., or whether the 'early' interneuronal linkages of the major reflex pathways are to a large extent separate in the brain stem. However, since it was not possible to identify the origins of the fibres exciting these neurons, this population of neurons is almost certainly mixed and could form part of the interneuronal pathways associated with a number of individual receptors and reflex responses. Clearly, before any effective analysis along these lines is attempted it would be necessary to have detailed information about the

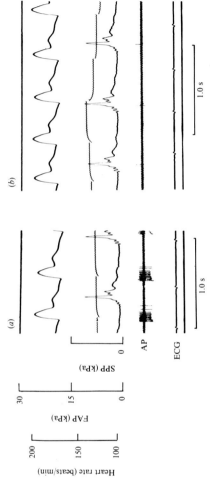

Fig. 19.6. Impulse activity of an interneuron which altered its time relationships during a period of electrical blockade of the intact ipsilateral vagal nerve. (*a*) was obtained during a control period when the spontaneous activity is related to each cardiac cycle. (*b*) was taken immediately after a period of high-intensity, low-frequency stimulation of the vagal nerve which produced blockade impulse transmission. The neuron now only responds to the perfused carotid sinus pressure pulse (**SPP**). AP = action potential; FAP = femoral arterial pressure (Fussey, Kidd & Whitwam, unpublished).

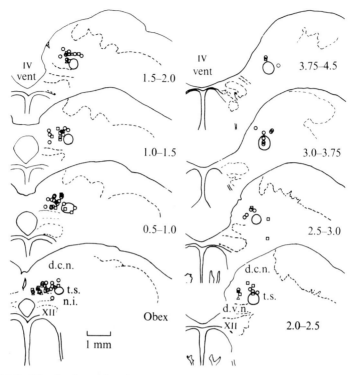

Fig. 19.7. Distribution of pulse-phased interneurons in brainstem. With few exceptions the neurons lie in the dorsal and medial sections of the nucleus tractus solitarius. Each section represents a 0.5-mm segment of the medulla and the distance (mm) rostral to the obex is indicated. The sites at which 40 single interneurons (□) were recorded are shown. For comparison, 55 sites at which multi-unit pulse phased activity (○) and eight sites (three presynaptic) at which recordings were made from pulse phased fibres (▵) are also shown. d.c.n., dorsal column nuclei; t.s., tractus solitarius; d.v.n., dorsal motor vagal nucleus; n.i., nucleus intercalatus; XII, Hypoglossal nerve nucleus; IV vent, IVth ventricle (Fussey, Kidd & Whitwam, unpublished).

location of the receptors. During this investigation the attempts to define the specific receptor area were consistently thwarted by the technical problems of identifying these receptor areas without inducing large oscillations in arterial and venous pressures. Thus, a selective stimulus to a known population of receptors is required which does not simultaneously introduce significant mechanical oscillations. Recently, Baetschi *et al.* (1975) have reported that neurons within the nucleus of the tractus solitarius and the pharahypoglossal area may be excited or inhibited by infusion or withdrawal of blood, or by 'atrial pulsation' induced by a perfusion system. However, each of these interventions is not specific for a known population of receptors; they can be expected to provide a non-specific stimulus to a wide variety of receptors within the cardio-aortic region. Since this investigation does not provide evidence of the pattern or the profile of receptors activated by these manoeuvres it is not possible to make a useful interpretation of the specific reflex pathways to which these neural changes may be related. Indeed, the possibility that some of the changes may be due to mechanical movements consequent on changes in venous pressure cannot be excluded.

You will not be surprised to hear that at Leeds we have been able to apply a selective stimulus to a discrete group of receptors, the left atrial receptors, without evoking large changes in vascular pressures and while carrying out a search for neurons responding to this stimulus (Keith, Kidd, Linden & Snow, 1973).

Previous studies from this laboratory have shown that inflation of small balloons in the pulmonary vein–atrial junctions induced reflex alterations in heart rate and urine volume (see Linden, 1975; this volume) and evidence shows that the balloons stimulated the left atrial receptors in the endocardial wall with fibres in the vagal nerves (Kidd, Ledsome & Linden, 1966). Inflation of the balloons with $1-1.5$ cm^3 of saline induced a brisk increase in the impulse activity, still with a cardiac rhythm from the receptors, which is maintained throughout the period of distension. Examination of these responses reveals that the frequencies of these action potentials are similar to those observed from left atrial receptors after infusion of saline or of blood (Fig. 19.8). Thus, inflation of the balloon provides a discrete stimulus to left atrial receptors which is similar to that observed under physiological conditions and these changes occur without concomitant changes in mean arterial blood pressure or left atrial pressure.

In dogs anaesthetised with chloralose small balloons were placed in the upper and middle pulmonary vein–atrial junctions. Stimulating and cooling thermodes were placed on the intact vagal nerves, and the medulla was explored for neural activity which changed after balloon distensions. Recordings were made from 40 sites at which the spontaneous activity of neurons increased when the balloons were inflated (Fig. 19.9). In each case the neurons responded repeatedly to periods of inflation and the responses were shown to be statistically significant increases over the pre- and post-stimulation spontaneous levels. At most sites the change in activity

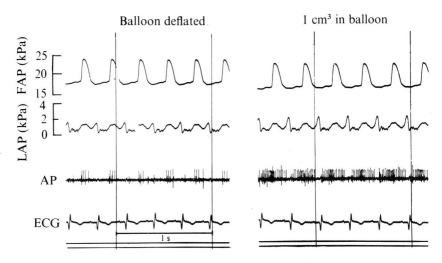

Fig. 19.8. Response of left atrial receptor to inflation (1 cm³ saline) of small balloons in left atrial–pulmonary vein junction. Note large increase in activity from the receptor during inflation which still retains a cardiac rhythm. FAP = femoral artery pressure; LAP = left atrial pressure; AP = action potential (Kidd, Ledsome & Linden, unpublished).

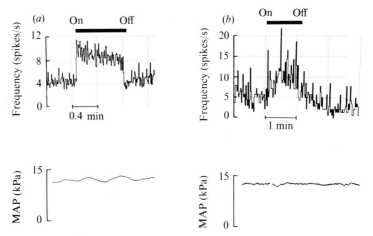

Fig. 19.9. Responses of brain stem neurons to inflation of left atrial balloons. *(a)* Activity of single neuron in nucleus tractus solitarius increased when the balloons are inflated. *(b)* Activity from single neuron in the medial reticular formation increased and decreased gradually following balloon inflation and deflation. Note the absence of concomitant alterations in mean arterial blood pressure (MAP) (Kappogoda, Keith, Kidd, Linden & Snow, unpublished).

was recorded from individual neurons whilst in some activity from several neurons was recorded simultaneously. For the purposes of this study these have not been separated since the objective of the investigation was to define the areas of the brain stem involved in the responses. Between the limits of 1 mm caudal to 3 mm rostral to the obex, there were two major areas within the brain stem in which the neural activity was increased: superficially, a number of sites were identified in the dorso-medial area of the nucleus tractus solitarius and, more deeply, in the medial reticular tissue adjacent to the paramedian reticular nucleus and the raphae nucleus, one neuron was located in the parahypoglossal area (Fig. 19.10). In the superficial area, the neurons responded quickly to distension of the balloons and the response was maintained throughout the period inflation (Fig. 19.9a). The spontaneous activity of most neurons had a cardiovascular rhythm, i.e. pulse phased (Fig. 19.11a); however, this was not invariable and the spontaneous activity of others consisted of an irregular, low-frequency (1–1.53) activity which appeared to bear no relationship to obvious cardiovascular or respiratory parameters. The spontaneous activity of the deep neurons never had a cardiovascular rhythm and frequently consisted of irregular bursting activity (Fig. 19.11b). Frequently, the impulse activity of these deep neurons increased more slowly after inflation of the balloons and subsided more slowly than those of the superficial group (Fig. 19.9b). We were satisfied that the changes in activity from neurons at both sites were not caused by mechanical artifacts, since movement of the electrode tip up to and beyond the neurons was not associated with alterations in spontaneous activity. In no case were the responses to balloon inflation associated with significant changes in left atrial pressure or arterial blood pressure.

Spontaneous and evoked activity from neurons belonging to both superficial and deep groups was observed during the cooling to 4 °C of one or both cervical vagal nerves. Fig. 19.12 shows the effect of vagal cooling on one neuron. It can be seen that cooling of the left vagal nerve (4 °C) abolished the response to balloon inflation and reduced spontaneous activity markedly; cooling of the right vagal nerves to the same temperature was without effect. At this time graded cooling of the vagal nerves to differentiate the effects of activity evoked in myelinated and non-myelinated afferent fibres from the left atrium (see Linden, this volume) has not yet been attempted. Evidence was also obtained which indicates that the increases in action potentials (recorded in pulses per second) which followed distension of the left atrial balloons were not solely the result of

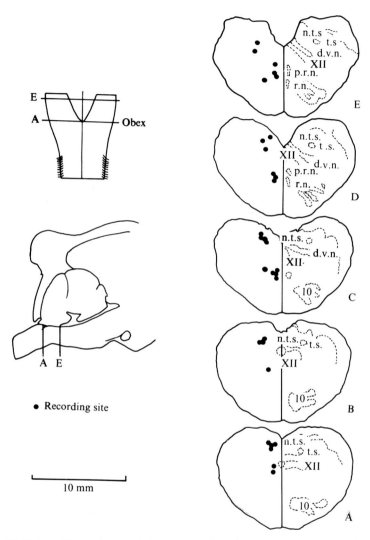

Fig. 19.10. Location of neurons in brain stem responding to stimulation of left atrial receptors. Each section is representative of a 1 mm slice of brain stem rostral to obex. Each site at which single and multi-unit activity responding to balloon inflation is shown. d.v.n., dorsal motor vagal nucleus, t.s., tractus solitarius; n.t.s., nucleus tractus solitarius; XII, hypoglossal nerve nucleus; 10, inferior olive; r.n, nucleus raphae; p.r.n., parmedian reticular nucleus (Kappagoda, Keith, Kidd, Linden & Snow, unpublished).

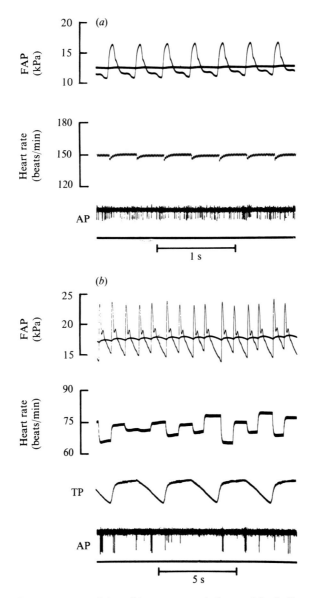

Fig. 19.11. Spontaneous activity of interneurons influenced by balloon inflation. *(a)* Pulse-phased activity from a neuron in nucleus tractus solitarius. *(b)* Spontaneous activity of a neuron in the medial reticular formation. FAP=femoral arterial pressure; AP=action potentials; TP=tracheal pressure (Kappagoda, Keith, Kidd, Linden & Snow, unpublished).

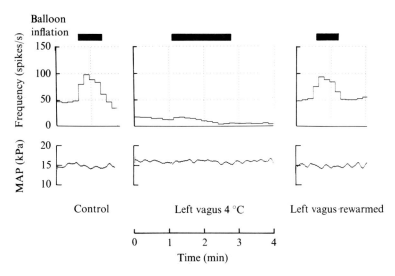

Fig. 19.12. Effect of cooling left vagal nerve to 4 °C on neuronal response to left atrial balloon inflation. *(a)* and *(c)* Show the responses with the vagal nerves at 37 °C. During *(b)* the left vagal nerve was cooled to 4 °C. Note abolition of response and reduction on spontaneous activity. MAP = mean arterial pressure (Kappagoda, Keith, Kidd, Linden & Snow, unpublished).

the concomitant increases in heart rate since the response was not affected by intravenous infusion of propranolol sufficient to block the increase in heart rate. Inflation of the balloons was never followed by inhibitory effects upon spontaneous activity of neurons.

We conclude that these neurons represent components of the neuronal pathways of the reflex responses evoked by stimulation of left atrial receptors (Fig. 19.13); that is, a sympathetically mediated increase in heart rate and an increase in urine flow postulated to be due to an inhibition of renal sympathetic nerve activity and to a humoral agent which is not the antidiuretic hormone (Linden, this volume; Kappagoda, this volume). The superficial group of neurons, located largely in the nucleus tractus solitarius, is likely to represent an early synaptic linkage on the pathways and there is some evidence of a transfer of the rhythmic cardiac modulation present in the vagal primary afferent fibres; it must be assumed that this modulation is lost later since there is no evidence for such a cardiac rhythmicity which does not originate from arterial baroreceptors in sympathetic efferent fibres (see Kirchheim, 1976). At present it is not possible to define whether the differences in spontaneous activity from

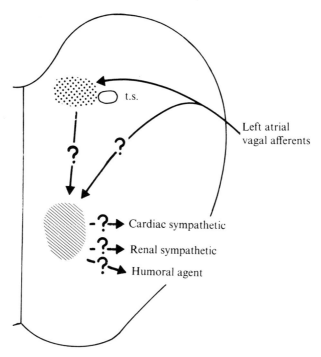

t.s.

Left atrial
vagal afferents

Cardiac sympathetic

Renal sympathetic

Humoral agent

Fig. 19.13. Diagramatic coronal section of dog brain stem indicating the major sites of neurons influenced by left atrial balloons. The question marks indicate connections not yet delineated. t.s. = tractus solitarius.

these neurons are reflecting further synaptic interconnections within the nucleus itself. The deep group may represent additional interneurons cascaded in the left atrial reflex pathway and may represent reticulospinal connections (Brodal, 1957), but further investigation will be required to define the nature of the linkages between the superficial and deep groups and the pathways to efferent neurons or effectors. The deep area is within the classical 'depressor' area and has also been implicated in the sympathetic components of baroreceptor pathways (e.g. Alexander, 1946; Humphrey, 1967; Miura & Reis, 1972; Coote & MacLeod, 1974). However, the general inhibitory effect on sympathetic efferent activity, albeit differing in magnitude for different sections, which is characteristic of the baroreceptor reflex response is quite different from the differential excitation of cardiac, inhibition of renal and absence of effect on splenic and lumbar sympathetic efferent activity evoked by left atrial receptor activation (Karim, Kidd, Malpus & Penna, 1972). The differences are likely to be accountable to the electrical stimulation techniques used in defining active medullary sites which will give discrepant results when a mixed population of neurons is excited. Further progress in defining central pathways for specific reflex pathways is only likely to be achieved by discrete activation of a known profile of receptors.

I would like to thank all my collaborators in these experiments who contributed so freely of both time and thought. The experimental work was supported by the Medical Research Council and the Wellcome Trust.

References

Agostoni, E;, Chinnock, J.E., Daly, M. de B. & Murray, J.G. (1957). Functional and histological studies of the vagus nerve and its branches to the heart, lungs and abdominal viscera in the cat. *J. Physiol.*, **135**, 182–205.

Alexander, R.S. (1946). Tonic and reflex functions of medullary sympathetic cardiovascular centres. *J. Neurophysiol.*, **9**, 205–17.

Allen, W.F. (1923). The origin and distribution of the tractus solitarius in the guinea pig. *J. comp. Neurol.*, **35**, 171–204.

Anderson, F.D. & Berry, C.M. (1956). An oscillographic study of the central pathways of the vagus nerve in the cat. *J. comp. Neurol.*, **106**, 163–81.

Åström, K.E. (1953). On the central course of afferent fibres in the trigeminal, facial, glossopharyngeal and vagal nerves and their nuclei in the mouse. *Acta physiol. scand.*, **29** (suppl. 106), 209–320.

Baetschi, A.J., Munzer, R.F., Ward, D.G., Johnson, R.N. & Gann, D.S. (1975). Right and left B-fiber input to the medulla of the cat. *Brain Res.*, **98**, 189–92.

Baumgarten, R.C., von & Arenda, C.L. (1959). Distribución de les aferencias cardiovascularis y respiratorias en les raices bulbanes del nervio vago. *Acto Neur. Latinoamer.*, **5**, 267–78.

Biscoe, T.J. & Sampson, S.R. (1970*a*). Field potentials evoked in the brain stem of the cat by stimulation of the carotid sinus, glossopharyngeal, aortic and superior laryngeal nerves. *J. Physiol.*, **209**, 341–58.

Biscoe, T.J. & Sampson, S.R. (1970*b*). Responses of cells in the brain stem of the cat to stimulation of the sinus, glossopharyngeal, aortic and superior laryngeal nerves. *J. Physiol.*, **209**, 359–73.

Bonvallet, M. & Sigg, B. (1958). Etude électrophysiologique des afférences vagales au niveau de leur pénétration dans le bulbe. *J. Physiol.* (Paris), **50**, 63–74.

Brodal, A. (1957). *The reticular formation of the brain stem. Anatomical aspects and functional correlations.* Oliver & Boyd: Edinburgh.

Chiba, T. & Doba, N. (1975). The synaptic structure of catecholaminergic axon varicosities in the dorso-medial portion of the nucleus tractus solitarius of the cat: possible roles in the regulation of cardiovascular reflexes. *Brain Res.*, **84**, 31–46.

Chiba, T. & Doba, N. (1976). Catecholaminergic axo-axonic synapses in the nucleus tractus solitarius (pars commissuralis) of the cat; possible relationship to presynaptic regulation of baroreceptor reflexes. *Brain Res.*, **102**, 255–65.

Coleridge, J.C.G., Hemingway, A., Holmes, R. & Linden, R.J. (1957). The location of atrial receptors in the dog: a physiological and histological study. *J. Physiol.*, **136**, 174–97.

Coote, J.H. & MacLeod, V.N. (1974). The influence of bulbospinal monoaminergic pathways on sympathetic nerve activity. *J. Physiol.*, **241**, 453–76.

Cottle, M.K. (1964). Degeneration studies of primary afferents of IXth and Xth cranial nerves in the cat. *J. comp. Neurol.*, **122**, 329–45.

Crill, W.E. & Reis, D.J. (1968). Distribution of carotid sinus and depressor nerves in the cat brain stem. *Am. J. Physiol.*, **214**, 269–76.

Culberson, T.L. & Kimmel, D.L. (1972). Central distribution of primary afferent fibres of the glossopharyngeal and vagal nerves in the oppossum, *Didelphis virginiana*. *Brain Res.*, **44**, 325–35.

Donoghue, S., Fox, R., Kidd, C. & Koley, B.N. (1977). The projection of vagal, cardiac C-fibres to the brain stem of the cat. *J. Physiol.*, **270**, 44–45P.

Fink, R.P. & Heimer, L. (1967). Two methods of selective silver impregnation of degenerating axons and their synaptic endings in the central nervous system. *Brain Res.*, **4**, 369–74.

Fussey, I.F., Kidd, C. & Whitwam, J.G. (1967). Single unit activity associated with cardiovascular events in the brain stem of the dog. *J. Physiol.*, **191**, 57–8P.

Fussey, I.F., Kidd, C. & Whitwam, J.G. (1970). The differentiation of axonal and soma-dendritic spike activity. *Pflügers Arch. ges. Physiol.*, **321**, 283–92.

Fussey, I.F., Kidd, C. & Whitwam, J.G. (1973). Activity evoked in the brain stem by stimulation of C fibres in the cervical vagus nerve of the dog. *Brain Res.*, **49**, 436–40.

Fuxe, K. (1965). Evidence for existence of monoamine neurons in the central nervous system. *Acta physiol. Scand.*, **64** (suppl. 247), 39–85.

Fuxe, K., Hokfeldt, T. & Nilsson, O. (1965). A fluorescence and electron microscopic study on certain brain regions rich in monoamine terminals. *Am. J. Anat.*, **117**, 33–46.

Gabriel, M. & Seller, H. (1970). Interaction of baroreceptor afferents from carotid sinus and aorta at the nucleus tractus solitarii. *Pflügers Arch. ges. Physiol.*, **318**, 7–20.

Harrison, F. & Bruesch, S.R. (1945). Intermedullary potentials following stimulation of the cervical vagus. *Anat. Res.*, **91**, 280.

Hellner, K. & Baumgarten, R. von (1961). Über rin Endigungsebiet afferentes, Kardiovascu-lärer Fasern des Nervus vagus im Rantenhirn der Katze. *Pflügers Arch. ges Physiol.*, **273**, 223–34.

Humphrey, D.R. (1967). Neuronal activity in the medulla oblongata of the cat evoked by stimulation of the carotid sinus nerve. In *Baroreceptors and hypertension*, ed. P. Kezdi, pp. 131–68. Pergamon Press: Oxford.

Ingram, W.R. & Dawkins, E.A. (1945). The intermedullary course of afferent fibres of the vagus nerve of the cat. *J. comp. Neurol.*, **82**, 157–68.

Jarisch, A. & Zotterman, Y. (1948). Depressor reflexes from the heart. *Acta physiol. scand.*, **16**, 31–51.

Karim, F., Kidd, C., Malpus, C.M. & Penna, P.E. (1972). The effects of stimulation of the left atrial receptors on sympathetic nerve activity. *J. Physiol.*, **227**, 243–60.

Keith, I.C., Kidd, C., Linden, R.J. & Snow, H.M. (1973). Modifications of neuronal activity in the dog medulla by stimulation of left atrial receptors. *J. Physiol.*, **238**, 17–18P.

Kerr, F.W.L. (1962). Facial, vagal and glossopharyngeal nerves in the cat. *Arch. Neurol. (Chicago)*, **6**, 264–81.

Kidd, C., Ledsome, J.R. & Linden, R.J. (1966). Left atrial receptors and the heart rate. *J. Physiol.*, **185**, 78–9P.

Kimmel, D.L. (1965). The terminations of vagal and glossopharyngeal afferent nerve fibres in the region of the commissural region in the cat, rat and guinea pig. *Anat. Res.*, **151**, 371.

Kirchheim, H.R. (1976). Systemic arterial baroreceptor reflexes. *Physiol. Rev.*, **56**, 100–76.

Koepchen, H.P., Langhorst, P., Seller, H., Polster, J. & Wagner, P.H. (1967). Neuronale Activität im unteren Hirnstamm mit Bezlehung zum Kreislauf. *Pflügers Arch. ges Physiol.*, **294**, 40–64.

Lam, R.L. & Tyler, H.R. (1952). Electrical responses evoked in the visceral afferent nucleus of the rabbit by vagal stimulation. *J. comp. Neurol.*, **97**, 21–36.

Linden, R.J. (1975). Reflexes from the Heart. *Prog. cardiovasc. Dis.*, **18**, 201–21.

Lipski, J., McAllen, R.M. & Spyer, K.M. (1975). The sinus and baroreceptor input to the medulla of the cat. *J. Physiol.*, **251**, 61–78.

Lipski, J. & Trzebski, A. (1975). Bulbo-spinal neurons activated by baroreceptor afferents and their possible role in inhibition of preganglionic sympathetic neurons. *Pflügers Arch. ges. Physiol.*, **365**, 181–92.

McAllen R.M. & Spyer, K.M. (1976). The location of cardiac vagal motoneurones in the medulla of the cat. *J. Physiol.*, **258**, 187–204.

Middleton, S., Woolsey, C.N., Burton, H. & Rose, J.E. (1973). Neural activity with cardiac periodicity in medulla oblongata of the cat. *Brain Res.*, **50**, 297–314.

Miura, M. & Reis, D.J. (1969). Termination and secondary projections of carotid sinus nerve in the cat brainstem. *Am. J. Physiol.*, **217**, 142–53.

Miura, M. & Reis, D.J. (1972). The role of the solitary and paramedian reticular nuclei in mediating cardiovascular reflex responses from carotid baro- and chemoreceptors. *J. Physiol.*, **223**, 525–48.

Öberg, B. & Thorén, P.N. (1972). Circulatory responses to stimulation of medullated and nonmedullated afferents in the cardiac nerve in the cat. *Acta. physiol. scand.*, **87**, 121–32.

Paintal, A.S. (1973). Vagal sensory receptors and their reflex effects. *Physiol. Rev.*, **53**, 195–27.

Porter, R. (1963). Unit responses evoked in the medulla oblongata by vagus nerve stimulation. *J. Physiol.*, **168**, 717–35.

Rhotan, A.L., O'Leary, J.L. & Ferguson, J.P. (1966). The trigeminal, facial, vagal and glossopharyngeal nerves in the monkey. *Arch. Neurol. (Chicago)*, **14**, 536–40.

Salmoiraghi, G.C. (1962). 'Cardiovascular' neurones in the brainstem of the cat. *J. Neurophysiol.*, **25**, 182–97.

Seller, H. & Illert, M. (1969). The localization of the first synapse in the carotid sinus baroreceptor reflex pathway and its alteration of the afferent input. *Pflügers Arch. ges. Physiol.*, **306**, 1–19.

Smith, R.S. & Pearce, J.W. (1961). Microelectrode recordings from the region of the nucleus tractus solitarius in the cat. *Can. J. Biochem. Physiol.*, **39**, 933–9.

Spyer, K.M. & Wolstencroft, J.H. (1971). Problems of the afferent input to the paramedian reticular nucleus, and the central connections of the sinus nerve. *Brain Res.*, **26**, 411–14.

Torvik, A. (1965). Afferent connections to the sensory trigeminal nuclei, the nucleus of the solitary tract and adjacent studies – an experimental study in the rat. *J. comp. Neurol.*, **106**, 51–141.

Trzebski, A., Peterson, L.H., Attinger, F., Jones, A. & Tempest, R. (1962). Unitary responses in the medulla oblongata related to carotid sinus receptors functions. *Physiologist*, **5**, 222.

Urabe, M., Yamazaki, S. & Araki, K. (1960. The central course of afferent fibres in the glossopharyngeal nerves and vagus nerves. *Neurologia, medic-chirurgica.*, **2**, 1–24.

Discussion

(a) Arndt noted that there were two responses to balloon inflation: one in which the response was rapid, and the other which had a slower time course. The slow ones would suggest that the central neurons have relatively slow time constants but was Kidd sure that the fast responding recordings were from second-order neurons rather than axons? Kidd replied that the group had been concerned about this point but the evidence indicated that with the micro-electrode and recording system in use the recordings were from interneurons rather than primary afferent axons.

(b) The comment was made that it is difficult to know whether the neuron from which recordings were made has anything to do with the physiological response. Kidd replied that that was not the point. The experiments were to show a functional linkage which related to reflex responses, i.e. functional anatomy: there was a good correlation between the stimulus which was specific and the response of the interneurons as well as the reflex effects.

(c) Kidd was asked whether it was possible, using electrical stimulation, to distinguish recordings from orthodromically conducting afferents or antidromically conducting efferents. Kidd pointed out that when recording multi-unit activity there could be a contribution from both antidromically excited efferent and synaptically activated neurons. It was not possible to distinguish these adequately in multi-unit recordings and no reliance should be placed on multi-unit activity for interpretive purposes.

(d) Gootman said she had done some experiments in which she had recorded from single neurons in the brain stem, looking at the outflow neurons. She had found units in the classical pressor area having activity which correlated with the cardiac cycle, efferent sympathetic discharge and the central respiratory cycle. There were some units with a tight double bursting which were markedly correlated with efferent sympathetic discharge and central respiratory drive. She wondered whether neurons in the tractus solitarius showed any correlation with central respiratory oscillations and whether Kidd had monitored phrenic discharge as an indicator of central respiratory activity. Kidd said that he had recorded from a small number of neurons with activity and there was a correlation with the cardiac cycle which was less strongly expressed and appeared to have a respiratory modulation. The animals were respired artificially and phrenic activity was not monitored.

20. The effects of medullary lesions on the reflex response to distension of the pulmonary vein–left atrial junctions

J. R. LEDSOME

Numerous investigators, utilising a variety of techniques, have provided evidence that the afferent fibres from cardiovascular receptors which run in the glossopharyngeal and vagus nerves, after entering the medulla, terminate in or close to the nucleus of the tractus solitarius (NTS) at the level of or rostral to the obex (reviewed by Kirchheim, 1976). In this area there is considerable overlap of the termination of fibres of the glosso-pharyngeal and vagus nerves. Biscoe & Sampson (1970a, b) studied field potentials generated in the brain stem and unit responses from cells in the immediate vicinity of the NTS following stimulation of the carotid sinus, glossopharyngeal, aortic and superior laryngeal nerves and concluded there was a considerable degree of overlap amongst the terminations of the nerves studied and a high degree of convergence of afferents upon the same unit. There may be significant modification of the pattern of impulses at the first synapse since during repetitive stimulation of the carotid sinus nerve the transmission across the synapse showed a sharp frequency limitation (Seller & Illert, 1969). Also, there is interaction between the inputs of the four major arterial baroreceptor nerves, the right and left aortic nerves and the right and left carotid sinus nerves in their effects on secondary neurons in the NTS near the obex (Gabriel & Seller, 1970). Functionally the depressor reflexes initiated by stimulation of the carotid sinus and aortic arch baroreceptors may be shown to interact in a manner producing an apparently similar non-linear summation (Angell-James & Daly, 1970). Thus the first synapse of the primary afferent fibres from the cardiovascular receptors appears to be an important site of convergence and overlap between inputs from a number of cardiovascular receptor areas.

Despite the electrophysiological evidence indicating close interaction between the inputs from the glossopharyngeal and vagus nerves anatomi-

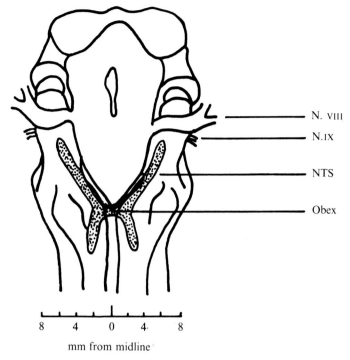

Fig. 20.1. Drawing of dorsal view of dog medulla showing approximate location of the tractus solitarius (NTS). Scale corresponds to average size of brain stem in mongrel dogs of 8–14 kg.

cal evidence from degeneration studies (Cottle, 1964), while showing considerable overlap of areas of input in the NTS, indicates that the afferent fibres from the carotid sinus are in preponderance rostral to the obex whereas the afferent fibres from the aortic nerve are preponderant in the intermediate portion of the NTS at the level of the obex. This pattern of distribution is in agreement with the work of Crill & Reis (1968) who studied potentials evoked antidromically in the carotid sinus and aortic nerves of the cat by stimulation of points in the NTS. Earlier, Oberholzer (1960) demonstrated that small, discrete lesions placed 1 mm rostral to the obex in the rabbit could abolish the depressor response to aortic nerve stimulation, leaving intact the response to distension of the isolated carotid sinus.

The pathways within the medulla which are followed by fibres from atrial receptors have been less extensively studied. It has been shown that vagal afferent fibres discharging with a cardiovascular rhythm enter the medulla in the more rostral rootlets of the vagus nerve (Bonvallett & Sigg, 1958) with those showing a pattern of discharge typical of atrial receptors being in the more caudal of the rostral rootlets (von Baumgarten, Koepchen & Aranda, 1959). Section of the rostral rootlets of the vagus nerve produced degeneration in the intermediate region of the NTS extending slightly caudal to the obex (Cottle, 1964). Unit discharge occurring during the P wave of the ECG and during the *T* wave and suggestive of input from atrial receptors was recorded by Hellner & von Baumgarten (1961) from points dorso-lateral to the tractus solitarius and 1–3 mm rostral to the obex in the cat. There is said to be a distinction between the medullary locations of the carotid pressoreceptors and the atrial receptors, the latter being slightly caudal (von Baumgarten, quoted by Gauer & Henry, 1963). More recently Baertschi *et al.* (1975) have investigated the distribution of medullary neurons responding (either activated or inhibited) to small changes in blood volume or to atrial pulsation, and presumed the changes to be caused by discharge from atrial receptors. Such neurons were found only in the NTS or close by in the underlying reticular formation. The connections of the atrial afferents beyond the secondary neurons are unknown. The hypothesis that stimulation of atrial receptors is associated with changes in the rate of release of antidiuretic hormone implies connections with the hypothalamus. Unit activity has been recorded from cells in the hypothalamus, the activity of which was affected by atrial stretch (Menninger & Frazier, 1972).

The experiments to be described were designed to test whether those areas of the medulla in which afferent fibres from the atria appear to terminate were in fact essential for the appearance of the reflex cardiovascular responses to atrial receptor stimulation. To stimulate the left atrial receptors the technique of distension of balloons in the left pulmonary vein–atrial junctions was used (Ledsome & Linden, 1964). This technique has been shown to be an effective stimulus to left atrial receptors (Linden, 1972). The experiments were carried out on dogs; it should be noted that almost all of the information regarding the central connections of cardiovascular receptors has been obtained from cats. However, preliminary investigations on cats showed that the reflex response to pulmonary vein–left atrial distension was small and not sufficiently reproducible to allow examination. To approach the medulla from the dorsal aspect it was necessary to remove a portion of the occipital bone and retract the cerebellum. It has been shown by Reis & Cuénod (1965) that cerebellectomy may change the response to carotid occlusion and carotid sinus stretch. To test the effect of removal of the cerebellum and pons on the reflex response to pulmonary vein distension, section of the brain stem was made through or just caudal to the inferior cerebellar peduncle. The section was made using a bank of stainless steel electrodes and a radio-frequency coagulator similar to that previously described for mid-collicular decere-bration in the dog (Albrook, Bennion & Ledsome, 1972). Section of the brain stem in the rostral medulla caused a decrease in heart rate and mean arterial pressure in all of ten dogs tested. There was some return of heart rate towards its previous value but arterial pressure remained lower than before the section. It was not possible to demonstrate any difference in the reflex increase in heart rate which occurred in response to pulmonary vein distension (Fig. 20.2). Also, the small transient decrease in mean arterial pressure observed within 20 s of pulmonary vein distension (Carswell, Hainsworth & Ledsome, 1970) was unaltered by the section. The reflex increase in heart rate was significantly reduced by injection of propranolol $(0.5 \text{ mg} \cdot \text{kg}^{-1})$ and was completely abolished by section of the cervical vago-sympathetic trunks. Thus the characteristics of the reflex reponse were similar to those seen in intact and decerebrate animals. Damage to the cerebellum during an approach to the dorsal medulla is unlikely to alter the magnitude of the reflex response to pulmonary vein distension.

To investigate the medullary pathways of the atrial receptor afferents lesions were made in the dorsal medulla, centred on the obex, with the intention of destroying the NTS bilaterally in the caudal part of that region associated with afferent cardiovascular fibres. Lesions were made using two techniques; in six dogs the lesion was made using a single electrode placed 1 mm deep at the obex and passing a current from a surgical coagulator (Birtcher, Blendtome). In nine subsequent animals a lesion was placed at the obex using a concentric array of stainless steel electrodes (23 gauge) consisting of one central electrode surrounded by a ring of six electrodes each 2 mm from the central electrode. The central electrode was placed at the obex and the tips of the electrodes were driven 1 mm below the surface of the medulla. A current of 50 μA was passed for 15 s from a Wyss Coagulator (J. Monti, Geneva) between the central electrode and each of the peripheral electrodes in turn. The lesions produced by the two techniques were similar; in all dogs the NTS and dorsal nucleus of the vagus were found to be destroyed over a length of about 2 mm at the level of the obex (Fig. 20.1). The hypoglossal nucleus was intact in some but damaged in other dogs.

Placing of the lesions as described invariably caused a marked increase in

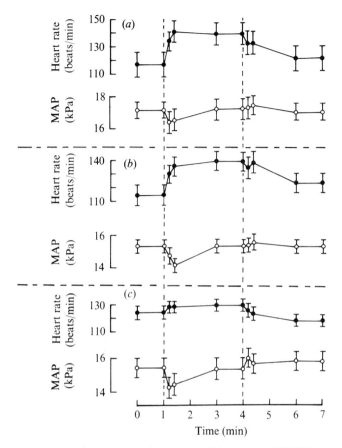

Fig. 20.2. Changes in heart rate and mean arterial pressure (MAP) in response to pulmonary vein distension. Pulmonary vein distension was applied between the vertical lines. *(a)* With brain stem intact; *(b)* after rostral medullary section; *(c)* after rostral medullary section plus propranolol 0.5 mg·kg^{-1}. Bars are ±s.e.m.

heart rate and mean arterial pressure. There was a gradual return to previous values over 10–15 min but the heart rate usually remained elevated by about 10 beats/min and mean arterial pressure remained increased by 2.0 kPa. In nine dogs the effects of pulmonary vein distension and carotid occlusion were compared before and after placing the lesion. In every case the increase in heart rate observed on distension of the pulmonary vein–atrial junctions was almost completely abolished as was the transient vasodilatation which accompanied pulmonary vein distension. In contrast, there were no significant differences in the response of

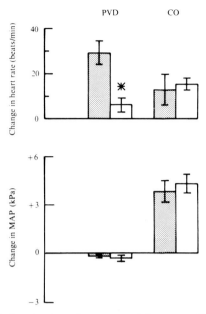

Fig. 20.3. Changes in heart rate and mean arterial pressure (MAP) in response to distension of the pulmonary vein–left atrial junction (PVD) and carotid occlusion (CO) before and after making a lesion at the obex. Nine dogs; three tests of each intervention averaged in each dog. Shaded columns, brain stem intact; non-shaded columns, after lesion at obex. * Indicates $P < 0.05$, bars are $\pm$ s.e.m.

heart rate or mean arterial pressure to carotid occlusion after placing of the lesion (Fig. 20.3).

In a further six dogs the effects of pulmonary vein distension and carotid occlusion were again compared but in addition the effects of stimulation of the left aortic nerve were examined. The left aortic nerve was identified in the neck as described by Edis & Shepherd (1971). The nerve was tied peripherally and the central end stimulated using low-intensity, high-frequency stimulation (2–4 V, 150 Hz, 0.2 ms) which induced a decrease in heart rate and a fall in arterial pressure. High-intensity, low-frequency stimulation of the nerve (6–10 V, 15 Hz, 2 ms) induced an increase in heart rate and arterial pressure in three dogs; in the other three dogs this response was not sufficiently reproducible to be examinable. Again in all six dogs, after placing a lesion in the vicinity of the obex, the reflex response to distension of the pulmonary vein–left atrial junctions was markedly reduced and a brisk response to carotid occlusion remained. The response to stimulation of the aortic nerve with low-intensity, high-frequency stimulation was totally abolished. The response to high-intensity, low-frequency stimulation of the aortic nerve was reduced but still present in each of the three dogs tested (Fig. 20.4). Decerebration before the making of the lesion did not affect the results.

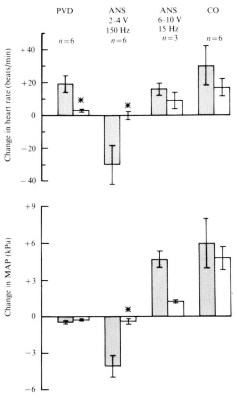

Fig. 20.4. Changes in heart rate and mean arterial pressure (MAP) in response to distension of the pulmonary vein–left atrial junctions (PVD), aortic nerve stimulation (ANS) and carotid occlusion (CO) before and after placing a lesion at the obex. Shaded columns, brain stem intact; non-shaded columns, after lesion at obex. Number of animals indicated by n; three tests of each intervention averaged in each animal. * Indicates $P < 0.05$; bars are $\pm$ s.e.m.

The observation that a bilateral lesion in the NTS abolished the reflex response to pulmonary vein distension while leaving intact the response to carotid occlusion is consistent with the hypothesis that fibres from the atrial receptors relay in or pass through the intermediate portion of the NTS close to the obex and that fibres from the carotid sinus baroreceptors relay more rostrally in the NTS. Numerous investigations in the cat have shown that lesions in the NTS 1–3 mm rostral to the obex completely prevent all responses to stimulation of the carotid baro- and chemoreceptors (e.g. Miura & Reis, 1972). The observation that the response to low-intensity stimulation of the aortic nerve, usually assumed to be caused by stimulation of baroreceptor afferents, was also abolished indicates that these vagal afferents also synapse more caudally than the glossopharyngeal afferents from the aortic baroreceptors in the medulla the question arises as to whether there is interaction between these inputs and at what level any able convergence and interaction between these inputs at this level. The fact that the response to carotid occlusion remained means only that sufficient fibres were left intact to mediate a response. The site of termination of aortic chemoreceptor afferents in the medulla is unknown; these are presumed to be responsible for the response to high-intensity, low-frequency stimulation of the aortic nerve. The results indicate that either a proportion of these fibres synapse in NTS rostral to the site of the lesion or that some fibres pass ventral or lateral to the site of the lesion before synapsing.

Modification of neuronal activity in the dog medulla by distension of the pulmonary vein–left atrial junctions has been described as occurring in two sites by Keith, Kidd, Linden & Snow (1975). Neurons in the region of the NTS had a definite cardiac rhythm and were closely associated with neurons activated by arterial baroreceptors. Other neurons were found in the ventral reticular formation down to the hypoglossal region and did not have a cardiac rhythm and discharged at lower frequencies. The latter site in the ventral reticular formation would not be affected by the lesions described here and possibly represents polysynaptic connections.

Since the afferents from atrial receptors appear to pass close to the afferents from the aortic baroreceptors in the medulla the question arises as to whether there is interaction between these inputs and at what level any such interaction occurs. Evidence has been presented that there is interaction at the first synapse between aortic and carotid baroreceptor afferents despite the apparent separation indicated by the results described here. It seems unlikely that there could be interaction between the atrial receptor input and the aortic baroreceptor input at the first synapse because of the strikingly different response patterns observed on activation of the two receptor areas. Also it was not possible to demonstrate interaction between the atrial receptor input and changes in carotid artery pressure (Carswell, Hainsworth & Ledsome, 1970). However, this point has not been examined in sufficient detail to allow firm conclusions and further work is needed.

There is increasing evidence for supramedullary modulation of baroreceptor evoked cardiac and vasomotor activity (Uther, Hunyor, Shaw & Korner, 1970; Kent, Drane & Manning, 1971; Korner, Shaw, West & Oliver, 1972; Adair & Manning, 1975). Until recently the functional significance of such modulation was not considered great because mid-collicular decerebration had little effect on baroreceptor reflexes in the anaesthetised animal. However, the experiments of Korner *et al.* (1972) on the conscious rabbit shows that stimulus–response curves for pontine and thalamic animals are displaced towards a lower heart rate compared with sham-operated animals. The results obtained by these workers are consistent with the hypothesis that both pentobarbital and chloralose anaesthesia modify the baroreceptor reflex response by affecting diencephalic regions influencing vagal efferents. They suggest that anaesthesia leads to an enhancement of baroreceptor-mediated sympathetic effects on the heart. Their results are interpreted by Kirchheim (1976) to imply an elevation of the threshold for baroreceptor activation of vagal efferents in anaesthetised animals. Supramedullary modulation of medullary reflexes might be expected to influence also the reflex responses to atrial receptor stimulation. In our experiments with rostral medullary section and in previously reported experiments with decerebration in dogs (Albrook *et al.*, 1972) brain stem section was associated with a decrease in heart rate demonstrated to be largely caused by an increase in cardiac vagal tone and a decrease in arterial pressure. An increase in vagal tone after brain stem section could best be explained as a removal of a supramedullary inhibitory drive to the medullary vagal efferents. Whilst these observations provide some support to the concept that vagal tone is usually low in anaesthetised animals this generalisation is certainly not always true of light chloralose anaesthesia in which there is often a slow heart rate with pronounced sinus arrhythmia (e.g. Carswell *et al.*, 1970, Fig. 3). Despite the increase in vagal tone induced by brain stem section we were unable to demonstrate any statistically significant change in the magnitude of the reflex increase in heart rate induced by distension of the pulmonary vein–left atrial junctions (Albrook *et al.*, 1972) before and after decerebration. It is of interest that after decerebration a combination of propranolol and bretylium tosylate was not effective in completely eliminating the reflex increase in heart rate associated with distension of the pulmonary vein–left atrial junctions (Burkhart & Ledsome, 1974) whereas in intact anaesthetised dogs

Kappagoda, Linden, Scott & Snow (1975) found that a combination of similar drugs usually completely prevented the response. We have interpreted these results as indicating that under conditions of moderate vagal tone the reflex response to distension of the pulmonary vein–left atrial junctions has an efferent vagal component. Kappagoda *et al.* (1975) have held firmly to the view that the efferent limb of the reflex is solely in the sympathetic system. In experiments in spinal animals (Burkhart & Ledsome, 1974) in which arterial pressure was maintained by infusion of noradrenaline and in which there was also moderate cardiac vagal tone, as indicated by the average heart rate of 84 beats/min, the magnitude of the reflex increase in heart rate induced by pulmonary vein distension was similar to that in the intact animal. Since the reflex increase in heart rate in these animals must have been due wholly to a decrease in vagal tone this implies an efferent vagal component to the reflex response not seen in the intact anaesthetised state. The suggestion of Kappagoda *et al.* (1975) that these changes were secondary to circulatory obstruction and a decrease in arterial baroreceptor stimulation and a decrease in arterial baroreceptor stimulation is unacceptable as in these experiments arterial pressure increased during distension of the pulmonary vein–left atrial junctions (Burkhart & Ledsome, 1974, Fig. 3). We believe this is strong evidence of a vagal efferent pathway for the reflexes and that the relative contributions of sympathetic and vagal efferent components of the cardiac response to distension of the pulmonary vein–left atrial junctions may vary according to the prevailing state of vagal and sympathetic tone of the animal. This implies a high degree of likelihood of modulation of the reflex response by supramedullary structures as has been observed for the arterial baroreceptor reflex responses. Arterial baroreceptor afferents converge on to sets of medullary neurons which also receive non-specific viscero-somatic afferents, the latter influencing not only the tonic autonomic outflow to the heart but also its reflex response to baroreceptor afferents (Kirchheim, 1976). Since the atrial receptor afferents appear to synapse in close association with these other afferents even if there is not interaction at the first synapse with other arterial baroreceptor afferents, it is likely that the reflex response to atrial receptor stimulation may also be modulated by the activity of viscero-somatic afferents. This is an area which has not yet been explored.

In summary, our observations indicate that in dogs the afferent fibres from atrial receptors pass through or synapse in the intermediate region of the NTS close to the obex. Here they are in close proximity to the afferent fibres from the aortic arch baroreceptors but are functionally separate from at least a proportion of the aortic chemoreceptor afferents and afferents from the carotid sinus baroreceptors. The possibility of interaction between atrial receptor afferent activity and that from the arterial baroreceptors has not been fully explored. Some evidence is discussed that the relative contributions of vagal and sympathetic efferent components of the reflex response to atrial receptor stimulation may vary, raising the possibility of modulation of the reflex response by supramedullary structures or viscero-somatic afferents.

Much of this work was carried out in cooperation with Sally M. Burkhart and with the assistance of Leilani Funnel whose help is gratefully acknowledged.

The work was supported by the Medical Research Council of Canada and the British Columbia Heart Foundation.

References

Adair, J.R. & Manning, J.W. (1975). Hypothalamic modulation of baroreceptor afferent unit activity. *Am. J. Physiol.*, **229**, 1357–64.

Albrook, S.M. Bennion, G.R. & Ledsome, J.R. (1972). The effects of decerebration on the reflex response to pulmonary vein distension. *J. Physiol. (London)*, **226**, 793–803.

Angell-James, J. & Daly, M. de B. (1970). Comparison of reflex vasomotor responses to separate and combined stimulation of carotid sinus and aortic arch baroreceptors by pulsatile and non-pulsatile pressures in the dog. *J. Physiol. (London)*, **214**, 51–64.

Baertschi, A.J., Munzner, R.F., Ward, D.G., Johnson, R.D. & Gann, D.S. (1975). Right and left atrial B-fibre input to the medulla of the cat. *Brain Res.*, **98**, 189–93.

Baumgarten, R. von, Koepchen, P. & Aranda, L. (1959). Untersuchungen zur Lokalisation der bulbaren Kreislanlzentren. *Vert. Dtsch. ges. Kreisl. Forsch.*, **25**, 170–2.

Biscoe, T.J. & Sampson, S.R. (1970*a*). Field potentials evoked in the brain stem of the cat by stimulation of the carotid sinus, glossopharyngeal, aortic and superior laryngeal nerves. *J. Physiol. (London)*, **209**, 341–58.

Biscoe, T.J. & Sampson, S.R. (1970*b*). Responses of cells in the brain stem of the cat to stimulation of the sinus, glossopharyngeal, aortic and superior laryngeal nerves. *J. Physiol. (London)*, **209**, 359–74.

Bonvallet, M. & Sigg, B. (1958). Etude électrophysiologique des afférences vagales au niveau de leur pénétration dans le bulbe. *J. Physiol. (Paris)*, **50**, 63–74.

Burkhart, S.M. & Ledsome, J.R. (1974). The response to distension of the pulmonary vein–left atrial junctions in dogs with spinal section. *J. Physiol. (London)*, **237**, 685–700.

Carswell, F., Hainsworth, R. & Ledsome, J.R. (1970). The effects of distension of the pulmonary vein–left atrial junctions upon peripheral vascular resistance. *J. Physiol. (London)*, **207**, 1–14

Cottle, M.K. (1964). Degeneration studies of primary afferents of IXth and Xth cranial nerves in the cat. *J. comp. Neurol.*, **122**, 329–45.

Crill, W.E. & Reis, D.J. (1968). Distribution of carotid sinus and depressor nerves in cat brain stem. *Am. J. Physiol.,* **214,** 269–76.

Edis, A.J. & Shepherd, J.T. (1971). Selective denervation of aortic arch baroreceptors and chemoreceptors in dogs. *J. appl. Physiol.,* **30,** 294–6.

Gabriel, M. & Seller, H. (1970). Interaction of baroreceptor afferents from carotid sinus and aorta at the nucleus tractus solitarii. *Pflügers Arch. ges. Physiol.,* **318,** 7–20.

Gauer, O.H. & Henry, J.P. (1963). Circulatory basis of fluid volume control. *Physiol. Rev.,* **43,** 423–81.

Hellner, K. & Baumgarten, R. von (1961). Uber ein Endigungsgebiet afferenter, kardiovascularer Fasern des Nervus vagus im Rautenhirn de Katze. *Pflügers. Arch. ges. Physiol.,* **273,** 223–34.

Kappagoda, C.T., Linden, R.J., Scott, E.M. & Snow, H.M. (1975). Atrial receptors and heart rate: the efferent pathway. *J. Physiol. (London),* **249,** 581–90.

Keith, I.C., Kidd, C., Linden, R.J. & Snow, H.M. (1975). Modification of neuronal activity in the dog medulla oblongata by stimulation of left atrial receptors. *J. Physiol. (London),* **245,** 80P.

Kent, B.J., Drane, J.W. & Manning, J.W. (1971). Suprapontine contributions to the carotid sinus reflex in the cat. *Circulation Res.,* **29,** 534–41.

Kirchheim, H.R. (1976). Systemic arterial baroreceptor reflexes. *Physiol. Rev.,* **56,** 100–76.

Korner, P.I., Shaw, J., West, M.J. & Oliver, J.R. (1972). Central nervous control of baroreceptor reflexes in the rabbit. *Circulation Res.,* **31,** 637–52.

Ledsome, J.R. & Linden, R.J. (1964). A reflex increase in heart rate from distension of the pulmonary vein–left atrial junctions. *J. Physiol.,* **170,** 456–73.

Linden, R.J. (1972). Function of nerves of the heart. *Cardiovasc. Res.,* **6,** 605–26.

Menninger, R.P. & Frazier, D.T. (1972). Effects of blood volume and atrial stretch on hypothalamic single unit activity. *Am. J. Physiol.,* **223,** 288–93.

Miura, M. & Reis, D.J. (1972). The role of the solitary and paramedian reticular nuclei in mediating cardiovascular reflex responses from carotid baro- and chemoreceptors. *J. Physiol. (London),* **223,** 525–48.

Oberholzer, R.J.H. (1960). Circulatory centers in medulla and midbrain. *Physiol. Rev.,* **40** (suppl. 4), 179–95.

Reis, D.J. & Cuénod, M. (1965). Central neural regulation of carotid baroreceptor reflexes in the cat. *Am. J. Physiol.,* **209,** 1267 79.

Seller, H. & Illert, M. (1969). The localization of the first synapse in the carotid sinus baroreceptor reflex pathway and its alteration of the afferent input. *Pflügers Arch. ges. Physiol.,* **306,** 1–19.

Uther, J.B., Hunyor, S.N., Shaw, J. & Korner, P.I. (1970). Bulbar and suprabulbar control of the cardiovascular autonomic effects during arterial hypoxia in the rabbit. *Circulation Res.,* **26,** 491–506.

Discussion

(a) In reply to a question Ledsome said that the transient fall in blood pressure and dilatation in the perfused hind limb and kidney following balloon distension of the pulmonary vein–atrial junctions was unaffected by β-receptor blockade but was abolished by α-blockade.

(b) It was noted that the lesion in the tractus solitarius abolished the responses and it was suggested that this might interrupt primary afferents and the deeper area in the region of the paramedian nucleus described in the preceding account might be the second link in the chain. Ledsome agreed with this interpretation but Kidd pointed out that it was not possible to know the sequence but only that these areas were links in a chain. Furthermore, it was important to be careful in the interpretation of results from lesion experiments because of the possibility that functional damage extended beyond the anatomical lesion.

(c) It was noted that the lesion made by Ledsome also abolished the effects of the depressor nerve stimulation and that fibres in the heart had been identified in the depressor nerve; e.g., according to Thorén, half the C fibres in the aortic nerve of the rabbit arose from the heart and lung. Coleridge added that in the dog also the aortic nerve contained C fibres and pulmonary stretch afferents.

(d) The possible efferent vagal component of the left atrial balloon reflex was discussed. Ledsome considered that the residual response after propranolol was not caused by inadequate block because the response to infused isoprenaline was 95% blocked. He considered that it was not caused by baroreceptors because the response occurred in spinal animals in which blood pressure usually increased slightly. The fact that it was vagal was inferred from the rapidity of the onset and decline of the response – within 10 s. He claimed that the vagal component was more marked in decerebrate animals in which there was a higher resting vagal tone. However, he had not tested the effects of bretylium in decerebrate dogs. Linden wondered what caused the increase in blood pressure and suggested that if it was thought to be caused by the increase in heart rate the effects of pacing the heart should be tested. Ledsome thought that it was not caused by spinal reflexes because it did not occur after vagal section.

Reflex responses during bleeding, infusion and infarction

21. Cardiopulmonary receptors with vagal afferents: location, fibre type and physiological role

D. E. DONALD & J. T. SHEPHERD

In unanaesthetised dogs subjected to hypotension induced by coronary occlusion the renal blood flow is much less affected than in dogs in which a similar decrease in arterial blood pressure and cardiac output is produced by haemorrhage (Gorfinkel, Szidon, Hirsch & Fishman, 1972). In analysing the different behaviour of the renal circulation in these two forms of shock, attention has been directed to receptors in the cardiopulmonary region since their activity has been shown to increase during coronary occlusion (Thorén, 1976b). Also, in cardiogenic shock there is a decrease in the postganglionic sympathetic nerve activity to the kidney which changes to an increase following section of the vagal nerves (Kezdi, Kordenat & Misra, 1974). During haemorrhage, withdrawal of the inhibition exerted by the cardiopulmonary receptors contributes to the increase in activity in the renal nerves (Clement, Pelletier & Shepherd, 1972).

Experiments such as these reflect the continued interest in those receptors in the heart and lungs which can cause reflex inhibition of the sympathetic outflow, and whose afferent fibres are in the vagi. This presentation will be confined largely to work conducted by the authors and their associates; the evidence for the reflex inhibition is reviewed and the possible role of these receptors in various abnormal circumstances is discussed.

Nature of inhibitory reflex

Interruption of afferent traffic

When the influence of the arterial baroreceptors is eliminated, block of cervical vagal traffic by cooling in the anaesthetised cat, rabbit and dog results in a rise in blood pressure, tachycardia and constriction of the resistance vessels of skeletal muscle, intestine and kidney and of the splanchnic capacitance vessels. There is, however, no change in the tone of the cutaneous veins. The vasoconstriction is unaffected by atropine or section of the vagi at the diaphragm and is caused by an increase in sympathetic adrenergic activity (Guazzi, Libretti & Zanchetti, 1962; Öberg & White, 1970; Mancia, Donald & Shepherd, 1973). Thus, receptors in the cardiopulmonary region subserved by vagal afferents exert a tonic inhibition on the central neurons controlling the sympathetic outflow to resistance and capacitance vessels, with the exception of the cutaneous veins.

Origin of inhibitory activity

To determine whether all or only parts of the cardiopulmonary region are involved in the tonic vasomotor inhibition the following conditions were instituted in anaesthetised, sino-aortic denervated dogs with diaphragmatic vagotomy. The caval venous return was oxygenated extracorporeally and returned to the aorta and the heart was removed leaving the ventilated lungs. Using the same extracorporeal circuitry, the lungs and the ventricles were removed leaving the beating atria perfused via the coronary arteries. With the venous return taken from the pulmonary arteries, oxygenated extracorporeally and returned to the left atrium, the lungs were removed and the atria were denervated leaving the working innervated ventricles. In each situation vagal cooling resulted in an increase in aortic blood pressure (Fig. 21.1). Thus, receptors in the lungs, the atria and the ventricles each are responsible for a vagally mediated tonic inhibition of the vasomotor centre (Mancia & Donald, 1975).

Fibre type responsible

When activity in medullated fibres from the cardiopulmonary region was blocked in seven anaesthetised, sino-aortic denervated rabbits by cooling the cervical vagi to 6 °C there was a mean increase in systemic arterial pressure of 3.3 kPa from a control value of 13.5 kPa. Subsequent interruption of non-medullated afferent traffic by cooling the vagi from 6 to 0 °C caused a further mean rise of 2.1 kPa. This demonstrates that receptors with non-medullated vagal afferents (C fibres) contribute to the tonic inhibition of the vasomotor centre. Because of overlap in sensitivity of different vagal fibres to cooling, the total contribution of these C fibres could not be evaluated (Thorén, Mancia & Shepherd, 1975).

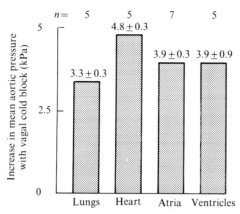

Fig. 21.1. Data from various studies of increases in aortic pressure during vagal cold block with only the lungs *in situ*, only the heart *in situ*, only the atria *in situ* and with the heart *in situ* but only the ventricles innervated. The relative importance of the tonic inhibition exerted separately by each area cannot be decided from these studies because the surgical techniques needed to isolate each area may not have preserved all afferent fibres. (Data from Mancia & Donald, 1975.)

In cats, anodal block of medullated fibres was induced by application of a direct current to the cervical vagus under circumstances in which there was no evidence for simultaneous blockade or activation of non-medullated fibres. In nine cats complete block of all afferent vagal traffic by cooling to 0 °C caused an increase in arterial pressure of 5.1 kPa; anodal block of medullated fibres resulted in a mean increase of 1.1 kPa. Thus, 4.0 kPa or 80% of the increase in arterial blood pressure can be ascribed to blockade of non-medullated fibres.

Interaction with other receptor systems

Cardiopulmonary vagal afferents and carotid baroreceptors

To examine the interaction between the cardiopulmonary and the carotid sinus reflexes the effect on aortic blood pressure of interrupting afferent vagal traffic from the cardiopulmonary region at different sinus pressures was studied in dogs. At carotid sinus pressures above 26.7 kPa vagal cold block did not cause an increase in aortic blood pressure either during normovolaemia or when the blood volume was increased to augment the inhibitory influence of the cardiopulmonary receptors. As sinus pressure was decreased an increase in aortic pressure with vagal cooling first became evident about 21.3 kPa, became progressively greater as sinus pressure was reduced and was maximal at sinus pressures between 13.3 and 6.7 kPa. At each level of carotid pressure between 21.3 and 6.7 kPa the responses to vagal block were greater during hypervolaemia. A similar pattern of interaction between the cardiopulmonary reflex and carotid sinus reflex was displayed by the hind limb and the renal vascular beds. Thus, the inhibitory influence exerted by the cardiopulmonary receptors varied inversely with that of the carotid baroreceptors, being maximal when carotid sinus inhibition was minimal and vice versa. Both systems appeared to influence the same pool of central neurons, since withdrawal of both normal and augmented cardiopulmonary inhibition was without effect when carotid sinus inhibition was maximal.

Cardiopulmonary vagal afferents, carotid baroreceptors and carotid chemoreceptors

The interplay between all three systems was examined in dogs by applying vagal blockade before and during chemoreceptor stimulation with different degrees of inhibition of the vasomotor centre by the carotid baroreceptors (Mancia, Shepherd & Donald, 1976). When the inhibition was maximal (carotid sinus pressure 29.3 kPa), vagal block and chemoreceptor stimulation either singly or in combination caused no or minimal increase in aortic blood pressure, or in perfusion pressure of a hind limb perfused at constant flow.

With intermediate inhibition (carotid sinus pressure 19.2 kPa) chemoreceptor stimulation resulted in a substantial increase in blood pressure. The modest response to vagal block at this sinus pressure was augmented six-fold by chemoreceptor stimulation. A possible explanation for this augmentation is that the increase in afterload caused by chemoreceptor stimulation changes cardiac behaviour and results in an increase in afferent vagal traffic from the cardiopulmonary area. Öberg & Thorén (1973) have shown that cardiac vagal afferents increase their frequency of discharge during the increase in resistance to left ventricular outflow, and in left ventricular end-diastolic pressure that results from carotid sinus hypotension or to injections of adrenaline.

When carotid sinus inhibition was absent (carotid sinus pressure 6.7 kPa), chemoreceptor stimulation was ineffective in increasing aortic blood pressure and hind-limb perfusion pressure. At this sinus pressure the vasomotor centre is inhibited strongly by the cardiopulmonary receptors and this may account for the absence of response to chemoreceptor stimulation. Whatever the final explanation of the various interactions the present studies demonstrate the dominant role of the carotid sinus reflex and the failure of chemoreceptor stimulation or of vagal afferents to evoke a vascular response when the carotid input is maximal.

Reflex in abnormal circumstances

Haemorrhage

In anaesthetised dogs section of the aortic nerves acutely abrogates the aortic baroreceptor and chemoreceptor reflexes (Edis & Shepherd, 1971; Mancia & Donald, 1975). With only the carotid baroreflex or the vagal reflex operative, haemorrhage causes similar degrees of constriction of the renal and mesenteric resistance vessels and of the spleen, whereas the cutaneous veins are unresponsive. The splenic contraction reflects the reflex response of the capacitance vessels throughout the splanchnic bed (Webb-Peploe, 1969).

The vasoconstrictions were mediated by an increase in sympathetic nerve activity rather than by circulating catecholamines, because the renal responses were abolished by denervation of the kidney and the splenic responses were recorded while this organ was isolated from the circulation. In anaesthetised rabbits with aortic depressor and carotid sinus nerves cut, a 10% decrease in blood volume increased renal sympathetic nerve activity by an average of 33%. This increase was unaffected by cutting the vagi at the diaphragm but was abolished or markedly attenuated by cooling or cutting the vagi in the neck. Thus the receptors mediating this reflex are situated in the cardiopulmonary area (Clement *et al.*, 1972).

By contrast to the similar effects of haemorrhage on the renal and splanchnic vessels with only the vagal or only the carotid baroreflex operative, the constriction of the hind-limb vessels is less with only the vagal reflex operative than with only the carotid baroreflex operation. This difference explains the greater decrease in blood pressure during haemorrhage in the former circumstances; in dogs with only the vagal reflex operative, only the carotid sinus reflex operative or neither of these reflexes operative, haemorrhage of 10% of the blood volume decreased the aortic blood pressure by 24, 18 and 42%, respectively (Pelletier, Edis & Shepherd, 1971).

In anaesthetised cats after section of the aortic and carotid sinus nerves a 10% decrease in blood volume caused increases in resistance in muscle and renal vascular beds, particularly in the latter. Since the responses were small or absent after avulsion of the cardiac nerves, it was concluded that cardiac receptors are involved (Öberg & White, 1970). The maximal reflex dilatation of the renal vessels obtained with right cardiac nerve stimulation was equal to that obtained by exposing the isolated carotid sinuses to a non-pulsatile pressure of 33.3 kPa: the maximal reflex dilatation of the skeletal muscle vessels when the cardiac nerve was stimulated averaged 66% of that resulting from maximal baroreceptor stimulation (Little, Wennergren & Öberg, 1975).

Thus, the cardiopulmonary receptors subserved by vagal afferents participate in the circulatory adjustments to haemorrhage but their importance in the support of blood pressure is less than that of the carotid baroreceptors because of the lesser influence of the former on the muscle resistance vessels.

Hypervolaemia

An increase in blood volume enhances the activity of medullated and non-medullated vagal afferents from the heart (Thorén, 1976a; Thorén, Donald & Shepherd, 1976). In rabbits with arterial baroreceptor denervation an estimated 10% increase in blood volume caused the renal nerve activity to decrease by an average of 41%; after interruption of the cervical vagi there was no significant change in activity (Clement et al., 1972). In anaesthetised, normovolaemic dogs with their aortic nerves cut, vagal cold block was carried out at carotid sinus pressures of 29.3, 20.0 and 5.3 kPa, during normovolaemia and after an estimated 15% increase in blood volume. The changes in blood flow of a hind limb and a kidney perfused at constant pressure were measured (Fig. 21.2). In both, the decrease in blood flow caused by vagal block at carotid sinus pressures of 20.0 and 5.3 kPa was augmented by hypervolaemia, that at 5.3 kPa being the greater. There was no response to vagal block at a carotid sinus pressure of 29.3 kPa in either normovolaemia or hypervolaemia. With this increase in blood volume there was a significant increase in renal but not in hind-limb blood flow at the control carotid pressure of 20.0 kPa. When the inhibition exerted by the carotid and the cardiopulmonary receptors at a carotid pressure of 20.0 kPa during normovolaemia was removed by reducing carotid pressure to 5.3 kPa and blocking the vagi, there was a 79% decrease in hind-limb blood flow, and a 35% decrease in renal flow. Interruption of afferent vagal traffic from cardiopulmonary receptors accounted for 63% of the total decrease in renal blood flow, but only for 26% of the total

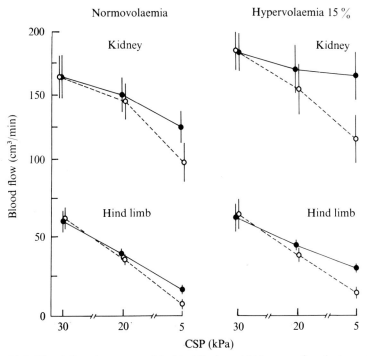

Fig. 21.2. Blood flow (mean ± s.e.) in hind limb and kidney perfused at constant pressure of 16.0 kPa in 13 dogs with aortic nerves cut. Observations made in control state (●) and during vagal cold block (○) at carotid sinus pressures (CSP) of 29.3, 20.0 and 5.3 kPa in normovolaemia and after blood volume had been increased by 15% with rheomacrodex. (Data from Mancia *et al.*, 1976).

decrease in hind-limb flow. Combined withdrawal of carotid and cardio-pulmonary inhibition during hypervolaemia caused a 66% decrease in hind-limb blood flow, and a 32% decrease in renal flow. Interruption of afferent vagal traffic now accounted for 90% of the decrease in renal blood flow and for 50% of the decrease in hind-limb flow.

These studies substantiate the observations made during haemorrhage that the influence of the cardiopulmonary receptors is greater on the renal than on the muscle resistance vessels, and suggest that the receptors can exert an inhibitory influence at normal carotid sinus pressures.

Coronary occlusion

In cats activity of left ventricular receptors with non-medullated vagal afferent fibres (mean conduction velocity 0.9 m/s) which normally have a low rate of discharge (mean 1.8 imp/s) can be greatly increased during a

1–1.5-min occlusion of a coronary artery (mean 15.4 imp/s). The receptors responding to coronary occlusion were located throughout the ventricle, including the ventricular septum. When a single coronary artery was occluded the receptors in the region normally supplied by that artery had the greatest increase in activity. This increase occurred during systole and in parallel with the systolic bulging of the ischaemic myocardium (Fig. 21.3). Receptors in other parts of the ventricle sometimes had a modest increase in activity which has been attributed to an increase in end-diastolic volume attending the occlusion of the major coronary artery. The maximal activity was not sustained; during occlusion of up to 40 min it decreased to 5 imp/s (mean value) after 5–10 min (Thorén, 1976*a*). Other studies have shown that ventricular receptors with vagal medullated fibres are also activated with experimental coronary occlusion and that depressor reflexes are induced (see Thorén, 1976*b*). That left ventricular receptors with afferent C fibres make the major or sole contribution to the vagally mediated depressor reflex seen in cats during coronary occlusion is suggested by the following. Electrical stimulation in cats of the afferent C fibres from the heart induces a depressor reflex whereas electrical stimulation of medullated fibres induces an excitatory response (Öberg & Thorén, 1973). In addition receptors with afferent C fibres greatly outnumber those with medullated fibres in the left ventricle (Coleridge, Coleridge & Kidd, 1964; Sleight & Widdicombe, 1965).

The findings of the above experimental studies have implications in the reflex control of the renal circulation in clinical situations. For example, during haemorrhagic shock, when the systemic arterial blood pressure and the central venous pressure decrease, input from the carotid and the cardiopulmonary receptors would be diminished, resulting in a strong constriction of the renal and muscular vascular beds. However, in cardiac failure accompanied by a decrease in arterial blood pressure and an increase in central venous pressure the decrease in carotid sinus pressure would act to constrict the renal vessels, but this change would be opposed by a simultaneous increase in activity of the cardiopulmonary receptors, thus serving to maintain renal blood flow. However, the constriction of the muscle resistance vessels caused by the reduced activity of the arterial baroreceptors would be less affected by the increased activity of the cardiopulmonary receptors.

Hypercapnia

In the anaesthetised dog (Mancia, Shepherd & Donald, 1975) and rabbit

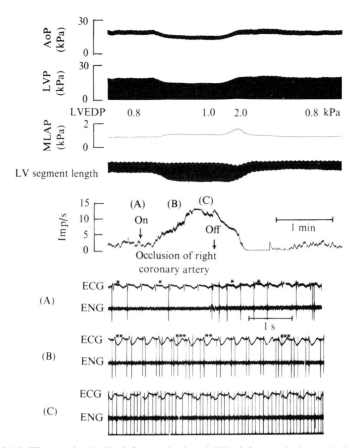

Fig. 21.3. The aortic (AoP), left ventricular (LVP), left ventricular end-diastolic (LVEDP), mean left atrial pressure (MLAP), left ventricular (LV) segment length (downward deflection means increased length) and discharge frequency in a left ventricular receptor are shown before and during occlusion of the right coronary artery. The ECG and neurograms (ENG) recorded at times indicated by (A), (B) and (C) are also shown. The receptor initially has a low frequency discharge (A) which is markedly increased about 20 s after onset of the coronary occlusion. In (B) the receptor discharge has a cardiac modulated rhythmicity with two or three impulses every cycle. From the total conduction time (220 ms) it is calculated that the receptor is activated in systole (* indicates the corrected position in the cardiac cycle of the receptor activation, which occurs 220 ms before the recorded spike). In (C) the receptor fires continuously. Note that the reciprocal S–T depression in the ECG is obvious already in (B). During the occlusion the increased firing occurs together with the bulging of the ischaemic area. The time lag between the release of occlusion and the decrease in systolic bulging was probably caused by an observed spasm of the coronary artery, which relaxed slowly on release of the ligature. The disappearance of the dyskinesia was accompanied by a rapid decrease in discharge frequency. (Data from Thorén, 1976. By permission of the American College of Cardiology.)

(Ott & Shepherd, 1973), mechanically ventilated and with sino-aortic denervation, the increase in renal vascular resistance due to cervical vagal cold block is two- to four-fold greater during hypercapnia than in normocapnia. In the dog the increase in the resistance of the muscle vessels during vagal block was similar in both situations. In the rabbit during hypercapnia and vagal block there was an increase in muscle resistance but it was much less than that seen in the kidney (Fig. 21.4). In the rabbit this increase in resistance with vagal block during hypercapnia was accompanied by a marked increase in activity in the renal sympathetic nerves (Ott, Lorenz & Shepherd, 1972).

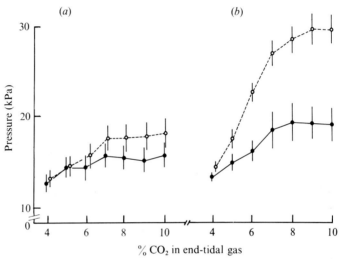

Fig. 21.4. Effect of increasing end-tidal carbon dioxide on increases in *(a)* hind-limb and *(b)* kidney perfusion pressures caused by vagal cold block. Means $\pm$ s.e. in seven rabbits ventilated at 15 cm$^3 \cdot$ kg^{-1}. Hind limb and kidney perfused at constant flow. ●, Control observations; ○, vagal cold block observations. (Data from Ott *et al.*, 1972.)

In additional studies in spontaneously breathing rabbits the effect of hypercapnia was studied with both increased ventilation (caused by inspiration of mixtures of carbon dioxide in oxygen) and decreased ventilation (caused by infusion of gallamine triethiodide during oxygen breathing). The increase in renal vascular resistance caused by vagal block during hypercapnic hyperventilation was six- to seven-fold greater than in normocapnic normal ventilation. A similar augmentation of the response to vagal block was seen during hypercapnic hypoventilation. Thus the powerful central constrictor effect of carbon dioxide can be counteracted in the renal vasculature by vagal afferents activated by lung movement; even minimal respiratory movements, which make little or no contribution to respiratory gas exchange, are sufficient to activate the pulmonary receptors responsible for the inhibitory reflex (Ott & Shepherd, 1975).

In the anaesthetised, mechanically ventilated rabbit with the carotid sinuses denervated and the vagi blocked, hypercapnia augments the vasomotor inhibition of renal and skeletal muscle vessels exerted by the aortic arch baroreceptors, the effect on renal resistance being the more pronounced (Ott & Shepherd, 1973). Thus, it seems that this phenomenon is a general property of reflexes that tonically inhibit the vasomotor centre. It seems most likely that carbon dioxide acts centrally to accentuate particularly the inhibitory control of the renal vessels by the vagal and other inhibitory afferents, thus counteracting the central excitatory effects of carbon dioxide. The preservation of renal blood flow through this mechanism would be of obvious benefit in contributing to the restoration of acid–base balance in respiratory acidosis.

Renin release

In addition to exerting a tonic control of the vasomotor centre, cardiopul-monary receptors with vagal afferents are involved in the neural control of renin secretion (Mancia, Romero & Shepherd, 1975). In anaesthetised dogs with aortic nerves sectioned and the carotid sinuses vascularly isolated and maintained at a pressure equal to the control arterial blood pressure, reversible cold block of the cervical vagal nerves resulted in a rapid and considerable increase in the secretion of renin (Fig. 21.5). The release of renin was abolished by denervating the kidney. That the control of renin secretion exerted by the cardiopulmonary receptors is sensitive to small changes in blood volume has been demonstrated recently in our laboratory. Thirteen dogs with intact carotid sinuses were anaesthetised with chlora-lose and subject to a non-hypotensive haemorrhage ($4 \mathrm{~cm}^3 \cdot \mathrm{kg}^{-1}$) before and after vagotomy. Six dogs did not release renin in either test situation. Six dogs released renin before vagotomy but not after. One dog released a large amount of renin before and a small amount after vagotomy (M. D. Thames, personal communication).

With renal perfusion pressure kept constant, cervical vagotomy in intact dogs maintained on a high salt diet caused increases in mean arterial blood pressure and in plasma renin activity (Yun, Delea, Bartter & Kelly, 1976). In dogs maintained on a low salt diet an increase in pressure in the left

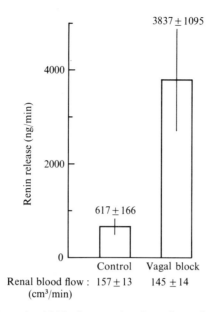

Fig. 21.5. Effect of vagal cold block on renin release from the left kidney in eight dogs (mean ±s.e.) with aortic nerves cut and carotid sinus pressure maintained at the level of mean aortic pressure prior to the block. Vagal block was maintained for 3 min and blood samples for renin measurement were drawn simultaneously from the aorta and left renal vein before block (control) and during the third minute of block. (Data from Mancia *et al.*, 1975. By permission of the American Heart Association.)

atrium and pulmonary veins caused by inflation of a balloon positioned at the mitral orifice depressed renin secretion from the innervated kidney to 58% of the control value, while secretion from the contralateral denervated kidney was unchanged (Zehr, Hasbargen & Kurz, 1976). In another study distension of the right atrium caused a decrease in renin secretion (Brennan, Malvin, Jochim & Roberts, 1971). The release of renin in response to haemorrhage is decreased in dogs with chronic cardiac denervation which left most of the atrial receptors innervated (Thames *et al.*, 1971).

These studies indicate that both atrial and ventricular receptors with vagal afferents are involved in the reflex control of renin secretion, and this can occur in animals with intact arterial baroreceptors. Pulmonary vagal afferents have been shown to affect renal nerve traffic. It is thus possible that pulmonary receptors with vagal afferents also are involved in the control of renin secretion.

Summary

Receptors situated in the atria, the ventricles and the lungs exert a tonic inhibition on the vasomotor centre. The inhibition is largely mediated by non-medullated vagal afferents and is most easily demonstrated when other inhibitory inputs are controlled or eliminated. Compared with the carotid baroreceptors the cardiopulmonary receptors have less influence on the muscle circulation, but have an equal influence on other vascular beds. In abnormal circumstances they may act to reinforce or oppose the influence of the carotid baroreceptors. Receptors in the atria and ventricles are involved in the neural control of renin secretion but the type of afferent vagal fibre has yet to be determined.

We wish to thank our collaborators in these studies, Drs D. L. Clement, A. J. Edis, G. Mancia, N. T. Ott, C. L. Pelletier, J. C. Romero, M. D. Thames, P. N. Thorén, M. M. Webb-Peploe and Mr R. R. Lorenz. We also thank Mrs J. Y. Troxell for typing the manuscript.

References

Brennan, L.A., Jr, Malvin, R.L., Jochim, K.E. & Roberts, D.E. (1971). Influence of right and left atrial receptors on plasma concentrations of ADH and renin. *Am. J. Physiol.*, **221**, 273–8.

Clement, D.L., Pelletier, C.L. & Shepherd, J.T. (1972). Role of vagal afferents in the control of sympathetic nerve activity in the rabbit. *Circulation Res.*, **31**, 824–30.

Coleridge, H.M., Coleridge, J.C.G. & Kidd, C. (1964). Cardiac receptors in the dog, with particular reference to two types of afferent ending in the ventricular wall. *J. Physiol.*, **174**, 323–39.

Edis, A.J. & Shepherd, J.T. (1971). Selective denervation of aortic arch baroreceptors and chemoreceptors in dogs. *J. appl. Physiol.*, **30**, 294–6.

Gorfinkel, H.J., Szidon, J.P., Hirsch, L.J. & Fishman, A.P. (1972). Renal performance in experimental cardiogenic shock. *Am. J. Physiol.*, **222**, 1260–8.

Guazzi, M., Libretti, A. & Zanchetti, A. (1962). Tonic reflex regulation of the cat's blood pressure through vagal afferents from the cardiopulmonary region. *Circulation Res.*, **11**, 7–16.

Kezdi, P., Kordenat, R.K. & Misra, S.N. (1974). Reflex inhibitory effects of vagal afferents in experimental myocardial infarction. *Am. J. Cardiol.*, **33**, 853–60.

Little, R., Wennergren, G. & Öberg, B. (1975). Aspects of the central integration of arterial baroreceptor and cardiac ventricular receptor reflexes in the cat. *Acta physiol. scand.*, **93**, 85–96.

Mancia, G. & Donald, D.E. (1975). Demonstration that the atria, ventricles and lungs each are responsible for a tonic inhibition of the vasomotor center in the dog. *Circulation Res.*, **36**, 310–18.

Mancia, G., Donald, D.E. & Shepherd, J.T. (1973). Inhibition of adrenergic outflow to peripheral blood vessels by vagal afferents from the cardiopulmonary region in the dog. *Circulation Res.*, **33**, 713–21.

Mancia, G., Romero, J.C. & Shepherd, J.T. (1975). Continuous inhibition of renin release in dogs by vagally innervated receptors in the cardiopulmonary region. *Circulation Res., 36,* 529–35.

Mancia, G., Shepherd, J.T. & Donald, D.E. (1975). Role of cardiac, pulmonary and carotid mechanoreceptors in the control of hind-limb and renal circulation in dogs. *Circulation Res., 37,* 200–8.

Mancia, G., Shepherd, J.T. & Donald, D.E. (1976). Interplay among carotid sinus, cardiopulmonary and carotid body reflexes in dogs. *Am. J. Physiol., 230,* 19–24.

Öberg, B. & Thorén, P. (1973). Circulatory responses to stimulation of medullated and non-medullated afferents in the cardiac nerve in the cat. *Acta physiol. scand., 87,* 121–39.

Öberg, B. & White, S. (1970). Circulatory effects of interruption and stimulation of cardiac vagal afferents. *Acta physiol. scand., 80,* 383–94.

Ott, N.T., Lorenz, R.R. & Shepherd, J.T. (1972). Modification of lung-inflation reflex in rabbits by hypercapnia. *Am. J. Physiol., 223,* 812–19.

Ott, N.T. & Shepherd, J.T. (1973). Modification of the aortic and vagal depressor reflexes by hypercapnia in the rabbit. *Circulation Res., 33,* 160–5.

Ott, N.T. & Shepherd, J.T. (1975). Modification of vagal depressor reflex by CO_2 in spontaneously breathing rabbits. *Am. J. Physiol., 228,* 530–5.

Pelletier, C.L., Edis, A.J. & Shepherd, J.T. (1971). Circulatory reflex from vagal afferents in response to haemorrhage in the dog. *Circulation Res., 29,* 626–34.

Sleight, P. & Widdicombe, J.G. (1965). Action potentials in fibres from receptors in the epicardium and myocardium of the dog's left ventricle. *J. Physiol., 181,* 235–58.

Thames, M.D., Ul-Hassan, Z., Brackett, N.C., Jr, Lower, R.R. & Kontos, H.A. (1971). Plasma renin responses to hemorrhage after cardiac autotransplantation. *Am. J. Physiol., 221,* 1115–19.

Thorén, P.N. (1976a). Atrial receptors with nonmedullated vagal afferents in the cat. *Circulation Res., 38,* 357–62.

Thorén, P.N. (1976b). Activation of left ventricular receptors with nonmedullated vagal afferent fibers during occlusion of a coronary artery in the cat. *Am. J. Cardiol., 37,* 1046–51.

Thorén, P.N., Donald, D.E. & Shepherd, J.T. (1976). Role of heart and lung receptors with nonmedullated vagal afferents in circulatory control. *Circulation Res., 38* (Suppl. II), 2–9.

Thorén, P.N., Mancia, G. & Shepherd, J.T. (1975). Vasomotor inhibition in rabbits by vagal nonmedullated fibers from cardiopulmonary area. *Am. J. Physiol., 229,* 1410 13.

Webb-Peploe, M.M. (1969). Isovolumetric spleen: index of reflex changes in splanchnic vascular capacity. *Am. J. Physiol., 216,* 407–13.

Yun, J.C.H., Delea, C.S., Bartter, F.C. & Kelly, G. (1976). Increase in renin release after sinoaortic denervation and cervical vagotomy. *Am. J. Physiol., 230,* 777–83.

Zehr, J.E., Hasbargen, J.A. & Kurz, K.D. (1976). Reflex suppression of renin secretion during distension of cardiopulmonary receptors in dogs. *Circulation Res., 38,* 232–9.

Discussion

(a) Some criticism was made of the technique used of studying reflex responses with the sino-aortic baroreceptors denervated, first, because some aortic afferents might not have been denervated and secondly,

because this procedure would result in an abnormal preparation with intense efferent sympathetic activity which might make venous responses difficult to obtain. Shepherd countered by pointing out that if any fibres from baroreceptors and chemoreceptors were left intact they did not produce any noticeable responses. The technique of denervation, he considered, was a classical physiological approach and other responses could be obtained using such a preparation.

(b) Goetz referred to a series of experiments in which it had been noted that moderate haemorrhage in conscious dogs resulted in renal vasodilatation, whereas the same procedure in anaesthetised dogs caused vasoconstriction. He wondered about the relative roles of cardiopulmonary receptors and atrial baroreceptors in these responses and felt that it was important to remember that anaesthetised animals and conscious man behave differently.

(c) There was a lengthy discussion about the problems associated with the use of anodal block. Some participants found that it was difficult to block myelinated fibres without stimulating C fibres or burning the nerve. This technique had been used successfully in the rabbits but Coleridge commented that 'there is physiology and there is the rabbit'. Neil added that he thought 'the rabbit isn't an animal – it's a flower!'

Thorén described the technique he had used successfully in the cat to block A fibres without exciting C fibres. (1) The sheath is completely removed from the nerve. This reduces the current required for block from about 300 μA to about 50 μA. (2) The temperature of the liquid paraffin is reduced to 32 °C. This reduces the sensitivity of C fibre to anodal block. (3) The current required for block is determined by increasing it gradually. This activates medullated fibres but if the appropriate current is subsequently switched on suddenly medullated fibres are blocked without initial activation.

Thorén added that the main drawback of the method is that C fibres would also be blocked to some extent. He said that the method was only useful for short periods and for C fibres of relatively low frequency. He conceded that the method would be almost impossible in the dog.

22. Cardiac baroreceptors in circulatory control in humans

F. M. ABBOUD & A. L. MARK

Although experiments in animals (Linden, 1973; Shepherd, 1974) and a few studies in humans (Roddie, Shepherd & Whelan, 1957; Roddie & Shepherd, 1958) have suggested that cardiopulmonary baroreceptors contribute to circulatory control the central role in reflex regulation of vascular resistance in humans has usually been attributed to arterial baroreceptors. However, during recent years we have conducted experiments in humans which implicate cardiopulmonary baroreceptors in control of vascular resistance in physiological states such as orthostatic stress and pathological states such as exertional syncope in aortic stenosis. The two general conclusions which emerge from these studies are first, that concepts of neurogenic control in humans should include an important contribution of cardiac as well as arterial baroreceptors and secondly, that there is substantial non-uniformity or selectivity in baroreceptor control of circulation in humans. Afferent impulses from baroreceptors are differentiated centrally to cause non-uniform sympathetic outflow to different vascular beds and segments. The experiments described here delineate important differences in the patterns of responses during activation of cardiopulmonary and carotid baroreceptor reflexes in humans.

Reflex adjustments during orthostatic stress

Low-pressure cardiopulmonary baroreceptors in reflex vasoconstrictor responses

When blood pools in leg veins during standing or lower body negative pressure (LBNP) there are decreases in venous return, central venous pressure and stroke volume, but arterial pressure is maintained by reflex

437

tachycardia and vasoconstriction. These reflex adjustments traditionally have been attributed to high-pressure arterial baroreceptors. However, Roddie *et al.* (1957) observed that vasodilator responses to increases in venous return do not correlate with the magnitude of increases in arterial pressure and suggested that low-pressure cardiopulmonary baroreceptors participate in reflex adjustments to changes in venous return. In addition we noticed several years ago that when blood is pooled in leg veins during LBNP, forearm vasoconstriction often occurred in the absence of an obvious fall in arterial pressure. These observations prompted us to evaluate systematically the role of low-pressure baroreceptors in the circulatory adjustments to venous pooling in humans.

We evaluated responses to graded LBNP in 11 healthy supine young men (Zoller *et al.*, 1972). Systemic arterial and central venous pressures were recorded with catheters introduced through a brachial artery and antecubital vein respectively. Forearm blood flow was measured by venous occlusion plethysmography using a Whitney mercury-in-silastic gauge. The lower part of the subject's body caudal to the iliac crest was enclosed in an airtight chamber. A commercial vacuum cleaner was connected to the chamber through an adjustable vent to produce graded LBNP.

LBNP at -1.3 kPa decreased central venous pressure and forearm blood flow and increased forearm vascular resistance (Table 22.1 and Fig. 22.1). The forearm vasoconstriction occurred without significant changes in mean arterial pressure, arterial pulse pressure, arterial dP/dt and heart rate (Table 22.1). Analysis of systemic arterial pressure before and at 4-s intervals after application of LBNP did not reveal even a transient decrease in systolic pressure (Zoller *et al.*, 1972), which might have triggered reflexes originating in arterial baroreceptors.

These findings during low levels of LBNP indicated that decreases in venous return and central venous pressure can produce reflex forearm vasoconstriction without significant changes in the determinants of carotid and aortic baroreceptor activity. This indicates that low-pressure baroreceptors, which are responsive to changes in cardiac filling pressures, contribute to forearm vasoconstriction.

During LBNP at -5.3 kPa, significant decreases in arterial pressure and increases in heart rate accompanied further decreases in central venous pressure (Table 22.1 and Fig. 22.1). Thus, LBNP at -5.3 kPa triggered both low- and high-pressure reflexes. LBNP at -5.3 kPa produced additional forearm vasoconstriction but the increment in forearm vascular resistance between LBNP at -1.3 kPa (low-pressure reflexes) and LBNP

Table 22.1. *Responses to lower body negative pressure (LBNP) in 11 normal subjects (from Zoller et al., 1972)*

	Control	LBNP −1.3	LBNP −5.3	Recovery
Central venous pressure (kPa)	0.97 ± 0.09	0.40 ± 0.08*	0.07 ± 0.08*	1.01 ± 0.08
Forearm blood flow (per 100 cm³ tissue) (cm³/min)	6.2 ± 1.0	4.4 ± 0.9*	3.5 ± 0.71*	6.2 ± 1.0
Forearm vascular resistance (per 100 cm³ tissue) (kPa·cm⁻³·min)	2.28 ± 0.25	3.38 ± 0.52*	4.00 ± 0.65*	2.36 ± 0.28
Systolic arterial pressure (kPa)	15.33 ± 0.45	15.25 ± 0.47	14.72 ± 0.53*	15.41 ± 0.45
Mean arterial pressure (kPa)	12.00 ± 0.49	12.04 ± 0.43	11.68 ± 0.45*	12.04 ± 0.37
Arterial pulse pressure (kPa)	5.43 ± 0.27	5.15 ± 0.27	4.39 ± 0.27*	5.37 ± 0.23
Arterial dP/dt (kPa/s)	100.24 ± 11.06	99.31 ± 13.33	86.91 ± 10.26*	99.18 ± 11.33
Heart rate (beats/min)	62.2 ± 2.2	64.7 ± 2.5	76.0 ± 4.3*	62.6 ± 2.1

Entries represent mean ±s.e. Values for arterial pressure were the average of two determinations approximately 15 and 45 s after onset of each level of LBNP.

* Values which are significantly different from control observations ($P < 0.05$).

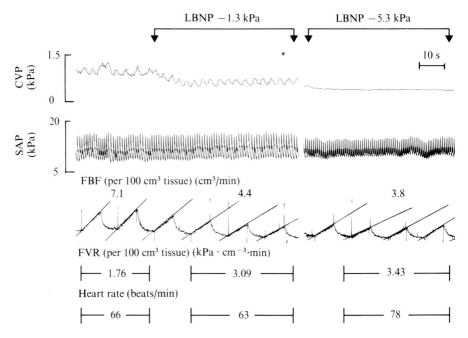

Fig. 22.1. Responses to graded lower body negative pressure (LBNP) in a normal human subject. LBNP at -1.3 kPa decreased central venous pressure (CVP) and produced forearm vasoconstriction without decreasing arterial pressure or pulse pressure or increasing heart rate. LBNP at -5.3 kPa decreased both central venous pressure and arterial pulse pressure and increased heart rate, but produced only slight additional forearm vasoconstriction. SAP = systemic arterial pressure; FBF = forearm blood flow; FVR = forearm vascular resistance. (Reprinted from Abboud *et al.*, 1976.)

-5.3 kPa (low- and high-pressure reflexes) was small (Table 22.1 and Fig. 22.1). This observation suggested that high-pressure baroreceptors might not have a major influence on forearm resistance in humans and that low-pressure reflexes might exert the predominant baroreceptor influence on forearm vasomotor tone in humans.

Cardiopulmonary and carotid baroreceptors in control of forearm resistance

To separate the relative contribution of low-pressure cardiopulmonary and high-pressure carotid baroreceptors in control of forearm resistance, we performed additional experiments (Mark, Eckberg, Abboud & Johannsen, 1974) utilising two different experimental approaches. In one approach we

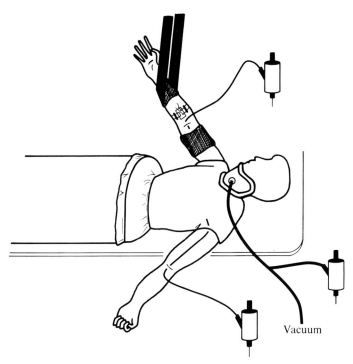

Fig. 22.2. Schematic diagram of methods used to produce neck negative pressure ('neck suction') and lower body negative pressure (LBNP). Neck suction is produced by creating negative pressure in the chamber beneath the neck collar. This stretches carotid baroreceptors and triggers reflex bradycardia and hypotension. LBNP was produced by enclosing the subject's body below the iliac crest in an airtight chamber and creating negative pressure in the chamber by using a commercial vacuum cleaner connected to the chamber through an adjustable vent. LBNP pools blood in veins of the lower extremities and decreases venous return and central venous pressure. Low levels of LBNP activate cardiopulmonary baroreceptor reflexes without decreasing systemic arterial pressure; high levels of LBNP activate both cardiopulmonary and arterial baroreceptor reflexes by decreasing central venous and systemic arterial pressures. Forearm blood flow is measured with a Whitney strain gauge plethysmograph and arterial and central venous pressures are recorded continuously. (Reprinted from Abboud *et al.*, 1976.)

compared responses to neck suction with responses to elevation of the legs and lower trunk. Neck suction was produced by applying a neck collar (Fig. 22.2) and creating negative pressure between the collar and the neck to distend the vascular structures in the neck (Eckberg, Cavanaugh, Mark & Abboud, 1975). This stretches carotid baroreceptors and causes bradycardia and hypotension (Ernsting & Parry, 1957; Bevegard & Shepherd, 1966; Eckberg *et al.*, 1975). Elevation of legs and lower trunk increases venous return and central venous pressure and stretches cardio-pulmonary baroreceptors.

Neck suction produced a significant lowering of heart rate and mean arterial pressure, but did not significantly decrease forearm vascular resistance (Fig. 22.3). One explanation for absence of striking reflex

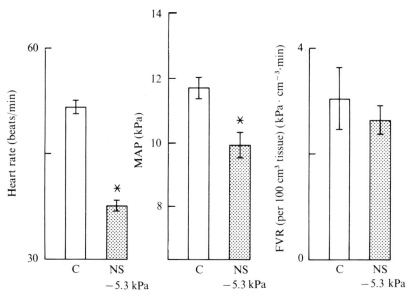

Fig. 22.3. Stretch of carotid baroreceptors with neck suction (NS) triggered reflex bradycardia and hypertension without producing significant forearm vasodila-tation. This suggests that carotid baroreceptors have minimal effects on forearm vascular resistance (FVR). Means ± s.e. C = control; MAP = mean arterial pressure; $n = 8$. * Indicates that $P < 0.05$. (Reprinted from Abboud *et al.*, 1976.)

vasodilatation could have been a low level of resting neurogenic constrictor tone in these healthy supine young men. However, in the same subjects in whom neck suction did not cause forearm vasodilatation, stretch of cardiopulmonary baroreceptors by elevation of the legs and lower trunk produced significant reflex vasodilatation (Fig. 22.4) as reported earlier (Roddie *et al.*, 1957).

In a second approach, we compared responses to LBNP at −5.3 kPa with responses to LBNP at −5.3 kPa plus simultaneous neck suction. By applying neck suction when the LBNP was −5.3 kPa, we could minimise

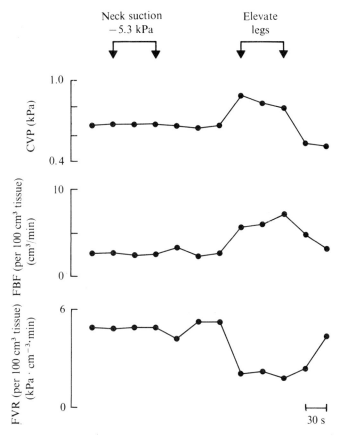

Fig. 22.4. Stretch of carotid baroreceptors with neck suction produced minimal effect on forearm vascular resistance (FVR) whereas stretch of cardiopulmonary baroreceptors from elevation of the legs and increases in central venous pressure (CVP) produced striking forearm vasodilatation. FBF = forearm blood flow. (Reprinted from Abboud *et al.*, 1975.)

the contribution to the carotid baroreceptor reflex and thereby separate the contribution of cardiopulmonary and carotid baroreceptor reflexes.

LBNP at −5.3 kPa decreased central venous pressure and arterial pulse pressure and produced reflex tachycardia and forearm vasoconstriction (Figs. 22.5 and 22.6). Simultaneous application of neck suction prevented the tachycardia during LBNP, but did not significantly attenuate the forearm vasoconstriction (Figs. 22.5 and 22.6).

These studies suggest that carotid baroreceptors regulate heart rate but have a minimal role in the regulation of forearm vascular resistance in

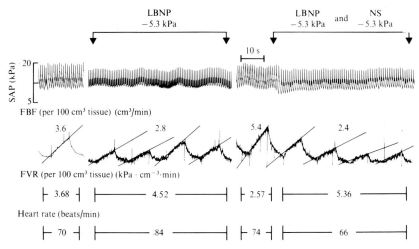

Fig. 22.5. Comparison of responses to lower body negative pressure and combined lower body negative pressure and neck suction. Simultaneous application of neck suction prevented tachycardia but did not inhibit the forearm vasoconstriction which occurred during LBNP alone. Values for forearm blood flow during interventions are average of the four determinations during the intervention. SAP = systemic arterial pressure; FBF = forearm blood flow; FVR = forearm vascular resistance.

humans. We conclude that forearm vasoconstriction during venous pooling results mainly from cardiopulmonary baroreceptors. We should, however, consider a possible role of aortic baroreceptors. Experiments in dogs suggest that aortic baroreceptors contribute to reflex adjustments during increases in arterial pressure, but do not participate appreciably in reflex adjustments during decreases in pressure (Hainsworth, Ledsome & Carswell, 1970; Dampney, Raylor & McLachlan, 1971; Donald & Edis, 1971; Edis, 1971). Specifically, aortic baroreceptors do not appear to be very responsive to changes in pulse pressure (Angell-James & Daly, 1970). This information would suggest that aortic baroreceptors do not make an important contribution during LBNP which decreases mainly pulse pressure not mean pressure. This conclusion is supported by the finding that the stretching of carotid baroreceptors by neck suction prevented the tachycardia during LBNP. If aortic baroreceptors made an important contribution to reflex responses during LBNP one might expect to observe some tachycardia during LBNP even when the contribution of carotid baroreceptors was minimised by application of neck suction during LBNP. Accordingly, we suggest that aortic baroreceptors probably do not make

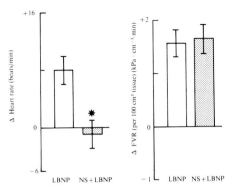

Fig. 22.6. Lower body negative pressure (LBNP) at −5.3 kPa increased heart rate and forearm vascular resistance. Simultaneous application of neck suction inhibited the tachycardia (*$P < 0.05$) induced by LBNP, but it did not inhibit the forearm vasoconstriction, suggesting that the carotid baroreceptor reflex modulates heart rate but not forearm vascular resistance (FVR) in humans. Mean ± S.E., $n = 8$. NS = neck suction.

an important contribution during LBNP and conclude that the forearm vasoconstriction results mainly from cardiopulmonary baroreceptors.

Carotid baroreceptor control of splanchnic resistance

Although the studies just described suggested that carotid baroreceptors are not very effective in modulating forearm vascular resistance in humans, it should not be concluded that they have a negligible influence on other regional circulations. Indeed, recent studies in our laboratories (Mark, Eckberg & Abboud, 1975) suggest that splanchnic vasoconstrictor responses to LBNP are mediated primarily through carotid baroreceptors. We measured splanchnic as well as forearm blood flow in seven normal young men during LBNP at −5.3 kPa and neck suction. Splanchnic blood flow was measured from the clearance of indocyanine green dye using the constant infusion method described by Rowell and his colleagues (Johnson, Rowell, Niederberger & Eisman, 1974; Rowell, 1976). LBNP at −5.3 kPa increased both forearm and splanchnic resistance (Fig. 22.7) as previously reported by Johnson *et al.* (1974).

Simultaneous application of neck suction did not alter the forearm vasoconstriction but it prevented most of the splanchnic vasoconstriction during LBNP (Fig. 22.7). This study suggests that stretch of carotid baroreceptors restrains adrenergic discharge to splanchnic vessels despite the fact that it does not exert major restraint on adrenergic discharge to

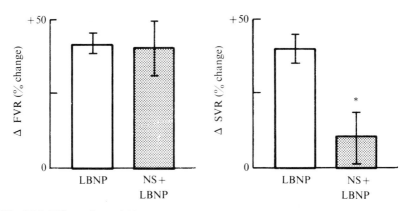

Fig. 22.7. Effect of carotid baroreceptors on splanchnic vascular resistance. Lower body negative pressure (LBNP) at -5.3 kPa increased forearm and splanchnic vascular resistance (FVR and SVR). Simultaneous application of neck suction (NS) did not modify the forearm vasoconstriction but significantly inhibited (*$P < 0.05$) the splanchnic vasoconstriction. (Reprinted from Abboud *et al.*, 1976.)

forearm vessels (Fig. 22.8). In contrast, cardiopulmonary baroreceptors have a major influence on adrenergic control of forearm vessels, but only a minor influence on splanchnic vessels (Fig. 22.8).

Non-uniform regional circulatory responses during activation of baroreceptor reflexes have been demonstrated previously in experimental animals (Abboud, 1972; Little, Wennergren & Öberg, 1975). Our study demonstrates non-uniformity of baroreceptor control of regional circulations in humans. Specifically, it delineates a profound difference in the pattern of non-uniformity between reflexes originating in cardiopulmonary and those emanating from carotid baroreceptors in humans.

Electrophysiological studies on the organisation of central vasopressor pathways by Gebber, Taylor & Weaver (1973) demonstrated that sympathetic outflow from the brain is distributed to postganglionic sympathetic nerves over two distinct types of pathways in cats, one pathway with a long latency and slow conduction and another pathway with short latency and rapid conduction. Responses evoked from the slowly conducting pathway were inhibited by arterial baroreceptors, whereas responses evoked from the rapidly conducting pathway were not inhibited. Taylor & Gebber (1975) recently demonstrated that there are differences in the representation of these two vasoconstrictor pathways to various vascular beds. Splanchnic nerves contained only the slowly conducting sympathetic pathway, i.e. that pathway which is influenced by arterial baroreceptors, whereas the external carotid nerves to skeletal muscle contained both slowly and rapidly conducting pathways. We speculate that adrenergic discharge during LBNP might be mediated through both slowly and rapidly conducting vasopressor pathways. If the forearm vasoconstriction were mediated through a more rapidly conducting pathway and the splanchnic vasoconstriction through a slowly conducting pathway, this might explain our finding that the forearm response was not blocked by stretch of carotid baroreceptors whereas the splanchnic response was inhibited.

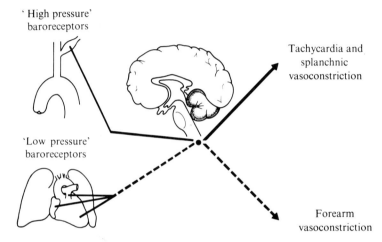

Fig. 22.8. Schematic diagram depicting the role of high pressure carotid and low pressure cardiopulmonary baroreceptors in circulatory control in humans. Carotid baroreceptors modulate heart rate and splanchnic vascular resistance and cardio-pulmonary baroreceptors modulate forearm vascular resistance. These observations indicate that there are striking contrasts in the pattern of regional sympathetic responses to activation of carotid and cardiopulmonary baroreceptor reflexes.

Baroreceptor control of venous tone

Non-uniformity of baroreceptor control in humans involves different vascular segments, as well as different vascular beds. This is illustrated by comparing responses of resistance and capacitance vessels in the extremities to activation of baroreceptor reflexes. Although earlier studies suggested that capacitance vessels or veins constrict during tilting or LBNP (Page *et al.*,1955; Sharpey-Schafer, 1961; Gilbert & Stevens, 1966), recent studies from several laboratories demonstrate that veins in the extremities do not participate appreciably in baroreceptor adjustments (Samueloff,

Browse & Shepherd, 1966; Epstein, Beiser, Stampfer & Braunwald, 1968, Abboud, Heistad & Mark, 1973). For example, LBNP produces striking arteriolar constriction in the forearm, but minimal venous constriction (Fig. 22.9). The absence of forearm venoconstriction during LBNP cannot be explained by lack of reflex responsiveness since a Valsalva manoeuvre or hyperventilation produces considerable reflex forearm venoconstriction. Thus it appears that baroreceptor activation of sympathetic pathways to

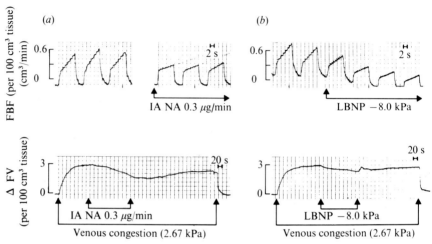

Fig. 22.9. Comparison of the effects of *(a)* intra-arterial noradrenaline (IA NA) and of *(b)* lower body negative pressure (LBNP) on forearm blood flow (FBF) (upper tracings) and forearm venous volume (FV) at a distending pressure of 2.67 kPa (lower tracings). Reduction in the slope of the flow curves indicates arteriolar constriction and reduction in the forearm volume indicates venous constriction. Intra-arterial noradrenaline caused marked venous as well as arteriolar constriction. LBNP also caused marked arteriolar constriction, but produced only minimal venoconstriction. (Reprinted from Abboud *et al.*,1976.)

arterioles during LBNP is not accompanied by activation of sympathetic pathways to veins in the forearm.

Since the leg veins are subjected to greater distending pressure than the forearm veins during upright tilt and contain more smooth muscle one might postulate that these veins might be more responsive to baroreceptor stimuli, but additional experiments have indicated that leg veins also fail to constrict neurogenically during standing (Samueloff *et al.*, 1966; Abboud *et al.*, 1973). The factor which opposes venous pooling in the legs during orthostatic stress is not neurogenic venoconstriction, but instead appears to be an increase in skeletal muscle tone which compresses the veins. It is possible, however, that veins in other regions, e.g. the splanchnic circulation, which lack the compressive force of skeletal muscle tone, might constrict reflexly during orthostatic stress. Indeed, splanchnic veins in animals constrict actively during activation of baroreceptor reflexes whereas cutaneous veins do not (Brender & Webb-Peploe, 1969; Iizuka *et al.*, 1970). Haemorrhage reduces splanchnic capacitance in humans, but it is not known whether this results from reflex constriction of capacitance vessels as opposed to humoral influences or passive decreases in splanchnic capacitance.

Baroreceptor control of plasma renin activity

During venous pooling and activation of baroreceptor reflexes there are striking increases in plasma renin activity. Recent studies indicate that low-pressure baroreceptors play an important role in the regulation of renin activity in animals (Brennan, Malvin, Jochim & Roberts, 1971; Thames *et al.*, 1971; Mancia, Romero & Shepherd, 1975; Zehr, Hasbargen & Kurz, 1976). We studied the relative contribution of low- and high-pressure baroreceptors to increases in plasma renin during venous pooling in humans.

We measured plasma renin activity (immunoassay) during graded LBNP in six healthy young men during *ad lib* sodium intake (Mark & Abboud, 1976). LBNP at -1.3 kPa decreased central venous pressure without decreasing arterial pressure and thus activated low-pressure but not high-pressure baroreceptor reflexes. LBNP at this level for 10 min produced forearm vasoconstriction, but did not increase plasma renin (Fig. 22.10). LBNP at -2.7 kPa also failed to increase plasma renin.

LBNP at -5.3 kPa decreased central venous pressure and arterial pressure and thereby activated both low- and high-pressure reflexes. LBNP at this level for 10 min produced tachycardia, forearm vasoconstriction, and three-fold increases in plasma renin (Fig. 22.10).

Three subjects were restudied after receiving a low sodium diet (10 mg/24 h) to increase renin responsiveness. The increase in renin with LBNP at -5.3 kPa (combined low- and high-pressure reflexes) was striking, but selective activation of low-pressure reflexes (LBNP at -1.3 kPa) again failed to produce substantial increases in renin.

The results suggest that selective activation of low-pressure baroreceptor reflexes does not produce significant increases in plasma renin in humans.

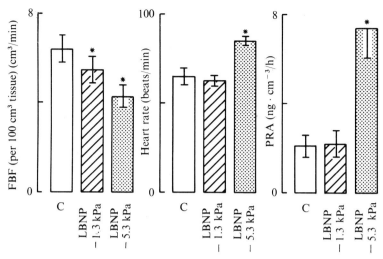

Fig. 22.10. Effects of graded lower body negative pressure (LBNP) on forearm circulation, heart rate, and plasma renin activity (PRA). Selective activation of low pressure baroreceptors with LBNP at −1.3 kPa produced forearm vasoconstriction, but did not increase heart rate or plasma renin activity. Although not shown here, LBNP at −2.7 kPa also failed to increase renin. LBNP at −5.3 kPa triggered both low and high pressure baroreceptor reflexes, produced forearm constriction, tachycardia and three-fold increases in plasma renin activity. Means ± s.e., C = control; $n = 6$. * Indicates $P < 0.05$, FBF = forearm blood flow.

These results could be interpreted to suggest that high-pressure baroreceptors play the predominant role in regulation of plasma renin during venous pooling. However, Zanchetti (1976) has suggested that selective activation of high-pressure baroreceptors (produced by decreasing carotid transmural pressure with positive pressure about the neck) also fails to increase renin. It would appear that physiological levels of tonic restraint from carotid baroreceptors may prevent reflex increases in renin activity during decreases in cardiopulmonary tonic restraint and vice versa. Accordingly, we conclude that there must be a concerted and not a selective decrease in tonic restraint from cardiopulmonary and carotid pressure baroreceptors to produce striking reflex increases in plasma renin during othostatic stress in humans.

Ventricular baroreceptors in aortic stenosis

Abnormal vascular responses to exercise

Since stretch of ventricular baroreceptors produces vasodilatation, brady-cardia and hypotension (Aviado & Schmidt, 1959; Salisbury, Cross &

Rieben, 1960; Ross, Frahm & Braunwald, 1961), we proposed that reflexes arising from stretch of left ventricular baroreceptors might contribute to exertional syncope in patients with aortic stenosis. We evaluated the possibility that exercise in patients with aortic stenosis increases left ventricular pressure, activates ventricular baroreceptors and promotes reflex vasodilatation (Mark *et al.*, 1973*b*).

During leg exercise reflex vasoconstriction normally occurs in a resting forearm and contributes to maintenance of arterial pressure (Bevegard & Shepherd, 1967). We tested the hypothesis that this normal forearm vasoconstrictor response to leg exercise is inhibited or reversed in patients with aortic stenosis, possibly because of activation of left ventricular baroreceptors (Fig. 22.11). Forearm vascular responses to leg exercise were measured in patients with aortic stenosis and in two control groups consisting of patients with mitral stenosis and patients without valvular heart disease.

Forearm vasoconstriction occurred during exercise in both control groups (Table 22.2 and Fig. 22.12). In contrast, forearm blood flow increased and forearm vascular resistance did not change during exercise in patients with aortic stenosis (Table 22.2 and Fig. 22.12). In the six patients with aortic stenosis and a history of exertional syncope there was vasodilatation in the forearm during leg exercise (Table 22.2 and Fig. 22.12). Inhibition or reversal of the expected forearm vasoconstrictor responses in aortic stenosis were associated with significant increases in left ventricular pressure (Table 22.2).

Three of the patients with aortic stenosis, a history of syncope and a forearm vasodilator response to leg exercise were restudied after aortic valve replacement. In all three vasoconstriction occurred in the forearm during leg exercise after operation (Fig. 22.13).

These observations indicated that reflex forearm vasoconstriction responses to leg exercise are inhibited or reversed in patients with aortic stenosis. We considered the possibility that the abnormal reflex responses might have resulted from non-specific depression of cardiovascular reflexes, but this was excluded by the finding that these patients displayed reflex vasoconstriction during application of ice to the forehead. We also considered the possibility that activation of arterial baroreceptors might have inhibited the vasoconstriction but arterial pressure did not increase significantly during the vasodilator response in the patients with aortic stenosis and a history of syncope. These observations make it unlikely that inhibition or reversal of vasoconstriction resulted from activation of

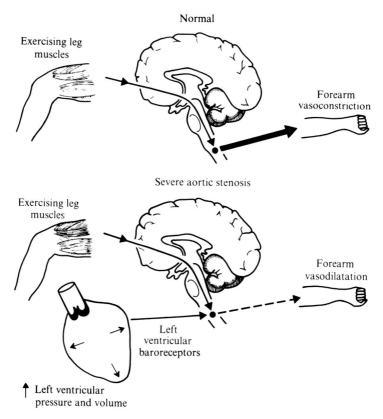

Fig. 22.11. Schematic diagram depicting the proposed effects of activation of left ventricular baroreceptors during exercise in patients with severe aortic stenosis. As shown at the top afferent impulses originating in exercising leg muscle normally produce reflex vasoconstriction in the non-exercising forearm. The hypothesis tested in these studies was that increases in left ventricular pressure and activation of left ventricular baroreceptors inhibit and reverse the forearm vasoconstriction.

arterial baroreceptors. Reflexes originating in low-pressure baroreceptors in left atrium and pulmonary vessels probably did not contribute to reversal of the vasoconstriction, since vasoconstrictor responses were not inhibited in patients with mitral stenosis in whom left atrial or pulmonary arterial wedge pressures increased by 1.2 ± 0.3 kPa during exercise.

We suggest that activation of left ventricular baroreceptors promoted the reflex vasodilatation during exercise in aortic stenosis. We cannot define the precise stimulus which might have activated ventricular baroreceptors, but four features characterised the patients with aortic

Table 22.2. *Responses to exercise in aortic stenosis*

Forearm blood flow (per 100 cm³ tissue) (cm³/min)		Forearm vascular resistance (per 100 cm³ tissue) (kPa·cm⁻³·min)		Mean arterial pressure (kPa)		Left ventricular systolic pressure (kPa)		Left ventricular end-diastolic pressure (kPa)	
R	E	R	E	R	E	R	E	R	E
Aortic stenosis: n=10									
3.9±0.5	4.9±0.7*	3.73±0.53	3.73±0.80	12.53±0.40	13.33±0.67‡	27.51±1.20	29.19±1.20*	2.27±0.27	2.80±0.53
Aortic stenosis with a history of exertional syncope: n=6[a]									
4.3±0.5	5.7±0.7*	3.20±0.53	2.53±0.40*	12.66±0.53	12.93±0.40	29.59±1.47	31.33±1.47*	2.27±0.40	2.93±0.80
Mitral stenosis: n=6									
3.4±0.4	2.7±0.5‡	4.53±0.80	6.67±1.73‡	13.86±0.80	14.01±0.01	18.93±1.87	19.33±1.73	1.73±0.53	1.73±0.13
No valvular heart disease: n=5									
5.6±1.5	4.2±1.2*	3.33±1.47	4.40±1.60*	12.00±0.93	12.26±0.80	15.06±0.93	14.86±0.93	1.20±0.13	1.20±0.13

Entries are mean ± s.e.

r = Resting values; E = values obtained during second minute of exercise.

* $P<0.05$.

‡ $P<0.10$.

[a] The patients with aortic stenosis and a history of syncope are a subgroup of the ten patients with aortic stenosis.

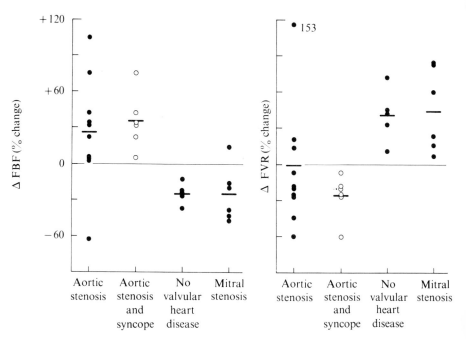

Fig. 22.12. Percentage change in forearm blood flow (FBF) and forearm vascular resistance (FVR) during the second minute of exercise. Dots represent the response of individual patients; the horizontal lines indicate the means of responses in each group. (Reprinted from Mark *et al.*, 1973*b*.)

stensosis who displayed the abnormal vascular responses. These features were: (1) elevated resting ventricular pressure; (2) increases in left ventricular systolic pressure during exercise; (3) increases in end-diastolic pressure during exercise; and (4) early or rapid increases in left ventricular pressures during exercise. Left ventricular systolic pressure increased in some of the control patients, but the increase was less abrupt and left ventricular end-diastolic pressure usually did not increase. The combination of changes which might have increased the activity of ventricular baroreceptors in aortic stenosis did not occur in the control patients.

The results of our study indicate that forearm vascular responses to exercise are abnormal in patients with severe aortic stenosis and are consistent with the hypothesis that activation of ventricular baroreceptors may promote reflex vasodilatation and syncope during exercise in aortic stenosis.

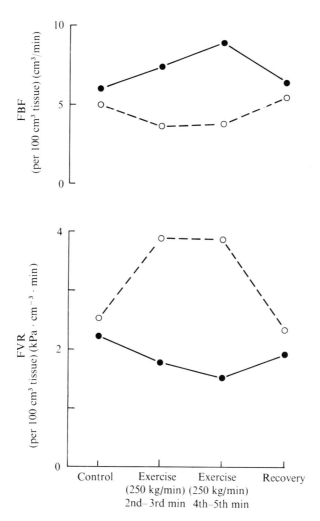

Fig. 22.13. Forearm vascular responses to supine leg exercise in a patient with severe calcific aortic stenosis before and after aortic valve replacement. Before operation (●) the patient displayed an abnormal forearm vasodilator response to leg exercise. After operation (○) the patient displayed a normal forearm vasoconstrictor response. FBF = forearm blood flow; FVR = forearm vascular resistance.

Reflex vascular responses to activation of ventricular baroreceptors in dogs

In a companion study in dogs (Mark, Abboud, Schmid & Heistad, 1973a), we explored further the mechanism by which aortic stenosis may promote reflex vasodilatation. Reflex vascular responses to acute left ventricular outflow obstruction were studied to compare the effects of activation of ventricular baroreceptors on vascular resistance in skeletal muscle (gracilis muscle) and skin (hind paw); to identify the afferent and efferent pathways which mediate the reflex vasodilatation; and to assess the relative contribution of ventricular baroreceptors and baroreceptors in left atrium and pulmonary vessels in responses to left ventricular outflow obstruction. The artery to the gracilis muscle and the cranial tibial artery to the paw were perfused separately at constant flow so that changes in perfusion pressure to each bed reflected changes in vascular resistance. Outflow obstruction was produced by inflating a balloon in the left ventricular outflow tract for 15 s while recording pressures in the left ventricle and aortic arch.

Inflation of the balloon increased left ventricular and atrial pressures and decreased pressure in the aortic arch (Fig. 22.14). Low and high levels of obstruction produced dilator responses averaging -0.7 ± 0.4 and -5.6 ± 0.1 kPa in muscle and -0.1 ± 0.1 and -0.4 ± 0.3 kPa in the paw.

The vasodilator responses to left ventricular outflow obstruction were blocked by bilateral vagotomy, sectioning the sciatic and obturator nerves and administration of phentolamine, but were not decreased by atropine or tripelennamine (Mark *et al.*, 1973*a*).

There are several points of interest in the responses to left ventricular outflow obstruction. The first is the difference in the vasodilator response in skin and skeletal muscle. Activation of ventricular baroreceptors produced only slight dilatation in skin compared with pronounced dilatation in muscle. Denervation and glyceryltrinitrate caused greater dilatation in the paw than did left ventricular outflow obstruction; decreases in perfusion pressure with denervation averaged -6.1 ± 1.3 kPa and with glyceryltrinitrate averaged -4.4 ± 0.8 kPa in paw whereas left ventricular outflow obstruction decreased perfusion pressure by only -0.4 ± 0.3 kPa. This indicates that reflex vasodilator response in the paw was not limited by a low level of resting neurogenic tone or by a negligible dilator capacity of these vessels. Instead, it suggests that activation of ventricular baroreceptors produces greater withdrawal of adrenergic drive to muscle than to skin and again emphasises the non-uniformity of baroreceptor control.

The second point is the role of left ventricular versus left atrial receptors in the reflex vasodilatation. This question is raised by the finding that left ventricular outflow obstruction increased both left ventricular and left atrial pressures. Obstruction to left ventricular inflow produced by

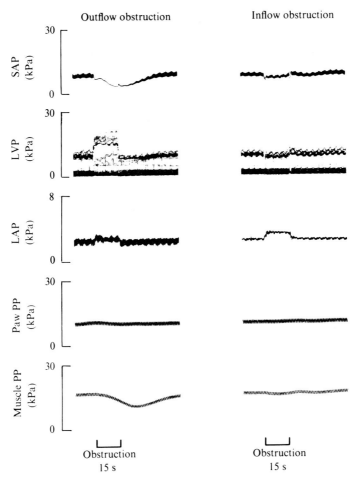

Fig. 22.14. Comparison of responses to left ventricular outflow obstruction and to left ventricular inflow obstruction produced by inflating a balloon at the mitral valve. SAP = systemic arterial pressure; LVP = left ventricular pressure; LAP = left atrial pressure; Paw PP and Muscle PP = paw and muscle perfusion pressure. (Reprinted from Mark *et al.*, 1973*a*.)

inflating a balloon at the mitral valve increased left atrial pressure to levels comparable with those during outflow obstruction but did not cause vasodilatation. This indicates that the reflex vasodilatation during ventricular outflow obstruction did not result from stretch of low-pressure baroreceptors in the atrial or pulmonary veins. These results are consistent with our finding that forearm vasoconstrictor responses to leg exercise are reversed in patients with aortic stenosis, but not in patients with mitral stenosis.

A third observation is that increases in left ventricular pressure produced vasodilatation despite a striking fall in arterial pressure which would be expected to trigger reflex vasoconstriction through arterial baroreceptors. This observation may be partly explained by experiments of Koike, Mark,

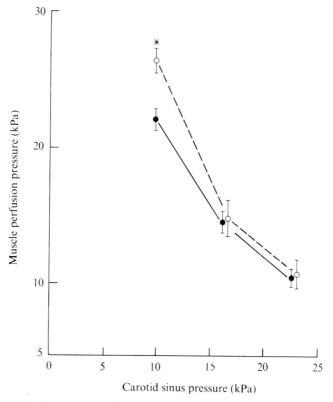

Fig. 22.15. Reflex vasoconstrictor responses during changes in carotid sinus pressure with cardiopulmonary vagal afferent activity intact (before vagotomy, ●) and with vagal afferent activity interrupted (after vagotomy, ○). Means ± s.e., $n = 14$. * Indicates $P < 0.05$. The reflex vasoconstrictor response in the muscle (i.e. increase in muscle perfusion pressure) with a fall in carotid sinus pressure was greater when cardiopulmonary vagal afferent activity was interrupted. The aortic depressor nerve had been cut and atropine had been administered at the beginning of these experiments so that when the vagotomy was performed it interrupted only cardiopulmonary afferents. These studies suggest that cardiopulmonary vagal afferent activity attenuates the gain of the carotid sinus reflex during hypotension. (Reprinted from Koike *et al.*, 1975.)

Heistad & Schmid (1975) in our laboratories which demonstrated that vagal afferent activity restrains the sympathetic discharge which is normally produced by carotid hypotension (Fig. 22.15).

The results of these studies in animals indicate that activation of left ventricular baroreceptors produces striking vasodilatation in skeletal muscle, but only slight vasodilatation in skin. The vasodilatation during activation of left ventricular baroreceptors occurs despite a fall in arterial pressure, apparently in part beause cardiopulmonary vagal afferent impulses inhibit the carotid baroreceptor reflex through a central interaction. The difference in the vasodilator responses in skeletal muscle and skin suggests a greater withdrawal of adrenergic constrictor tone from vessels in skeletal muscle than in skin. Activation of sympathetic cholinergic or histaminergic dilator pathways does not contribute to the dilatation.

The studies of the authors and their colleagues which are described in this manuscript were supported by a Clinical Investigatorship, a Research and Education Associateship, and a grant (MIRS 5462) from the Veterans Administration and by Program Project Grant HL-14388, Research Grants HL-16149 and RCDA HL-00041 from the National Heart, Lung and Blood Institute, and Grant M01 RR59 from the General Clinical Research Centers Program, Division of Research Resources, National Institutes of Health.

References

Abboud, F.M. (1972). Control of the various components of the peripheral vasculature. *Fedn Proc. Fedn Am. Socs exp. Biol.*, **31**, 1226–39.

Abboud, F.M., Heistad, D.D. & Mark, A.L. (1973). Effect of lower body negative pressure on capacitance vessels of forearm and calf. *Fedn Proc. Fedn Am. Socs exp. Biol.*, **32**, 396.

Angell-James, J.E. & Daly, M. de B. (1970). Comparison of the reflex vasomotor responses to separate and combined stimulation of the carotid sinus and aortic arch baroreceptors by pulsatile and nonpulsatile pressures in the dog. *J. Physiol.*, **209**, 257–93.

Aviado, D.M., Jr & Schmidt, C.S. (1959). Cardiovascular and respiratory reflexes from the left side of the heart. *Am. J. Physiol.*, **196**, 726–30.

Bevegard, B.S. & Shepherd J.T. (1966). Circulatory effects of stimulating the carotid arterial stretch receptors in man at rest and during exercise. *J. clin. Invest.*, **45**, 132–42.

Bevegard, B.S. & Shepherd, J.T. (1967). Regulation of the circulation during exercise in man. *Physiol. Rev.*, **47**, 178–213.

Brender, D. & Webb-Peploe, M.D. (1969). Influence of carotid baroreceptors on different components of the vascular system. *J. Physiol.*, **205**, 257–74.

Brennan, L.A., Jr, Malvin, R.L., Jochim, K.E. & Roberts, D.E. (1971). Influence of right and left atrial receptors on plasma concentrations of ADH and renin. *Am. J. Physiol.*, **221**, 273–8.

Dampney, R.A.L., Raylor, M.G. & McLachlan, E.M. (1971). Reflex effects of stimulation of carotid sinus and aortic baroreceptors on hindlimb vascular resistance in dogs. *Circulation Res.*, **29**, 119–27.

Donald, D.E. & Edis, A.J. (1971). Comparison of aortic and carotid baroreflexes in the dog. *J. Physiol.*, **215**, 521–38.

Edis, A.J. (1971). Aortic baroreflex function in the dog. *Am. J. Physiol.*, **221**, 1352–7.

Eckberg, D.L., Cavanaugh, M.S., Mark, A.L. & Abboud, F.M. (1975). A simplified neck suction device for activation of carotid baroreceptors. *J. Lab. clin. Med.*, **85**, 167–73.

Epstein, S.E., Beiser, G.D., Stampfer, M. & Braunwald, E. (1968). Role of the venous system in baroreceptor mediated reflexes in man. *J. clin. Invest.*, **47**, 139–52.

Ernsting, J. & Parry, D.J. (1957). Some observations on the effects of stimulating the stretch receptors in the carotid artery in man. *J. Physiol.*, **137**, 45–6.

Gebber, G.L., Taylor, D.G. & Weaver, L.C. (1973). Electrophysiological studies on organization of central vasopressor pathways. *Am. J. Physiol.*, **224**, 470–81.

Gilbert, C.A. & Stevens, P.M. (1966). Forearm vascular responses to lower body negative pressure and orthostasis. *J. appl. Physiol.*, **21**, 1265–72.

Hainsworth, R., Ledsome, J.R. & Carswell, F. (1970). Reflex responses from aortic baroreceptors. *Am. J. Physiol.*, **218**, 423–9.

Iizuka, T., Mark, A.L., Wendling, M.G., Schmid, P.G. & Eckstein, J.W. (1970). Differences in response of saphenous and mesenteric veins to reflex stimuli. *Am. J. Physiol.*, **219**, 1066–70.

Johnson, J.M., Rowell, L.B., Niederberger, M. & Eisman, M.M. (1974). Human splanchnic and forearm vasoconstrictor responses to reductions of right atrial and aortic pressures. *Circulation Res.*, **34**, 515–24.

Koike, H., Mark, A.L., Heistad, D.D. & Schmid, P.G. (1975). Influence of cardiopulmonary vagal afferent activity on carotid chemoreceptor and baroreceptor reflexes in the dog. *Circulation Res.*, **37**, 422–9.

Linden, R.J. (1973). George E. Brown Memorial Lecture: Function of cardiac receptors. *Circulation*, **48**, 463–80.

Little, R., Wennergren, G. & Öberg, B. (1975). Aspects of the central integration of arterial baroreceptor and cardiac ventricular receptor reflexes in the cat. *Acta. physiol. scand.*, **93**, 85–96.

Mancia, G., Romero, J.C. & Shepherd, J.T. (1975). Continuous inhibition of renin release in dogs by vagally innervated receptors in the cardiopulmonary region. *Circulation Res.*, **36**, 529–35.

Mark, A.L. & Abboud, F.M. (1976). Effects of low and high pressure baroreceptor reflexes on plasma renin in humans. *Circulation*, **54** (suppl. II), 95.

Mark, A.L., Abboud, F.M., Schmid, P.G. & Heistad, D.D. (1973*a*). Reflex vascular responses to left ventricular outflow obstruction and activation of ventricular baroreceptors in dogs. *J. clin. Invest.*, **52**, 1147–53.

Mark, A.L., Eckberg, D.L. & Abboud, F.M. (1975). Selective contribution of cardiopulmonary and carotid baroreceptors to forearm and splanchnic vasoconstrictor responses during venous pooling in man. *Physiologist*, **18**, 305.

Mark, A.L., Eckberg, D.L., Abboud, F.M. & Johannsen, U.J. (1974). Relative contribution of low and high pressure baroreceptors in circulatory adjustments to venous pooling in man. *J. clin. Invest.*, **53**, 50A–51A.

Mark, A.L., Kioschos, J.M., Abboud, F.M., Heistad, D.D. & Schmid, P.G. (1973*b*). Abnormal vascular responses to exercise in patients with aortic stenosis. *J. clin. Invest.*, **52**, 1138–46.

Page, E.B., Hickam, J.F., Sieker, H.O., McIntosh, H.D. & Pryor, W.R. (1955). Reflex venomotor activity in normal persons and in patients with postural hypotension. *Circulation*, **11**, 262–70.

Roddie, I.C. & Shepherd, J.T. (1958). Receptors in the high-pressure and low-pressure vascular systems: their role in the reflex control of the human circulation. *Lancet*, **1**, 493–6.

Roddie, I.C., Shepherd, J.T. & Whelan, R.F. (1957). Reflex changes in vasoconstrictor tone in human skeletal muscle in response to stimulation of receptors in a low-pressure area of the intrathoracic vascular bed. *J. Physiol.*, **139**, 369–76.

Ross, J., Jr, Frahm, C.J. & Braunwald, E. (1961). The influence of intracardiac baroreceptors on venous return, systemic vascular volume and peripheral resistance. *J. clin. Invest.*, **40**, 563–72.

Rowell, L.B. (1976). Measurement of hepatic-splanchnic blood flow in man by dye techniques. In *Dye curves: the theory and practice of indicator dilution*, ed. D. A. Bloomfield, 1st edn, pp. 209–29. University Park Press: Baltimore.

Salisbury, P.F., Cross, C.E. & Rieben, P.A. (1960). Reflex effects of left ventricular distention. *Circulation Res.*, **8**, 530–4.

Samueloff, S.L., Browse, N.L. & Shepherd, J.T. (1966). Response of capacity vessels in human limbs to head-up tilt and suction of lower body. *J. appl. Physiol.*, **21**, 47–54.

Sharpey-Schafer, E.P. (1961). Venous tone. *Br. med. J.*, **2**, 1589–95.

Shepherd, J.T. (1974). Lewis A. Conner Memorial Lecture: The cardiac catheter and the American Heart Association. *Circulation*, **50**, 418–28.

Taylor, D.G. & Gebber, G.L. (1975). Baroreceptor mechanisms controlling sympathetic nervous rhythms of central origin. *Am. J. Physiol.*, **228**, 1002–13.

Thames, M.D., Ul-Hassan, Z., Brackett, N.C., Jr, Lower, R.R. & Kontos, H.A. (1971). Plasma renin responses to hemorrhage after cardiac autotransplantation. *Am. J. Physiol.*, **221**, 1115–19.

Zanchetti, A. (1976). Renin release: experimental evidence and clinical implications in arterial hypertension. *Circulation*, **54** (suppl. II), 1.

Zehr, J.E., Hasbargen, J.A. & Kurz, K.D. (1976). Reflex suppression of renin secretion during distention of cardiopulmonary receptors in dogs. *Circulation Res.*, **38**, 232–9.

Zoller, R.P., Mark, A.L., Abboud, F.M., Schmid, P.G. & Heistad, D.D. (1972). The role of low pressure baroreceptors in reflex vasoconstrictor responses in man. *J. clin. Invest.*, **51**, 2967–72.

Discussion

(a) It was felt that there was a difference in the baroreceptor reflex in man compared with experimental animals. In man the main efferent component was a vagal bradycardia. However, the comparison was complicated by the fact that most of the manoeuvres in man (Valsalva, tilting, etc.) would also change the stimulus to cardiopulmonary receptors. The baroreceptor reflex in man was noted to be inhibited in both isometric and rhythmic exercise. A further point, following on from the observation that the baroreceptor reflex was inhibited in patients with aortic stenosis which was attributed to left ventricular distension, was whether baroreceptor inhibition occurred during exercise in patients with angina who would also develop ventricular distension. No answer was given to this.

(b) A question was raised as to whether the technique of neck suction could be regarded as stimulating only carotid baroreceptors. Possible reflexogenic areas which might be affected were tracheal receptors and chemoreceptors. Mark replied that responses resembled those obtained using other techniques. They were not affected by breathing 100% oxygen and they did not appear to resemble the usual type of respiratory response.

(c) Another question was whether the lower body suction would reduce cardiac output and whether the resulting small fall in arterial pressure would be corrected by arterial baroreceptors before it could be observed. Mark thought this was unlikely because Shepherd had shown that much larger decreases in carotid pressure produced by direct compression did not give vasoconstriction in the forearm. Also by simultaneously applying neck suction to counteract this change, lower body suction still caused tachycardia. This was regarded as evidence that baroreceptors were inhibited.

The Bezold–Jarisch reflex in man

F. M. ABBOUD, D. L. ECKBERG, J. M. KIOSCHOS &
C. W. WHITE

from Cardiovascular Center and Department of Medicine, Veterans Administration
and University Hospitals, University of Iowa, Iowa City, Iowa 52242, USA

Diagnostic coronary arteriography affords an opportunity to study reflexes triggered by cardiac receptors in intact, unanaesthetised man. Meglumine diatrizoate (Renografin-76), 6–8 cm^3, was injected selectively into the left coronary arteries of 14 patients. Arterial pressure, forearm blood flow (plethysmography), and heart rate were measured after coronary arteriography: *(a)* in the control state; *(b)* during atrial pacing at 100 beats/min; and *(c)* after intravenous atropine, 2.0 mg. Following arteriography, mean arterial pressure fell 2.9 ± 0.7 kPa, forearm blood flow fell 1.3 ± 0.3 cm^3/min $\times$ 100 cm^3, and heart rate fell 21 ± 6 beats/min. Reductions of arterial pressure and forearm blood flow caused by coronary arteriography were comparable before and during atrial pacing. Atropine significantly reduced the changes of mean arterial pressure (to -1.3 ± 0.1, $P < 0.025$), forearm blood flow (to -0.3 ± 0.3, $P < 0.01$), and heart rate (to -4 ± 1, $P < 0.01$). Injections of contrast medium into the ascending aorta did not provoke significant haemodynamic changes.

Our results suggest that intracoronary injection of contrast medium in man elicits a vasodepressor reflex which is not explained on the basis of bradycardia. The attenuation of this response by atropine suggests that it is mediated, in part, by an efferent cholinergic mechanism.

Miscellaneous short papers

Cardiovascular reflex mechanisms during acute haemorrhage in neonatal pigs

N. GOOTMAN & P. M. GOOTMAN

from Departments of Physiology, SUNY Downstate Medical Center and Albert Einstein College of Medicine; and Division of Pediatric Cardiology, Long Island Jewish Hillside Medical Center of SUNY at Stony Brook, NY, New York, USA

Compensatory mechanisms involved in cardiovascular responses to haemorrhage in the newborn were evaluated in piglets ranging in age from day of birth to two weeks old. Animals lightly anaesthetised with 0.25% halothane in N_2O-O_2 were haemorrhaged by sequential removal of 5 cm^3 of arterial or venous blood totalling 20–30 $cm^3 \cdot kg^{-1}$. Arterial pH, pCO_2 and pO_2 were monitored and adjusted within normal ranges. Aortic pressure, heart rate, femoral, renal and carotid flows (with electromagnetic transducers) were recorded continuously. The capacity for central neural cardiovascular control was tested before and after haemorrhage and after re-infusion by evaluating responses to electric stimulation of hypothalamic vasoactive sites. Haemorrhage led to increased femoral and renal resistances and tachycardia in animals of all ages, of shorter duration in the youngest. Changes were greater after arterial than after venous haemorrhage. Heart rate, blood pressure and peripheral resistance responses to hypothalamic stimulation were diminished or lost after haemorrhage in the youngest piglets; only responses to low-intensity stimulation were lost in older ones. Partial recovery of responses to low-intensity stimulation occurred after re-infusion in younger animals. These observations, combined with those of our earlier studies demonstrating age-dependence of cardiovascular responses to baroreceptor activity, exogenous catecholamines and vagotomy, suggest that reflex mechanisms are less effective in acute haemorrhage in the neonate as a consequence of immaturity of central control.

The authors would like to thank Dr B. J. Buckley for help and advice.

Supported in part by Nassan Heart Association Grant No. 443 and NIH. NS. 12031.

Is the elevated renal function in patients with acute heart failure a homeostatic mechanism?

E. D. BENNETT, N. BROOKE, Y. LIS & A. WILSON

from Medical Unit and Department of Radiology, St George's Hospital, London, UK

Renal function has been measured in patients with a presumptive diagnosis of acute myocardial infarction. Patients with radiological evidence of moderate left ventricular failure had a statistically higher creatinine clearance (124 ± 9.6 cm^3/min) ($P < 0.01$) and urine volume (2245 ± 286 cm^3/24 h) than the patients with normal chest X-rays, whose creatinine clearance was 87.0 ± 4.1 cm^3/min and urine volume 1652 ± 145 cm^3/24 h and also those patients with pulmonary oedema, whose creatinine clearance was 67.0 ± 8.1 cm^3/min and urine volume 1495 ± 378 cm^3/24 h. By the third day these differences had disappeared.

Of patients with moderate left ventricular failure 48% had a creatinine clearance greater than 120 cm^3/min and excreted significantly more sodium (140 ± 15.6 mmol/24 h) ($P < 0.01$) than patients in the same group with a creatinine clearance less than 120 cm^3/min and a sodium excretion of 57 ± 8.3 mmol/24 h and also significantly more sodium than patients with normal chest X-rays, and a sodium excretion of 114 ± 15.5 mmol/24 h ($P < 0.01$). It has been suggested that acute elevation of left-sided filling pressure may activate atrial and ventricular receptors which produces reflex renal vasodilatation. This in turn leads to an elevation of glomerular filtration and results in increased sodium and water excretion. It is postulated that this is a homeostatic mechanism, the function of which is to protect an overloaded heart. It is of interest that two patients with acute pericarditis, but without evidence of heart failure, showed a similar elevation of creatinine clearance which fell on resolution of the clinical condition. This suggests that activation of superficial epicardial receptors can also lead to elevation of renal function.

Autonomic reflex responses during the initial stage of canine endotoxin shock

M. O. K. HAKUMÄKI, M. O. HALINEN & H. S. S. SARAJAS

from Departments of Physiology, University of Kuopio, Kuopio, and College of Veterinary Medicine, Helsinki, Finland

Endotoxin induces bradycardia in dogs anaesthetised with sodium pentobarbitone (Levy & Blattberg, 1967). However, from plasma catechol responses it has been argued that endotoxin increases the sympathetic activity (Archer, Black & Hinshaw, 1975), while the reactions of the heart rate have been interpreted to imply vagal activation (Chien *et al.*, 1966).

Autonomic and cardiovascular reaction patterns to endotoxin (*E. coli*, 1 mg·kg^{-1} i.v.) were studied in 30 dogs under morphine–chloralose anaesthesia. In 14 of the dogs the possible modulation of the reactions by small doses of pentobarbitone was explored.

In both groups of dogs there was an immediate cardiac acceleration (from 137, 140–189, 200 beats/min) in response to endotoxin. Concurrently, a profound hypotension ensued. With falling aortic pressure the aortic arch baroreceptor activity practically ceased. There was also an associated decrease in the left atrial pressure and in the left atrial B-type receptor discharges. Subsequently, chemoreceptor-like firing appeared in several afferent vagal recordings and spontaneous respiration commenced. On the efferent side the initial hypotension was accompanied by a rise in the cardiac sympathetic and cervical vagal discharge rates. The vagal response was particularly striking in those dogs which were also given pentobarbitone.

The results outlined, supported by the previous indirect evidence *(vide supra)*, strongly suggest that the bradycardial reactions elicited by endotoxin, especially in dogs under pentobarbitone anaesthesia, result from increased vagal efferentation.

References

Archer, L.T., Black, M.R. & Hinshaw, L.B. (1975). Myocardial failure with altered response to adrenaline in endotoxin shock, *Br. J. Pharmac.*, **54**, 145–55.

Chien, S., Chang, C., Dellenback, R.J., Usami, S. & Gregerson, M.I. (1966). Hemodynamic changes in endotoxin shock. *Am. J. Physiol.*, **210**, 1401–10.

Levy, M.N. & Blattberg, B. (1967). Changes in heart rate induced by bacterial endotoxin. *Am. J. Physiol.*, **213**, 1485–92.

Response of skeletal muscle vasculature to coronary artery occlusion

C. M. MONTEFUSCO, J. B. NOLASCO & D. F. OPDYKE

from College of Medicine and Dentistry of New Jersy, Newark, New Jersey, USA

The possibility that the failure of an increase in vascular resistance contributes to hypotension after myocardial ischaemia induction was examined in skinned dog hind limbs following occlusion of the circumflex coronary artery. Animals anaesthetised with pentobarbital sodium (30 mg·kg^{-1} i.v.) and with intact vagus nerves exhibited reduced cardiac output and greatly increased total peripheral resistance. Femoral blood flow, femoral vascular resistance, and mean arterial blood pressure were maintained at or near control levels. Bilateral cervical vagotomy performed before coronary artery ligation produced decreases in cardiac output as before. Mean arterial pressure exhibited sustained, serial decreases, and only moderate increases in total peripheral resistance were observed. Unlike the results obtained in vagotomised dogs with myocardial ischaemia, atropinisation performed before coronary artery occlusion resulted in smaller and slower reductions of cardiac output, maintenance of mean arterial pressure and femoral vascular resistance at control values and increased total peripheral resistance. These results indicate that: (1) vagal afferent nerves serve a protective function by mediating normal compensatory manoeuvres and (2), an additional mechanism, activated by ischaemia but masked by the vagi, may cause inappropriate vasodilatation when vagal afferent tone is lost.

Supported in part by USPHS NIH-HLI Grant No. 14563.

Invited lecture
Relevance of cardiovascular reflexes

B. FOLKOW

To come to Leeds on this occasion to lecture on reflex cardiovascular control is like bringing coals to Newcastle at a time when the world's leading coal producers are there for a sales convention. However, Professor Linden was optimistic enough to ask, I unrealistic enough to accept and you unlucky enough to be exposed, so we had best get on rapidly.

On facing so many experts on receptor biophysics, bulbar interneuronal connections, fibre discharge patterns and what not, I had best abstain from such electrophysiological complexities and rather look at the reflex circulatory control from a pragmatic plumber's point of view. In other words, what principal information reaches the central command, what sweeping orders does the latter send out and how are they executed, without caring too much about the coding, and the exact headquarter organisation or precisely how its switchboard functions. What, above all, does it mean for the sensitive balance between flow, pressure, flow resistance, capacitance and filling, and how are these functional entities brought together into an efficiently running system? Thus, the present survey by no means aims at any detailed review of the enormous literature, but is rather intended as a discussion of some general principles that seem to be of particular interest for overall *cardiovascular* function. For this reason only a few references that provide further access to the literature are given.

To start with the important Poiseuille relationship $Q = P/R$ (Q = flow; P = pressure; R = resistance), cardiovascular experts are traditionally inclined to consider P more than Q, and reflex specialists are probably no exception in this respect. Still, the final purpose of the system is to deliver a decent flow, Q, to allow nutritional exchange in the all-important capillary compartment. The level of pressure, P, to achieve this would have been of only subordinate relevance had not complex dimensional and logistical

473

relationships set a lower limit for P, preferentially to ensure the ultrafiltrate needed by the kidneys for their vital cleansing function. On the other hand, too high a pressure simply means a waste of energy which biological systems usually shun and, furthermore, which ultimately leads to lesions in both pump and tubing. Thus, there are indeed good reasons to keep around an optimal P level, at least as a daily average. Nevertheless, in perfectly normal individuals P, here representing *mean* arterial pressure, may range from 8 to 9 kPa during deep sleep, up to at least 16 to 17 kPa during a fulminating display of rage, while Q, here meaning cardiac output, is surprisingly well tailored for the nutritional demands as long as blood oxygen content is normal.

One basic question that arises from such simple deliberations is whether the cardiovascular receptors might not somehow be able to 'sense' Q, and report about it, besides their established ability to inform abundantly about the stretch offered by pressure at the strategic input ('volume') and output ('pressure') sites and about the composition of the blood pumped out for consumption. Later we will consider such possibilities as we get closer to the problems involved.

Another important question that arises is, of course, concerned with the general competence of reflex cardiovascular control, i.e. what extra of performance does it provide and at which points are these extra margins of particular relevance? As always, a sensible starting point for answering such questions is to see what happens when the reflex control is *not* there.

Now, most cardiovascular systems still manage in situations of acute denervation, but without doubt their margins are considerably narrowed. There is, e.g., no longer any real adjustment of the filling of the system, nor any appropriate tightening around the content. This becomes particularly precarious where wall compliance is the highest, i.e. at the cardiac filling site, leaving this all-important process to the mercy of incidental hydrostatic influences. Happily enough, however, most species only trust the four-legged position, implying that most of the returning blood runs almost downhill along the highly compliant 'filling side' to the pump. Thus there is little trouble at this point upon denervation (unless the total contents were to be reduced as well) because the pump is still reasonably well primed. It then manages quite well thanks to its inbuilt automaticity and heterometric–homeometric autoregulation. As long as Q is thus upheld, so is R in most circuits because of their pronounced precapillary myogenic tone during rest and, in fact, neurogenic contributions to R are in the resting equilibrium trivial. Even an exercise-induced R reduction may be reasonably tolerated by boosting simultaneously venous return thanks to the muscle pump and hence enhancing Q also, hopefully in reasonable proportion to the R reduction.

It is man only (and the improbable giant kangaroo deprived of its nervous control) that gets into serious trouble by insisting on spending long periods upright and thus placing the major capacitance fraction below the heart. The capacitance vessels which, in contrast to heart and resistance vessels, have little or no wall autoregulation to fall back upon, now simply yield and down goes the filling of the pump. Cardiac and resistance vessel autoregulation are admirable mechanisms but are *not* meant to face such a situation alone. In fact, according to autoregulation principles both Q and R now fall, and this in a situation where the reverse is badly needed. Collapse is thus inescapable unless the intelligent subject, if given time to use his wits before fainting, decides to either lie down or to combat the venous pooling by his muscle pump.

Thus, without reflexes it is no longer possible to integrate pressure, flow, resistance, capacitance and filling into a smooth and efficient machinery and, above all, the ability to adjust properly to extra demands or accidental shortcomings is markedly impaired. Furthermore, for freeing his anterior extremities for purposes more useful than mere locomotion, man pays an extra price which makes him particularly dependent on efficient compensatory adjustments. There are thus good reasons to spend some time on the cardiovascular reflexes, exploring their principal arrangement, competence level and interactions – mutually, 'downstream' with the effector control mechanisms, and 'upstream' with the highest command centres.

Principal organisation of the reflex centres

Let us start, not with the afferent or efferent links, but with their central connections, i.e. with the *reflex centres,* and briefly define their characteristics. This is justified in this case, since the influence of a given piece of incoming information is highly dependent on what the centres can accomplish and what they are doing to start with. For decades it was believed that the medullary cardiovascular centres (or conveniently but not quite correctly the vasomotor centre (VMC))* constituted a largely

---

* It has sometimes been debated whether the vasomotor centre deserves the status designation of 'centre', since it is so intimately influenced by still higher command offices, but in the end this boils down largely to semantics. For similar reasons the command headquarters of an army corps might be called, e.g., 'relay station', being no doubt subordinated both to the general high command and the government. The general in charge of the corps would, however, most likely 'blow his top' at such a designation of his command centre, and for understandable reasons.

undifferentiated meshwork of neurons, delivering tonic activity along a likewise diffuse meshwork of efferent sympathetic fibres. The well-established divergence and convergence arrangement of this peripheral autonomic innervation, both at ganglionic and peripheral sites, seemed to favour strongly such a view, with the hormonal component thought to come in as a likewise diffusely acting reinforcer.

However, there were several quite early studies pointing in another direction. For example, the cutaneous vasoconstrictor fibres appeared to be commanded as a more or less separate unit, as a link involved in thermoregulation. Furthermore, as early as the 1930s Hermann Rein and his colleagues in Germany, applying his 'Thermostromuhr' to various vascular beds, noted that the muscle circuit was usually more affected than the renal one upon, e.g., unloading of the baroreceptors. However, almost nothing was known at that time about the relationship between sympathetic discharge frequency and effector response in these circuits so the renal vessels might simply have responded less extensively than the muscle ones to a given enhancement of constrictor fibre discharge. For technical reasons early electrophysiological studies were of limited help since they allowed only for qualitative analyses, i.e. whether activity increased or decreased, but could not clear up whether a possible differentiation in terms of unequal degrees of excitation or inhibition existed.

For a more quantitative analysis of the patterns of autonomic fibre discharge it was of advantage that systematic explorations of the frequency–response characteristics of the various cardiovascular neuro-effector systems became available (Folkow, 1955, 1960; Mellander, 1960). Such studies revealed a steeply hyperbolic relationship between frequency of discharge and effector response, implying that only minor shifts in discharge rate in the lower, physiological range usually resulted in marked changes in effector response. The 'resting' physiological discharge rate to, e.g., the muscle circuit, appeared to be quite low, at most 1–2 Hz, hardly ever increasing beyond 6–8 Hz even at intense reflex excitation, and this rule seemed to hold for autonomic control as a whole (Folkow, 1960). The background of this tonic activity is still not fully understood, but it is basically induced locally in the bulb, and local chemosensitive structures may well be of great importance for triggering the activity in analogy with the organisation of bulbar respiratory control (Öberg, 1976).

Against such a background it became clear that even fairly modest degrees of quantitative differentiation in 'inherent' VMC discharge to the various circuits, and/or in its reflex modulation, would imply a major differentiation with respect to the ensuing neurogenic effector adjustments, which in the final reckoning is the key point. A comparison of such results of direct, graded sympathetic stimulations with, e.g., Rein's earlier recordings of reflex vascular adjustments made it likely that some type of differentiated VMC organisation existed.

This was tested in experiments where VMC was unloaded of essentially all cardiovascular 'proprioceptor' control while recording pressure and flow in several vascular beds simultaneously. It was then found that the renal and cutaneous vessels (paw, dominated by the cutaneous A–V shunts) were usually only little affected, despite the fact that their constrictor fibres exerted most powerful effects even at low rates of discharge, while the circuits of skeletal muscle and (perhaps to a lesser extent) the gastro-intestinal (GI) tract were in this situation markedly constricted (Löfving, 1961). However, if only the excitability of the bulbar centres was boosted by the local chemical changes caused by slight hypoventilation, the renal circuit could be as massively constricted as the muscle or GI circuits upon baroreceptor unloading. Upon slight hyperventilation, on the other hand, the constrictor neuron pools governing the skeletal muscle circuit were still considerably engaged but those governing the renal vessels were hardly active at all (Fig. 1*a*). In contrast, stimulation of 'somatic pressor' afferents (probably mainly, but not only, pain fibres; see below) engaged the renal vessels to a much greater extent than the muscle ones (Fig. 1*b*). Furthermore, activation of the hypothalamic defence area strongly engaged the renal vasoconstrictor fibres but inhibited, if anything, the constrictor fibres to the precapillary muscle vessels, where instead, dilator fibres are activated (Folkow & Neil, 1971).

These findings imply, first, that the bulbar neuron pools controlling the vasoconstrictor fibres to the various systemic circuits may, during 'rest', have separate levels of tonic discharge, secondly, that they may be separately engaged by incoming central or peripheral neurogenic influences and thirdly, that the level of excitability of these neuron pools may be reset in relation to each other by, e.g. local chemical changes in the brain stem itself, thereby greatly modifying the balance of the pattern of discharge. In more recent analyses, employing modern neurophysiological techniques also, the principles outlined above have been largely confirmed and considerably extended (Öberg, 1976). These characteristics of the neuron pools constituting the VMC provide an adequate explanation of the apparently contradictory results that have been reported in the

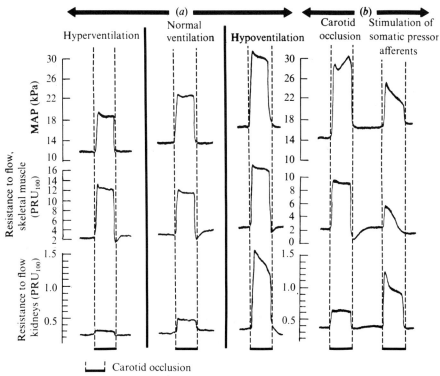

Fig. 1. Illustration of the differentiated vasomotor centre (VMC) control of vasoconstrictor fibre discharge: *(a)* alterations in the balance of pressor and resistance responses in skeletal muscle and kidney, brought about by standardised carotid occlusions in anaesthetised, curarised and vagotomised cat under circumstances where VMC excitability is simultaneously changed by a variation of respiration. Note the negligible engagement of the renal vascular bed during slight hyperventilation, but with an increasingly powerful 'neurogenic recruitment' of this circuit along with stepwise reductions in respiration. (Modified from Folkow, Johansson & Löfving, 1961.) MAP = mean arterial pressure. *(b)* Same technique and parameters, but here illustrating the balance in reflex responses to carotid occlusion and to activation of somatic pressor afferents, respiration being here constant. Note the greatly differing extents of reflex vasoconstriction in kidney and skeletal muscle at these two procedures. (Modified from Johansson, 1962, Fig. 14.)

literature over the years concerning the patterns of central or reflex cardiovascular changes in experimental animals. Clearly, differences in anaesthesia, respiratory state, pH, induction of pain fibre activation or of hypothalamic discharge may well so greatly modulate the excitability of the central neuron pools and their mutual relationships that almost *any* pattern of response may be obtained and yet truly reflect the current situation. Thus, rather than reflecting any shortcomings in the conflicting studies, their results may illustrate the remarkable versatility of the VMC organisation. Species differences are also likely to play a role here. Man, e.g., may well have been forced to organise the functional balance of his VMC differently from cats or dogs, considering the problems with an upright posture.

It appears that the organisation of the VMC has much in common with the nearby, likewise tonically active, respiratory centres: the latter can control the various respiratory muscles as separate entities, and the corresponding neuron pools no doubt have different excitability levels. Thus, some are already active in the undisturbed resting equilibrium (like the diaphragm and external intercostals), others become engaged in a graded fashion first when the excitatory drive on the centres is enhanced (the auxiliary respiratory muscles). In analogy, at an undisturbed resting equilibrium, with no extra 'chemical drive' and/or excitation via pain fibres or higher 'alerting' centres, the VMC in its supervision of overall cardiovascular homeostasis sends out tonic activity mainly to circuits like those in the muscle and GI tract, leaving the renal and cutaneous circuits little affected by *this* type of neurogenic direction. On closer inspection such an arrangement seems most appropriate since the latter two circuits are intimately involved in two other homeostatic mechanisms of the highest priority, i.e. the water–electrolyte–pH balance and the temperature equilibrium, and must in the long run be integrated in such controls. Only more episodically, e.g. during pain, increased alertness, exercise, or when the cardiovascular homeostasis becomes compromised (as in man, some-times, in the upright position) the latter two circuits are increasingly recruited neurogenically into overall circulatory homeostasis, and then simply by raising the excitability level of their bulbar neuron pools.

It may be questioned how a differentiated neurogenic control such as that outlined above is possible in the face of the extensive divergence–convergence principle that no doubt characterises the sympathetic innerva-tion. The answer may be a very simple one: this divergence–convergence principle is probably merely used to 'bind together' a given vascular circuit, or part of it (e.g. into pre- and postcapillary segments), to form efficient 'functional units' for central command, almost as skeletal muscle control is based on a number of motor units of suitable extension. This creates relatively independent groups of central neurons, but mutually linked together within each group, which command the various circuits (or appropriate parts of them) thus serving as a central keyboard for remote control of the cardiovascular 'functional units'. There seems to be no real need for any central regulation of *individual* arterioles or veins in, say, a skeletal muscle; the 'community rule' exercised by the local vascular control machinery is more competent to serve such needs.

Electrophysiological recordings of single sympathetic fibres (see, e.g., Öberg, 1976) reveal that in skeletal muscle some 90% of them exhibit tonic activity, the remaining ones being presumably the sympathetic dilator fibres that lack tonic activity. In some other tissues the proportion of tonically active sympathetic fibres seems to be somewhat lower, but it should be realised that many sympathetic efferents have other destinations than the blood vessels and are then, of course, *not* linked up with the integrated neuron pools controlling the vessels. For example, in the intestine there is a separate sympathetic inhibitory innervation to the cholinergic intramural ganglionic cells and in the skin there are fibres to piloerectors, sweat glands, etc. Evidently, only those neurons which are destined to the same effector system, and thus having the same functional significance, are likely to be 'bound together' by the convergence–divergence arrangement. Therefore, there is no inherent disagreement, so far, between the arrangement of VMC control proposed here and electrophysiological findings concerning sympathetic nerve fibres.

By such a proposed arrangement the VMC can most efficiently control its subordinated circuits, or sections of circuits, by a limited number of neurons, almost like a general commands his troops, or a central government rules local communities. Thus, in ordinary, undisturbed circumstances quite a lot of 'home rule' is allowed in vascular control, based on the smooth interaction between myogenic tonic activity and local chemical–mechanical feedback mechanisms, but this local control may in times of emergency be almost completely overruled by sweeping orders from central levels, on the basis of their far more multifacetted information about the overall situation. The smooth operation of this command organisation is therefore intimately dependent on the messages from the (1) cardiovascular proprioceptors, but is also considerably influenced by information from (2) other parts of the organism (various intero- and exteroreceptors) and, particularly important, information from its (3) external environment (telereceptors) also, after integration at the highest central nervous system (CNS) levels. Some problems associated with this reflex control will now be taken up for consideration.

Myelinated versus unmyelinated cardiovascular afferents

Some general aspects of the cardiovascular mechanoreceptors will first be considered, including some principles of their reflex effects. Mainly for technical reasons, current knowledge about the unmyelinated C fibre afferents is but a tiny fraction of that available concerning the myelinated ones, despite the fact that they constitute quite a large fraction of the cardiovascular afferents; in some cases the majority of fibres belong to the former category. In a way they represent 'the submerged part of the iceberg' but in this particular case a part most worthy of closer exploration.

The very fact that there *are* two such sets of afferents (Paintal, 1973; Kirchheim, 1976), where the myelinated fibres presumably represent a phylogenetically younger and more sophisticated edition, is in itself an indication of a functional differentiation and then probably both on the afferent and the efferent sides. In exploring these with electrophysiological techniques it must be easy to become spellbound by the impressive trains of high-frequency, large spikes in the myelinated afferents, particularly when contrasted with the occasional, humble bumps that travel up in the unmyelinated ones. This is most deceptive, however, because the low-key mumbling in the many C afferents often seems to result in more dramatic reflex effects, if anything, than that which the 'fluent and articulate' myelinated fibres accomplish. For example the high-pressure receptor C fibres, active at only 1–2 Hz, produce substantial depressor effects, reaching a maximum (profound reflex blood pressure fall and bradycardia) at only 10–20 Hz. These effects are more pronounced than those induced by the myelinated ones in the same nerve at their maximum frequency of around 100 Hz. Further, the balance between pressure reduction and bradycardia is different in the two sets with more emphatic bradycardia arising from the C fibre group, at least in the cat (Fig. 2).

On the low-pressure, 'volume' side the differences between the two sets of afferents is even more dramatic with regard to their reflex effects on the heart and vessels, since here the responses go in *opposite* directions, as schematically illustrated in Fig. 3. By exerting a modest excitatory influence on cardiac sympathetic activity and a weak one on the vasoconstrictor fibres, the myelinated cardiac afferents form an exception from the general rule that vagal and glossopharyngeal cardiovascular mechanoreceptors produce generalised sympathetic inhibition in association with vagal excitation. The cardiac C fibre afferents exert a profound depressor influence, especially vagal bradycardia and inhibition of renal sympathetic activity (see Öberg, 1976), and at very low frequencies of discharge.

With regard to the mode of receptor activation, the two sets of afferents convey typical mechanoreceptor messages both on the high-pressure and

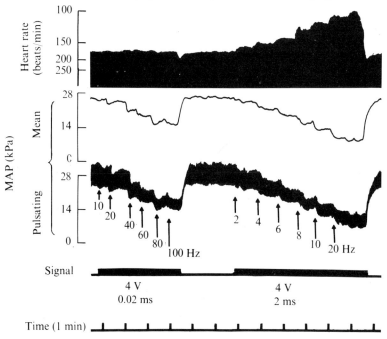

Fig. 2. Frequency–reflex response relationships and quantitative differentiation of depressor and bradycardia responses to stimulation at increasing rates of either *myelinated* (left part) or *unmyelinated* (right part) baroreceptor afferents in the aortic nerve of cats. In the first case the aortic nerve is stimulated at a strength–pulse duration relationship that excites myelinated afferents only, but over a range of frequencies, 10–100 Hz, that largely covers their full range of reflex responses, with only very weak depressor responses in the lower range. In the second case the stimulation strength excites *all* afferents (as also checked by a maximal appearance of the C fibre wave in the neurogram). Note that substantial depressor responses now occur already at 2–4 Hz and at only 10–20 Hz a maximal reflex response occurs with unmyelinated afferent stimulation which also produces far more vagal bradycardia than stimulation of the myelinated ones. Therefore, this type of stimulation may be considered as 'semi-selective' with respect to the reflex depressor responses to C fibre activation, although chemoreceptors with both myelinated and unmyelinated afferents are stimulated too. However, their reflex sympathetic activation is here overwhelmed by the depressor responses and their slight reflex bradycardia response adds to the baroreceptor reflex (B. Öberg, E. Kendrick, P. Thorén & G. Wennergren, unpublished observations).

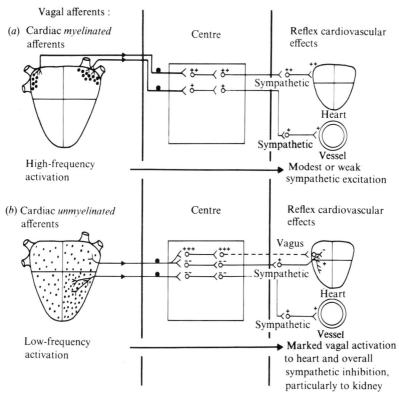

Fig. 3. Schematic illustration of reflex cardiovascular responses to: *(a)* selective electrical stimulation of cardiac (presumably mainly atrial) *myelinated* afferents; *(b)* electrical stimulation of cardiac (mainly left ventricular) *unmyelinated* afferents, in principle accomplished as in Fig. 2. Note that the former afferents, even at high rates of discharge, produce only weak to modest sympathetic excitation. The unmyelinated cardiac afferents, on the other hand, produce vagal bracycardia and overall sympathetic inhibition which is already evident at 2–4 Hz and maximal at 10–20 Hz, and where the renal vascular circuit particularly is influenced. (Mainly based on findings by Öberg & Thorén, 1973.)

low-pressure sides but they may well, at least in part, reflect different *types* of wall deformation. While the receptor-wall relationships appear to be relatively straightforward for the high-pressure myelinated afferents, with an adventitial 'parallel-coupled' arrangement, it may be a more complex arrangement for the C fibres (cf. Kirchheim, 1976). Their finest arborisations might penetrate into the vascular media, in which case there is the chance that a 'series-coupled' arrangement is also present as was originally proposed by Landgren (1952). The question, however, is still an open one.

An arrangement of this type is important as these fibres would then sense the degree of smooth muscle activity, providing them with another dimension of control, in addition to merely reflecting the degree of stretch and hence the pressure level. As the media coat at the receptor sites is exposed to both nervous and hormonal influences it is clear that a functional 're-setting' peripherally of *both* types of receptors is a mechanism to be seriously considered, though exactly how important it is remains to be seen.

On the low-pressure, 'volume' side, on the other hand, the atrial A and B fibres represent both types of arrangement (Paintal, 1973), but what this means in terms of a possible differentiation of their *reflex* effects is not known. Some of them seem to be situated in the atrial wall so as to serve both as series- and parallel-coupled receptors, depending on the current situation of stretch and inotropism – this is evidently often the case for the C fibres, particularly those arborising in the thick-walled left ventricle. Several of these fibres, e.g., sense not only the degree of end-diastolic stretch, but their discharge is also modified by the degree of inotropism at a given level of pre-stretch (Fig. 4). It is likely that these receptors, depending on the extent of arborisation and penetration into the muscle layers, may in fact cover a wide spectrum of stretch and distorsion, ranging in their functional characteristics from a pure 'parallel-coupling' to a pure

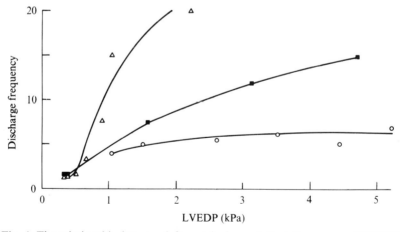

Fig. 4. The relationship between left ventricular end-diastolic pressure (LVEDP) and discharge frequency in a left ventricular C fibre afferent, first at 'normal' myocardial inotropism (■), secondly at enhanced inotropism (induced by a β-receptor agonist; △) and thirdly at lowered inotropism (induced by β-receptor blockade; o). (P. Thorén, unpublished.)

'series-coupling', with a variety of combinations of these two extremes. As a result the ventricular C fibre afferents may be expected to provide the centres with multifacetted information of ventricular events both in systole and diastole, though still belonging to the fairly homogenous and superficially 'dull' family of afferent unmyelinated fibres (cf. Thorén, Donald & Shepherd, 1976).

The interesting question then arises whether these ventricular C fibre afferents, depending on whether they preferentially sense diastolic stretch or systolic tension, may also exert accordingly differentiated *reflex* effects. They without doubt all appear to be classical 'depressor fibres', but some observations (B. Öberg, personal communication) might point in the above-mentioned direction. For example, intense systolic distortions, whether aortic pressure is high or low, produce not infrequently an unusually powerful reflex bradycardia, perhaps sometimes in connection with the peculiar reflex vagal activation to the GI tract associated with vomiting, as illustrated, e.g., by Abrahamsson and Thorén (cf. Thorén *et al.*, 1976; Öberg, 1976). On the other hand, even intense overfilling of the system, which is likely to activate markedly ventricular afferents with preferentially 'parallel-coupled' arrangements, does not usually produce reflex effects with such an intense bradycardia component. Of course, these are very difficult things to quantify so that the differences mentioned might be quantitative rather than qualitative in nature, i.e. a matter of intensity of afferent discharge where high activation leads to recruitment of additional efferent links. The possibility should, however, be considered that C fibre afferents, which sense somewhat different events in the cardiac cycle because of their arborisation in the wall, may also be relatively differentiated in their contacts with the various neuron pools in the bulbar centres.

If nothing else, considerations of this nature amply serve to illustrate that we have barely scratched the surface of the submerged parts of this particular iceberg. It should also be stressed that problems of this nature are by no means only for receptor superspecialists, or reflections of a *l'art pour l'art* attitude, but are likely to have important implications also for down-to-earth cardiovascular physiologists and cardiologists, which is a good enough excuse for taking them up in this context.

Reflex effects of 'volume' receptors

Since this symposium has dealt abundantly with every conceivable aspect of the cardiac receptors, and by experts in the field at that, only a few aspects of their reflex influences, as seen from a plumber's point of view, will be considered here. The interest concerning volume control has been very much concentrated on the *atrial* section of the low-pressure side, and presumably so because it is here that the particularly well-studied myelinated A and B afferents originate. However, in reality the most important part of the low-pressure side is of course its 'functional endpoint', the left ventricle in diastole. The dominating receptor afferents in this part of the heart is the 'gray mass' of C fibres, which on closer inspection may be even more interesting and important than the atrial myelinated ones, as discussed above.

Good volume control calls for receptors of suitable placement and sensitivity, with appropriate connections centrally and a variety of efferent links and effectors, including those to the functionally different cardiovascular units, so as to affect intake and losses, the equilibrium between intra- and extravascular fluid compartments and the balance between wall tension and content within the cardiovascular system. The important concept of such a control system began to take substantial shape mainly thanks to the studies by Henry, Gauer, Linden and their colleagues (cf. Gauer, Henry & Behn 1970; Linden, 1973, 1975). These studies revealed in essence the presence of a nervous afferent link for which the atrial myelinated afferents with their receptors were considered to be the most likely candidate, acting mainly through hormonal efferent link(s), the renal losses of water and salt. There is no need here to go into details since these interesting and complex events have been excellently covered in other lectures. It should be stressed, however, that both the antidiuretic hormone (ADH) and the 'diuretic substance' of Linden and his colleagues appear to function as reciprocal regulators mainly of renal *water* losses, while the long-term volume control calls for a modulation of primarily the renal excretion of sodium chloride. As pointed out above, there are good reasons to consider in this context also the ventricles and their receptors, and here the C fibre afferents emerge as the most likely candidates, beside the atrial myelinated afferents (cf. Öberg, 1976; Thorén *et al.,* 1976). The vasovagal cardiac C afferents can exert considerable reflex effects on renal sodium chloride excretion, partly by modulating renal function via the sympathetic fibres, partly via the renin–angiotensin system and, hence, via modulations of the aldosterone secretion (Wennergren, Henricksson, Weiss & Öberg, 1976). Moreover, an overall volume control has, of course, more to it than the important renal excretory function *per se* and some of these other aspects will also be briefly discussed.

It seems clear from the studies mentioned that stretch of the atria at sites where the myelinated afferents originate induces the above-mentioned neuro-hormonal reflex changes in renal excretion. However, from this fact alone it does not necessarily follow that these myelinated afferents only, or even mainly, serve as the initiator of such reflexes. In general the entire myocardium contains arborisations of vagal C afferents, though particularly the left ventricle (cf. Thorén *et al.*, 1976). It is therefore likely that local atrial stretch activates such afferents as well, and the question arises as to which group of afferents is the most important one for inducing the renal effects.

It is technically possible to stimulate, electrically, only the myelinated afferents in the cardiac nerve, since their threshold is substantially lower than that of the C fibre afferents. As mentioned above this technique was utilised to explore the cardiovascular reflex (neurogenic) effects of the myelinated afferents, which cause a mild sympathetic activation, as outlined earlier and illustrated in Fig. 3*a*. Atrial stretch at the site of the myelinated fibre arborisations also produces a sympathetically conveyed tachycardia (Linden, 1973, 1975). Incidentally, these neurogenic cardiovascular effects are in the *opposite* direction of what an orthodox volume receptor, excited by increased filling, would be expected to induce. It would be worthwhile to investigate whether such selective, electrical activation of the myelinated cardiac nerve afferents mimic the reflex (hormonal) changes in renal excretion produced by atrial stretch since the atria contain unmyelinated vagal afferents as well. *A priori* the chances appear very good that this will be the case but if not the unmyelinated afferents come into focus here also, and then the question arises as to what the functional significance of the vivid A and B fibre activity is.

In all likelihood they are involved in several important types of processes, including aspects of volume control. Their induction of reflex tachycardia (cf. Linden, 1973, 1975) has given the old controversy concerning the Bainbridge reflex new vitality, and such a role seems quite appropriate, the more so since this type of reflex influence appears to be particularly vivid in awake animals (Vatner, Boettcher, Heyndrickx & McRitchie, 1975). This, in turn, may imply an important involvement of suprabulbar centres for this type of reflex and – in general – the addition of the more specialised, myelinated cardiovascular afferents might imply their particular involvement in suprabulbar, more sophisticated cardiovascular adjustments.

Among many suggestions concerning their functional importance, it has also been proposed that they may inform the centres about the rate of cardiac pumping (Arndt, Brambring, Hindorf & Röhnelt, 1971). Beyond doubt the message, particularly of the A fibres, is so coded as to be most suitable for this purpose. The question then arises whether some receptor type might inform the centres also about the *stroke volume,* in which case full information about cardiac output would be at hand. Some of the left ventricular C fibres discharge in close proportion to the degree of end-diastolic stretch and, moreover, for a given pre-stretch 'they further increase their activity upon increased inotropism as shown in Fig. 4 (P. Thorén, unpublished). Others may reflect either presystolic filling or the contractile force, depending on the type of arborisation in the wall, independently of other C fibres. This implies that the two main parameters (beside the afterload, which is sensed and signalled by the high-pressure receptors) determining stroke volume may both be signalled by the same type of ventricular receptors, with other receptors providing different information.

Theoretically, therefore, the bulbar centres might receive coded messages that reflect *both* rate and stroke volume, and hence cardiac output, provided the central computer has the wits to integrate the information in the right way. If so, the problem is 'what to do with it'; are perhaps some reflex adjustments performed on the basis of such complex information and primarily aimed at controlling cardiac output *per se*? As mentioned in the introduction, Q is, after all, more relevant than P for the organism. It may be justified to ask whether the highly complex and sophisticated receptor-reflex machinery only concentrates on keeping P on the ingoing and outgoing sides in check, leaving the control of Q to the interaction between peripherally induced resistance changes and the central control of pressures at the input and output sides. This may all seem a bit far-fetched but it should be stressed that biological control systems have repeatedly revealed a competence and sophistication that have surpassed the wildest expectations of investigators. For example, the tiny brain of a bat in flight handles perfectly a vastly more complex environmental information.

Whatever the case the unmyelinated cardiac afferents seem admirably fitted to do good service in reflex volume control. Many of them are tonically active and hence continuously contribute to adjust, e.g. the tightening of the system around its content, should the filling of the system be altered. Since they also reflexly modulate in the appropriate direction (like the arterial baroreceptors) the pre- and postcapillary resistance ratio, reflex transcapillary 'auto-transfusion' of tissue fluid (Öberg, 1964) at reduced filling of the system, and the reverse at blood loss (see below).

An interesting question here is whether unmyelinated cardiac afferents might exert a relatively more powerful modulation of those neuron pools that control the venous 'capacitance vessels', just as they seem to exert a preferential influence on the heart and the renal vascular bed (see, e.g., Öberg, 1976). The venous 'low-pressure' side is particularly important as a dynamic forechamber to the pump and functionally includes the pulmonary circuit and its influence on the left heart. To 'bind together' the low-pressure system with the ventricles by means of some preferential reflex modulation originating from this part of the system would seem appropriate. So far, available results concerning the left ventricular receptors and their reflex effects on systemic veins (cf. Öberg, 1976; Thorén *et al.,* 1976) hardly support this possibility. However, it might be worthwhile to investigate whether the *pulmonary* blood volume is possibly thus modulated by left ventricular receptors; it seems to be the most readily adjusted 'capacitance section' in the organism (Aarseth, 1971) and is immediately available for the left ventricle.

Further, the ventricular afferents reflexly reduce, with particular efficiency, the sympathetic discharge to the kidney, which directly (by neurogenic haemodynamic means) enhances urine formation and sodium loss, and perhaps does so also by a more roundabout way (neurogenic–renin–angiotensin–aldosterone) (cf. Wennergren, 1975). The reverse will, of course, happen when these tonically active receptors are unloaded by a reduced filling.

Ventricular afferents might, together with the baroreceptors, even modulate the fluid–salt uptake from the intestine, since evidence was quite recently presented by Lundgren and his group (Brunsson *et al.,* 1976) that the vasoconstrictor fibre innervation of the gastro-intestinal mucosa markedly affects the water–salt uptake in the villous countercurrent exchanges. This uptake increases at a raised activity in the regional sympathetic discharge and, since gastro-intestinal vasoconstrictor fibre discharge is intimately controlled both by the high-pressure receptors and by the unmyelinated low-pressure ones, there may thus exist a volume receptor control even of intestinal fluid–salt uptake. A receptor-modulated influence on the thirst mechanism has been repeatedly documented and it thus appears as if virtually *all* aspects of the organism's water–salt and volume house-keeping is closely guarded by cardiac receptors, though with good cooperation from the baroreceptors (see below). Among the cardiac receptors the unmyelinated ventricular variety appear to be the most 'all-round ones' with respect to the various reflex adjustments involved. The various aspects of receptor control of volume, and of some other key cardiovascular parameters as well, are tentatively outlined in Fig. 5 which is no doubt incomplete and partly wrong but perhaps anyhow justified.

To give a final touch to these considerations about *cardiac* receptors and their importance for overall cardiovascular control, it should be mentioned that the C fibre cardiac afferents seem to play an important role in one of the most spectacular displays of cardiovascular reflex control within the whole field of biology, i.e. the diving 'reflex' (see also below). In this reflex they are to a great extent responsible for the profound bradycardia seen during submersion in expert divers such as the duck (Blix, Wennergren & Folkow, 1976). The intense sympathetic discharge then 'packs' venous blood into the heart where end-diastolic pressures rise accordingly to 2.0–3.0 kPa. This brings about strong distension, an activation of the ventricular C fibre afferents and hence a marked reflex intensification of the 'primary' vagal bradycardia. The sympatho-inhibitory reflex component normally induced by these afferents is, however, in *this* situation completely 'occluded' centrally by the intense chemoreceptor activation of the vasoconstrictor fibres (cf. Wennergren, 1975; Öberg, 1976). What is left is a strong reflex brake on the pump, to match its output to the now enormously increased systemic resistance.

Reflex effects of the baroreceptors

It appears as if the unmyelinated high-pressure afferents display a somewhat higher, and perhaps also wider, range of receptor thresholds than the myelinated ones (cf. Kirchheim, 1976; Thorén & Jones, 1977). However, if their mode of activation is in part also different and dependent on, e.g., the level of smooth muscle activity at the receptor sites as outlined previously, selective re-settings of even *opposite* directions might be accomplished. It may therefore be too early to judge all the functional implications of observed differences in threshold for the two receptor types. Further, it is not yet possible to see fully all the consequences of the quantitatively different reflex patterns induced by the myelinated high-pressure afferents versus the corresponding unmyelinated ones (Fig. 2). The latter show some similarities to the unmyelinated low-pressure afferents in so far as both induce particularly pronounced reflex vagal

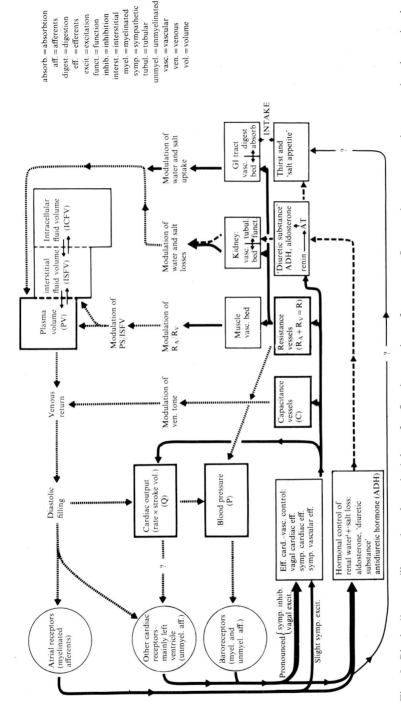

absorb. = absorbtion
aff. = afferents
digest. = digestion
eff. = efferents
excit. = excitation
funct. = function
inhib. = inhibition
interst. = interstitial
myel. = myelinated
symp. = sympathetic
tubul. = tubular
unmyel. = unmyelinated
vasc. = vascular
ven. = venous
vol. = volume

Fig. 5. **Block diagram** illustrating the principles of reflex integration of pressure, flow, resistance, capacitance and volume in the cardiovascular system, with particular emphasis on volume control.

activation, but whether the high-pressure afferents also affect the renal vascular bed in a similar preferential direction as the cardiac C fibre afferents is not clear.

In general, the important problem as to how the various groups of cardiovascular 'depressor' afferents differ in their reflex contacts with the functionally different neuron pools that control the heart and various compartments deserves much more attention. Recent studies have provided interesting new information (see Öberg, 1976) and more is likely to emerge, though the technical difficulties are considerable. As mentioned, perhaps the nearest question at hand in this respect concerns the precapillary resistance and postcapillary capacitance compartments and whether their constrictor fibres might be differently engaged by the various 'depressor afferents', in this particular case the high-pressure C fibre afferents versus the high-pressure myelinated ones. Also, a quantitative difference here would be of great functional importance, but so far no clear differences have been observed, though the final word may not yet have been said since all aspects are not fully explored.

However, even if there are no such differences in fibre discharge, a number of peculiarities in neuro-effector characteristics may nevertheless result in response patterns that appear different in their haemodynamic consequences, which, from the cardiovascular point of view, is just as important. First, on the *pre*capillary side the constrictor fibre supply differs in extent between circuits, secondly, it is mainly a remote modulator of a (usually) pronounced inherent vascular tone and, thirdly, this modulating influence, particularly in some circuits, is counterbalanced by powerful 'autoregulatory' adjustments (cf. Fig. 6). As a result, even in situations where sympathetic discharge is uniform it still implies a considerable redistribution of flow and, moreover, the overall changes in systemic resistance are often only modest, except at pronounced increases of sympathetic discharge. In contrast, on the *post*capillary capacitance side, which from a functional point of view serves as a 'unit' in securing cardiac filling, inherent effector activity is usually negligible so that here constrictor fibre discharge dominates events, while at the same time the neurogenic responses are better maintained than on the precapillary side. Furthermore, the hyperbolic frequency–response curve of the postcapillary compartment is clearly more steep and displaced to the left compared with that of the precapillary one (Mellander, 1960).

As a net result the reflex adjustments of the capacitance side may be *relatively* more pronounced than those on the precapillary side in the

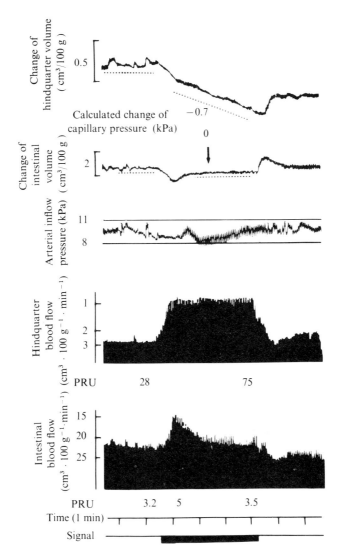

Fig. 6. Consequences of tissue fluid absorbtion by reflex increases in vasoconstrictor fibre activity, induced by blood loss in the cat. In skeletal muscle the enhanced discharge leads to a sustained and, with time, considerable mobilisation of interstitual fluid. This 'autotransfusion' is a consequence of the well-maintained reflex increase in the pre- and postcapillary resistance ratio, which keeps mean capillary pressure considerably lowered and hence shifts the Starling equilibrium towards absorbtion in this large tissue mass. In the intestine, on the other hand, an autoregulatory precapillary adjustment rapidly ensues in response to the reflex vasoconstrictor fibre activation. Consequently the flow reduction is here only transient with no sustained shift in mean capillary pressure or in transcapillary fluid balance (from Öberg, 1964).

low-range register of fibre discharge, while the reverse is true as constrictor fibre activity is further intensified. Depending on the prevailing level of sympathetic discharge, given receptor activations may therefore result in a *variable* balance between resistance and capacitance adjustments, where the latter tend to dominate in the ordinary, lower range of tonic activity. However, all haemodynamic signs of neurogenic capacitance adjustments may vanish in situations where venous pressures are so low that veins begin to collapse (below approximately 0.7–1 kPa). If veins are nearly collapsed beforehand it does not matter much whether constrictor fibres are activated or not, since no real expulsion of blood can then take place. A lot of apparent controversies in results, and interpretations of results, have for such reasons occurred during the years. Further, these circumstances emphasise that the full impact and haemodynamic importance of reflex venous control is revealed first in the erect position of man, when dependent veins are kept well distended by the raised transmural pressure.

It is quite important for overall haemodynamics that the unrestrained VMC discharge appears to be the most intense to the muscle vascular bed, and here the high-pressure afferents exert an especially powerful damping effect, supported by the low-pressure C fibre afferents. It follows that these reflex adjustments particularly affect the muscle circuit, preferentially on its *pre*capillary side, which also considerably modulates the pre- and postcapillary resistance ratio in this tissue. Other things being equal, this implies a reflexly governed re-setting of mean capillary pressure, and hence of the Starling transcapillary exchange between vessels and interstitual fluid, which is particularly voluminous in this large tissue mass (Öberg, 1964, 1976). A sensitive reflex device for 'autotransfusion' from the muscle intestitial fluid depots thus occurs whenever blood volume is reduced, and vice versa. Paradoxically, therefore, the reflex *resistance* adjustments are to a great extent used as a major device also in the highly important control of *volume* on which cardiac output finally depends. This special aspect of volume control is under the particular guidance of the high-pressure receptors, but they act, of course, in close cooperation with the low-pressure stretch receptors, as outlined above.

On the whole the reflex control of volume and hence, in the end, of cardiac output is secured by both low-pressure and high-pressure receptors and is, as to its time sequence, arranged as 'three lines of defence'. The first is based on the immediate adjustments of cardiac and vascular effectors, tightening the walls around a dwindling content and when this is not enough, adjusting R to Q, but in a 'rationing' fashion where the more vitally important tissues are favoured. The second is a slower consequence of the first, in so far as the reflexly lowered mean capillary pressure in the muscle circuit starts a fairly slow but finally most efficient autotransfusion of interstitial fluid. The third line is directed towards reducing losses by reflex kidney adjustments, and increasing uptake both by eliciting thirst and by facilitating fluid–electrolyte transfer across the intestinal wall (Fig. 5).

In this complex machinery it should be realised that the high-pressure receptors are by no means unable to sense volume deficits. Even if a volume reduction is so small that only a slight fall in stroke volume ensues (with perhaps no significant cardiac output reduction in case the ventricular C afferents have compensated efficiently by a reflex heart rate increase) the combined set of high-pressure receptors may sense the consequent reduction in pulse amplitude. In reverse, the low-pressure receptors may to some extent also sense an increased arterial pressure, partly because the high-pressure receptors then reduce reflexly ventricular contractility, from which a rise in end-diastolic volume and hence in ventricular C fibre discharge may ensue. Thus, each of the two sets of receptors may reflexly influence the haemodynamic situation 'seen' by the other set, such that both become involved and cooperate concerning both volume and pressure control. The possibility that they might together even inform about cardiac output has been discussed above.

Thus, even if the vagal–glossopharyngeal high- and low-pressure stretch receptors in most normal situations act in harmonious concert, it is not just an academic question whether their reflex inhibitory patterns are in part differently balanced. This becomes increasingly obvious in pathophysio-logical situations, e.g. when a failing heart leads to overstimulation of the low-pressure receptors and a reduced stimulation of the high-pressure ones. The presence of differently balanced reflex patterns may then result in a considerable imbalance in haemodynamics, of a nature that is not met with otherwise. One example is the intense ventricular C fibre firing from bulging sections of a partially ischaemic myocardium. This elicits profound reflex bradycardia, overall sympathetic inhibition and even reflex vomiting (cf. Thorén, 1972; Öberg, 1976), and does so in a critical situation where other adjustments are badly needed. The concomitant unloading of the high-pressure receptors may in this particular situation have little say, as the ventricular ones have not only largely silenced sympathetic discharge but also produced maximal bradycardia because of their particularly intense vagal activation. A state of cardiogenic 'shock' may ensue where acute collapse seems imminent and in fact not unfrequently occurs.

Abolition, e.g. by atropine injection of the particularly intense bradycardia, which here stands almost unopposed by other reflex mechanisms, is empirically found to be quite successful in many of these cases. There are other such examples, e.g. the attacks of severe bradycardia in some cases of aortic stenosis where, again, a sudden imbalance in low-pressure/high-pressure receptor activity and in reflex effects may be responsible.

A most important pathophysiological aspect of the high-pressure receptors is their gradual but fairly rapid 're-setting' along with the development of hypertension. This re-setting seems, at least to the greater part, to be a consequence of an adaptive change of the arterial wall itself, in terms of an increasing thickness and stiffness at the receptor sites (cf. Aars, 1969; Jones, 1976). Thereby the receptors so to say 'accept' the hypertensive state and by their reflex modulations they try to maintain the high pressure rather than counteract it, as is their normal response to acute pressure rises. This re-setting is well documented by neurophysiological techniques concerning the myelinated baroreceptor afferents, but whether the unmyelinated ones also are re-set to the same extent is not clear. Here, an ongoing investigation in the Department of Physiology at the University of Göteburg (P. Thorén and J. V. Jones) is likely to provide further information. Indirect evidence may indicate that all restraining reflex mechanisms are perhaps not fully re-set in hypertension, since there are frequently signs of a remaining 'brake' tending to keep sympathetic discharge relatively damped, even when hypertension has reached its established phase. However, at present the information is still too vague to allow for any definitive conclusions.

'Sympathetic' cardiac afferents

In recent years another type of cardiac stretch afferents, running in the sympathetic trunks, has been explored by Uchida, Brown, Malliani and their colleagues (see, e.g., Malliani, Recordati & Schwartz, 1973, Öberg, 1976). Apart from the pain afferents from the heart, long since acknowledged to run here, the sympathetic trunks carry other sets of afferent fibres, both myelinated and unmyelinated, for which the adequate stimulus appears to be cardiac mechanical events and possibly also local chemical changes associated with the normal range of cardiac activity. It is not yet quite clear in which respects the information sensed by these 'sympathetic' afferents differs from that conveyed by the 'specific' vagal mechanoreceptors, but the reflex effects of the former fibre group is of an *excitatory* nature, causing tachycardia and vasoconstriction. Their reflex influence is thus of a positive feedback nature and is evidently in part conveyed at spinal levels. The functional significance of this set of cardiac afferents is not yet clear, but it is possible that they constitute a myocardial counterpart to the thin afferents originating in skeletal muscles and here sensing mechanical and chemical events during activity (see below). Together with the C pain fibres these latter fibres belong to the 'somatic pressor afferents', and during, e.g., exercise they seem to convey an additional excitatory 'drive' to the bulbar cardiovascular and respiratory centres.

The arterial chemoreceptors

This is not the place to deal extensively with this important type of receptor and its reflex cardiovascular effects and the reader is referred to the excellent work by Korner and Daly and their groups (cf. Korner, 1971; Angell-James & Daly, 1972; Öberg, 1976). In principle, the 'primary' chemoreceptor reflex implies a sympathetic vasoconstrictor fibre activation (except to the cutaneous vessels or at least their A–V anastomoses) but in association with a reflex vagal restraint on heart rate, a pattern that forms a basis for the complex 'diving reflex' as briefly discussed above (cf. Folkow, 1968; Angell-James & Daly, 1972; Shepherd & Pelletier, 1975). However, a concomitant chemoreceptor activation of respiration, and hence of pulmonary stretch receptors, adds an important reflex modulation in so far as tachycardia now ensues, with appropriate modulations of the sympathetic discharge to the vascular bed. In any case, the pattern is basically one of sympathetic activation and the question arises as to what happens when this pattern is confronted by the various 'depressor' reflexes, elicited by the above-mentioned sets of mechanoreceptors. Dr Öberg has dealt in more detail with this interesting competition between opposing reflex patterns, and one example occurring during diving has already been discussed above. In principle it appears as if an intense chemoreceptor drive fairly efficiently 'occludes' the reflex vasodepressor influence from the various mechanoreceptors while they often act in *synergism* with regard to the reflex bradycardia.

'Somatic pressor and depressor' afferents

These fibre groups are composed of mainly unmyelinated C fibres and thin myelinated afferents, respectively (cf. Johansson, 1962). Both contain an important contribution of pain fibres, where, e.g., superficial pain or ischaemic pain tends to elicit reflex sympathetic activation, while 'deep' pain of the type induced by distorsion, blunt trauma to deeper tissues, etc., tends to evoke an overall depressor reflex associated with, e.g., gastro-intestinal activation, nausea, and also with inhibition of somatomotor activity. These depressor influences are in part relayed in, or in close vicinity to, the so-called 'depressor area' of the bulbar cardiovascular control centres, which the various depressor reflexes discussed above are also dependent on (Johansson, 1962; Öberg, 1976).

However, apart from various sets of pain fibres it is increasingly clear that particularly the somatic pressor afferents also contain fibres that in skeletal muscle, e.g., are activated by the mechanical–chemical changes accompanying normal muscle activity – an idea championed by Russian physiologists for decades (cf. Khayutin, 1966) but only recently more widely accepted (see, e.g., McCloskey & Mitchell, 1972; Clement & Shepherd, 1974). This type of input is likely to provide an additional excitatory drive to the cardiovascular centres in exercise, in cooperation with similar influences from suprabulbar centres and also from the arterial chemoreceptors, once these are affected by the increasing changes in the pH of blood, etc. It appears that the combined influence of such excitatory influences tends to suppress the reflex bradycardia component elicited whenever the baroreceptors and cardiac receptors are activated by rising pressures in and beyond the heart, in a comparable manner to the way the hypothalamic defence area is activated (see below).

Cortico-hypothalamic influences on cardiovascular reflex control

It is clear that although the basic framework of the reflexes emanating from the cardiovascular proprioreceptors may be organised at lower bulbar levels, the various sets of receptors also markedly affect higher centres. An interesting question here is whether such reflex influences occurring at the more sophisticated higher levels of the CNS are primarily effected by means of the myelinated afferents which phylogenically presumably represent a later addition to allow more specialised reflex control.

To take a few examples, Zanchetti (1976) has shown that chemorecep-tors can facilitate the induction of a hypothalamic 'sham rage', while an intense baroreceptor activation has the opposite effect. Such intense baroreceptor activation also affects EEG in the direction of reduced alertness and sleep. Furthermore, the studies by Korner and his colleagues on chemoreceptor control in awake rabbits illustrate that part of the reflex pattern is conveyed at suprabulbar levels (see Korner, 1971). Moreover, a cardiovascular receptor modulation of, e.g., ADH secretion in connection with volume control simply must be conveyed via hypothalamic centres.

So much for the *afferent* input from cardiovascular receptors to higher CNS levels where particularly the limbic system–hypothalamus and the ascending reticular formation seem to be the main targets and where the modulations induced may well be conveyed via the various sets of ascending monoamine neuron systems originating in the brain stem.

Interesting interactions occur concerning the *efferent* effects upon reflex cardiovascular control, induced from particularly limbic–hypothalamic centres governing emotional behaviour and instincts. On the basis of telereceptor information the external environment is scanned by the CNS computer which evaluates the information, disregards the trivia and concentrates on the more interesting and alerting signals. Depending on whether such environmental elements are interpreted as danger and threat, or represent something appealing like food, sexual partners, etc., the limbic system–hypothalamus induces an 'emotionally charged' pattern of behaviour that is appropriate to deal with that particular external stimulus. Especially when the situation implies immediate danger or challenge for the individual, this type of emotionally charged behaviour pattern is temporarily given absolute priority, with suppression of most other events that involve somatomotor behaviour, autonomic control of inner organs and the hormonal control of metabolism, water–salt balance, etc.

Two drastic examples along these lines will be given to illustrate what then happens concerning reflex homeostasis of the circulation. These are the *defence reaction,* elicited in situations where attack, defence or outright flight are imminent, and the *'playing dead' reaction* (in man, emotional fainting is the probable parallel) which seems to occur particularly in such situations of danger or embarrassment where the organism is, or feels, 'cornered', with no chance of escape (cf. Folkow & Neil, 1971; Zanchetti, 1976).

Concerning the circulation of blood, the defence reaction transforms the cardiovascular system into a preferential 'heart–skeletal muscle–brain pump system', implying an anticipatory pressure rise and cardiac output increase. In the periphery, constriction occurs in most vascular beds except those of skeletal muscle, myocardium and brain where flow increases. Clearly, such a sudden superimposed command from 'the ministry of defence' must imply a considerable interference with the subordinate VMC

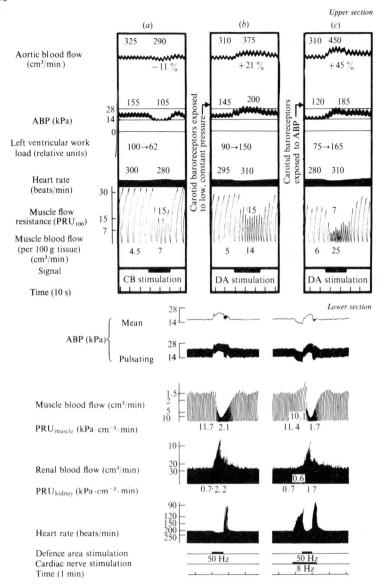

Fig. 7. *Upper section:* Vagotomised, curarised, anaesthetised cat, with recordings of aortic flow, arterial pressure (ABP), heart rate and muscle flow resistance. *(a)* The reflex sympatho-inhibitory response to carotid baroreceptor (CB) activation; *(b)* the haemodynamic changes to a standardised defence area (DA) stimulation when the carotid baroreceptors are 'uncoupled' by being exposed to a low, constant pressure; *(c)* the haemodynamic changes to the same defence area stimulation but when the carotid baroreceptors are fully exposed to the pressure changes and

centre and its homeostatic control based on the steady stream of information from the various cardiovascular receptors. These receptors immediately sense the centrally induced rise in arterial mean and pulse pressures, in heart rate, in myocardial filling and inotropism, etc. and normally they would by their appropriate reflex effects damp such a superimposed excitatory influence. However, in this emergency situation such an interference cannot be permitted and an interesting interaction takes place. The hypothalamic discharge produces a virtually complete suppression centrally of that part of the reflex pattern that would normally produce vagal bradycardia (cf. Hilton, 1965). The parts of the reflex patterns that modulate sympathetic discharge to the vascular bed, on the other hand, are less, if at all, suppressed by the centrally induced defence reaction (Fig. 7).

This differentiated interaction has an interesting haemodynamic consequence because, as mentioned above, the baroreceptors have an especially powerful inhibitory influence on the constrictor fibre control of the skeletal muscle vessels. Therefore, the net result of this interaction between the central drive and the reflex modulation of vasoconstrictor fibre activity serves to *enhance* muscle blood flow and further allows the heart to deliver the increased output at a somewhat lower pressure load. In other words, this example of a neurogenic interference between the excitatory defence reaction and the inhibitory reflex influence from the cardiac and arterial mechanoreceptors results in a 'haemodynamic facilitation' of the centrally induced adjustment by further favouring the supply of blood to those regions now needing it, but doing so at a total lower cost for the pump (cf.

(caption continued)
allowed to modulate reflexly the defence reaction. Note from the differences between *(b)* and *(c)* that the interaction between the excitatory defence reaction and the inhibitory baroreceptor reflex influence implies a 'haemodynamic facilitation' that *enhances* cardiac output and muscle blood flow increase but at a lower pressure load, as compared with the situation when the defence reaction is unaffected by the baroreceptor reflex (from Lisander, 1970). *Lower section:* Anaesthetised, curarised cat. Effect of defence area stimulation alone (left) and when such a stimulation is superimposed upon a continuous cardiac nerve stimulation which, via the unmyelinated afferents produces intense vagal bradycardia and sympathetic inhibition (right). Note that the defence reaction virtually completely 'occludes' the reflex bradycardia, while the extent of muscle vasodilatation is *enhanced* and the extent of renal vasoconstrictor response moderately attenuated by the ongoing reflex effects of the unmyelinated cardiac afferents (from Wennergren, 1975).

Lisander, 1970; Wennergren, 1975). A similar suppression centrally of the
reflex bradycardia mechanism seems to occur in connection with the
central and peripheral excitatory drive on the cardiovascular system during
exercise.

The overall depressor 'playing dead' reaction, in all probability corresponding to emotional fainting in man, is also induced from limbic–
hypothalamic structures which in the cat at the hypothalamic level are
situated only a few millimetres from the defence area. At the bulbar level
the efferents seem to have a relay in the medullary 'depressor area' which,
as mentioned earlier, also serves as a relay for the cardiac, arterial and
'somatic depressor' reflexes (cf. Löfving, 1961). This area therefore seems
to be a final common link in eliciting any type of overall sympathetic
inhibition and vagal bradycardia, which on the cardiovascular side seem to
represent the key changes in 'playing dead' and emotional fainting.

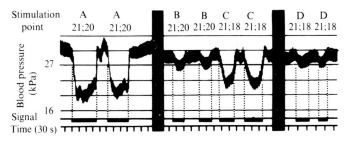

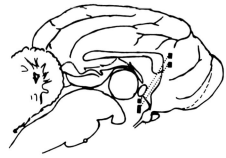

Fig. 8. Curarised, lightly anaesthetised cat with baroreceptor reflexes excluded.
Strong depressor responses elicited from anterior parts of the cingulate gyrus
(normally associated with bradycardia and, usually, somatomotor inhibition). This
depressor response emanating from the limbic system is almost entirely abolished
by electrolytic lesions of the medially placed hypothalamic 'sympatho-inhibitory'
area and is in the bulb conveyed via the medullary 'depressor area' (from Löfving,
1961).

If very intense, this centrally induced depressor pattern may by itself be so powerful that the normal buffer reflex influences of the various cardiovascular receptors cannot compensate for it, and a vasovagal syncope then occurs as a direct response to the central situation. If the central drive is less intense, however, the buffer reflexes may partly compensate and produce a more graded reduction in sympathetic activity and increased vagal activity to the heart. However, if the net sympathetic inhibition results in enough venous pooling in an individual in the erect position the heart may become so poorly filled that the left ventricle, in particular, contracts around a reduced content resulting in such distorsion in the wall that a secondary activation of ventricular receptors occurs as first suggested by Henry and his group. It has been illustrated experimentally in cats how such 'squeezing' of the left ventricle may, indeed, strongly activate the ventricular C afferents which would then superimpose their typical reflex bradycardia and sympathetic inhibition on the prevailing inhibitory pattern, perhaps precipitating an outright fainting (see Thorén, 1972; Öberg, 1976).

In other words, there is here an interesting cooperation between the central structures involved in the emotionally induced sympathetic inhibition–vagal bradycardia response and the reflexly induced pattern, of similar direction, initiated from the ventricles.

These few examples may suffice to illustrate the interesting and often considerably differentiated interactions that occur between the autonomic centres at telen- and diencephalon levels of the brain and those present at lower bulbar levels, with their functional role mainly directed towards daily cardiovascular homeostasis.

References

Aars, H. (1969). *Effects of arterial hypertension on aortic baroreceptor activity*. Universitetsforlaget: Oslo.

Aarseth, P. (1971). Nervous influence on the pulmonary capacitance vessels in the rat. *Acta physiol. scand.*, **83**, 60–9.

Angell-James, J. & Daly, M. de B. (1972). Some mechanisms involved in the cardiovascular adaptations to diving. In *Symposia of the Society for Experimental Biology*, pp. 313–41. Cambridge University Press: Cambridge.

Arndt, J.O., Brambring, P., Hindorf, K. & Röhnelt, M. (1971). The afferent impulse traffic from atrial A-type receptors in cats. Does the A-type receptor signal heart rate? *Pflügers Arch. ges. Physiol.*, **326**, 300–15.

Blix, A.S., Wennergren, G. & Folkow, B. (1976). Cardiac receptors in ducks – a link between vasoconstriction and bradycardia during diving. *Acta physiol. scand.*, **97**, 13–19.

Brunsson, I., Eklund, S., Jodal, M., Lundgren, O. & Jovall, H.S. (1976). The influence of the sympathetic nerves on intestinal net water absorption. *Acta physiol. scand.*, **98** (supp. 440), 73.

Clement, D.L. & Shepherd, J.T. (1974). Influence of muscle afferents on cutaneous and muscle vessels in the dog. *Circulation Res.*, **35**, 177–83.

Folkow, B. (1955). Nervous control of the blood vessels. *Physiol. Rev.*, **35**, 629–63.

Folkow, B. (1960). Range of control of the cardiovascular system by the central nervous system. *Physiol. Rev.*, **40** (suppl. 4), 93–9.

Folkow, B. (1968). Circulatory adaptations to diving in ducks. In *Proc. int. union of physiological sciences*, vol. VI. xxiv International Congress of Physiological Sciences: Washington.

Folkow, B., Johansson, B., & Löfving, B. (1961). Aspects of functional differentiation of the sympatho-adrenergic control of the cardiovascular system. *Med. Exp.*, **4**, 321–8.

Folkow, B. & Neil, E. (1971). *Circulation*. Oxford University Press: Oxford.

Gauer, O.H., Henry, J.P. & Behn, C. (1970). The regulation of extracellular fluid. *A. Rev. Physiol.*, **32**, 547–95.

Hilton, S.M. (1965). Hypothalamic control of the cardiovascular responses in fear and rage. *The scientific basis of medicine annual reviews*. pp. 217–38. The Athlone Press, London.

Johansson, B. (1962). Circulatory responses to stimulation of somatic afferents. *Acta physiol. scand.*, **57** (suppl. 198), 1–91.

Jones, J.V. (1976). Time-course and extent of carotid sinus baroreceptor resetting during experimental renal hypertension in rats. *Acta physiol. scand.*, **98** (suppl. 440), 125.

Khayutin, V.M. (1966). Specific and non-specific responses of the vasomotor centre to impulses of spinal afferent fibres. *Acta physiol. Acad. Sci. hung.*, **29**, 131–43.

Kirchheim, H.R. (1976). Systemic arterial baroreceptor reflexes. *Physiol. Rev.*, **56**, 100–76.

Korner, P.I. (1971). Integrative neural cardiovascular control. *Physiol. Rev.*, **51**, 312–67.

Landgren, S. (1952). On the excitation mechanism of the carotid baroreceptors. *Acta physiol. scand.*, **26**, 1–35.

Linden, R.J. (1973). Function of cardiac receptors. *Circulation*, **48**, 463–80.

Linden, R.J. (1975). Reflexes from the heart. *Prog. cardiovasc. Dis.*, **28** (3), 201–21.

Lisander, B. (1970). Factors influencing the autonomic component of the defence reaction. *Acta physiol. scand.*, **80** (suppl. 351), 1–42.

Löfving, B. (1961). Cardiovascular adjustments induced from the rostral cingulate gyrus. *Acta physiol. scand.*, **53** (suppl. 184), 1–82.

Malliani, A., Recordati, G. & Schwartz, P.J. (1973). Nervous activity of afferent cardiac sympathetic fibres with atrial and ventricular endings. *J. Physiol.*, **229**, 457–69.

McCloskey, D.I. & Mitchell, J.H. (1972). Reflex cardiovascular and respiratory responses originating in exercising muscle. *J. Physiol.*, **224**, 173–86.

Mellander, S. (1960). Comparative studies on the adrenergic neuro-hormonal control of resistance and capacitance blood vessels in the cat. *Acta physiol. scand.*, **50** (suppl. 176), 1–86.

Öberg, B. (1964). Effects of cardiovascular reflexes on net capillary fluid transfer. *Acta physiol. scand.*, **62** (suppl. 229), 1–98.

Öberg, B. (1976). Overall cardiovascular regulation. *A. Rev. Physiol.*, **38**, 537–70.

Öberg, B. & Thorén, P. (1973). Circulatory responses to stimulation of medullated and nonmedullated afferents in the cardiac nerve in the cat. *Acta physiol. scand.*, **87**, 121–32.

Paintal, A.S. (1973). Vagal sensory receptors and their reflex effects. *Physiol. Rev.*, **53** (1), 159–227.

Shepherd, J.T. & Pelletier, C.L. (1975). Carotid chemoreflex and circulatory control. In *The peripheral arterial chemoreceptors*, ed. M. J. Purves, pp. 463–79. Cambridge University Press: Cambridge.

Thorén, P. (1972). *Studies on left ventricular receptors signalling in nonmedullated vagal afferents*, Ph.D. thesis. Elanders Boktryckeri: Kungsbacka.

Thorén, P., Donald, D.E. & Shepherd, J.T. (1976). Role of heart and lung receptors with nonmedullated vagal afferents in circulatory control. *Circulation Res.*, **38**, (Suppl. II), 2–9.

Thorén, P. & Jones, J.V. (1977). Characteristics of aortic baroreceptor C-fibres in rabbits. *Acta physiol. scand.*, **99**, 448–56.

Vatner, S.F., Boettcher, D.H., Heyndrickx, G.R. & McRitchie, R.J. (1975). Reduced baroreflex sensitivity with volume loading in conscious dogs. *Circulation Res.*, **37**, 236–42.

Wennergren, G. (1975). Aspects of central integrative and efferent mechanisms in cardiovascular reflex control. *Acta physiol. scand.*, **96** (suppl. 428).

Wennergren, G., Henricksson, B.Å., Weiss, L.G. & Öberg, B. (1976). Effects of stimulation of nonmedullated cardiac afferents on renal water and sodium excretion. *Acta physiol. scand.*, **97**, 261–3.

Zanchetti, A. (1976). Hypothalamic control of circulation. In *The nervous system in arterial hypotension*, ed. S. Julius & M.D. Ester, pp. 397–428. S.C. Thomas Publishing: Springfield, Illinois.

Closing address

O. H. GAUER

I have no more than 900 000 ms to integrate the many substantial papers presented at this symposium. This is nearly impossible so I must confine myself to a random selection of a few annotations.

The meeting started appropriately with demonstrations of the histology of the various receptors found in the heart. This was done in three beautiful papers. Considerable attention was given to the unmyelinated end-nets. Many problems related to a separation of the fibres into their various functions are still unsolved. I wonder whether a closer cooperation between histologists and electrophysiologists would or could lead to faster progress in this important field. While the distinction between afferent and efferent end-nets is possible with Falck's method the question of the recording characteristics of what may possibly be different species of afferent end-nets remains unanswered. This uncertainty contrasts with our knowledge of the recording properties of the myelinated unencapsulated endings. The B receptors have long been known to yield information on cardiac filling. Newer investigations furnish strong evidence that the discharge activity of the A receptors during atrial contraction is fixed regardless of changing atrial mechanics. Therefore, the integrated signal rate per time unit indicates heart rate. It is intriguing to speculate that A and B receptors together furnish the central nervous system with information on cardiac output.

As to the afferent neural pathways there can be no doubt that the vagus deserves major interest. The work of the Milan group reminds us, however, that sympathetic afferent pathways may play an important part in the so-called reflex activity within the vegetative system. Lynne Weaver has evidence that information carried in cardiac sympathetic afferent fibres may even be transmitted directly to the kidney without passing up to the

medulla oblongata. The existence of so many previously unknown
pathways explains why the distinction between a reflex and a miracle is not
so easy. It was assumed that we were dealing with a reflex when a certain
effect disappeared after cutting the vagus, while we were confronted with a

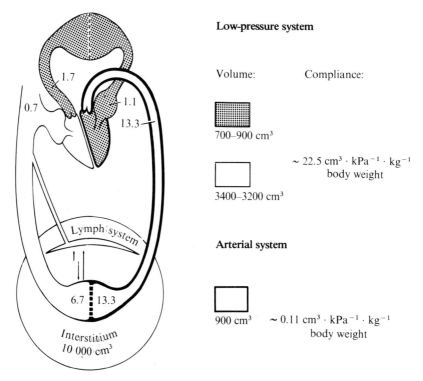

Low-pressure system

Volume: Compliance:

700–900 cm³

~ 22.5 cm³ · kPa^{-1} · kg^{-1}
body weight

3400–3200 cm³

Arterial system

900 cm³ ~ 0.11 cm³ · kPa^{-1} · kg^{-1}
body weight

Fig. 1. Distribution of blood volume and distensibility of the major compartments
of the circulation. The numbers in the various sections indicate the mean pressures
in kPa. On the right side the estimated volumes and compliances of the main
compartments are indicated. The compliance of the arterial system is given at its
working pressure of 13 kPa. The arterial system holds approximately 15% of the
estimated blood volume, and the arterial pressure is caused entirely by cardiac
dynamics and total peripheral resistance. The low pressure system consists of the
systemic veins, the right heart, the pulmonary circulation and the left ventricle in
diastole. It holds 85% of the blood volume and the pressure is mainly determined by
the distensibility of the total vascular bed and the blood volume, the latter being a
static parameter. Note that the shaded area, representing the filling reservoir of the
left ventricle, is studded with mechanoreceptors. A reflex effect of these receptors on
the arterial system can readily be signalled by a change in heart rate. In contrast, the
effect on the low pressure system is hidden in the response of numerous hormonal
mechanisms (Fig. 2) and therefore not immediately obvious (Gauer, 1972).

miracle when the effect persisted. How complicated things are may be elicited from the amazing observation of P. D. Gupta and M. Singh that the tachycardia of the Bainbridge reflex can no longer be elicited after T1 has been sectioned.

The paramount importance of the cardiac receptors is because of their location in the strategically most important transition zone between the low-pressure system and the arterial system (Fig. 1). The low-pressure system holds 85% of the total blood volume and comprises the systemic capacitance vessels, the right heart and the pulmonary circulation. The left ventricle is Janus-headed, facing in both directions. It is the pump of the arterial system in systole and a reservoir belonging to the low-pressure system in diastole. While the atrial B receptors are ideally suited to record cardiac filling pressure the left ventricular receptors have been found to indicate changes in diastolic volume of the ventricle (Thorén and his colleagues) as well as changes in mechanical parameters related to ventricular dynamics.

The effects of stimulation of cardiac receptors are to be found in both systems. Those on heart rate and blood pressure, that is on the arterial system, are easy to measure but often difficult to interpret. Thus, we find the paradox of an increased heart rate with a reduced filling of the heart, yet a distension of the right atrium initiates a tachycardia (Bainbridge reflex). By contrast it is easy to see that the goal of the effects of a change in the filling of the low-pressure system and the ensuing receptor response is volume regulatory. But the details of how this comes about are obscure because they do not involve an as immediate, easily measured, adjustment as on the high-pressure side. They are hidden in subtle shifts of venous tone and in the delayed responses of a number of hormone systems (Fig. 2). This is the reason why the physiology of the low-pressure system including volume control has only recently received adequate attention by circulatory physiologists.

The experiments of the Leeds group, distending extra small balloons in the orifice of the veins leading into the left atrium, showed that the tachycardia to be expected according to Bainbridge does indeed occur. This is a response of arterial haemodynamics. On the other hand, the accompanying reduction in antidiuretic hormone activity is a clear-cut expression of the mechanisms of homeostatic control of the low-pressure system.

These and similar experiments described at this meeting together with histological studies and observations of the effects of stimulating and

510

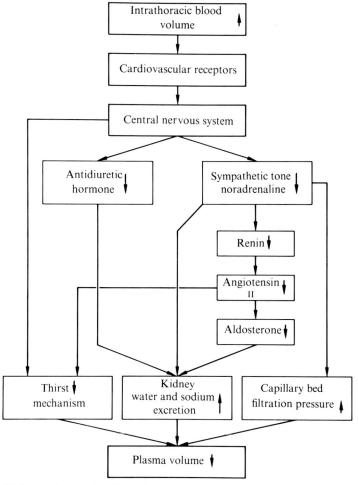

Fig. 2. Major pathways for the reflex control of plasma volume following an expansion of the intrathoracic blood volume. An increase in intrathoracic blood volume will stimulate cardiovascular receptors thus conveying information to the central nervous system (CNS). The water and sodium regulatory mechanisms are activated via decreased antidiuretic hormone (ADH) levels on the one hand and sympathetic and renin–angiotensin control of kidney function on the other. Fluid uptake is diminished because of the action in the CNS and via decreased angiotensin effect on thirst. Restraint from fluid uptake and increased fluid excretion together decrease the extracellular fluid volume and hence plasma volume; furthermore, it is possible to affect plasma volume by changing the capillary filtration pressure via the sympathetic outflow. (For the sake of clarity several important points have been omitted, e.g. a possible diuretic and saliuretic hormonal factor, the interaction between the ADH and sympathetic, renin–aldosterone system as well as the possible effects of the various hormones on vascular tone (Gauer & Henry, 1976).)

interrupting neural pathways and, more recently, of the effects on various hormonal systems serve to pinpoint the location and recording properties of the various receptors. The attempts to come to reasonable (teleological) interpretations have been partially successful although many unsolved problems remain. The many new facts presented and the vivid discussion have certainly contributed towards a better understanding of the overall regulation of the circulation.

As the last speaker I again want to thank Drs Linden, Kidd and Hainsworth for the great effort they put into the task of bringing this group together for a successful and stimulating meeting.

References

Gauer, O.H. (1972). Kreislauf des Blutes. In *Physiologie des Menschen*, ed. O. H. Gauer, K. Kramer & R. Jung, p. 234. Urban & Schwarzenberg: Munich, Berlin and Vienna.

Gauer, O.H. & Henry, J.P. (1976). Neurohormonal control of plasma volume. *International Review of Physiology II*, **9**, 145–90.

INDEX